T0143470

Graph Searching Games and Probabilistic Methods

DISCRETE
MATHEMATICS
AND
ITS APPLICATIONS

Series Editors

Miklos Bona
Patrice Ossona de Mendez
Douglas West

R. B. J. T. Allenby and Alan Slomson, How to Count: An Introduction to Combinatorics, Third Edition

Craig P. Bauer, Secret History: The Story of Cryptology

Jürgen Bierbrauer, Introduction to Coding Theory, Second Edition

Katalin Bimbó, Combinatory Logic: Pure, Applied and Typed

Katalin Bimbó, Proof Theory: Sequent Calculi and Related Formalisms

Donald Bindner and Martin Erickson, A Student's Guide to the Study, Practice, and Tools of Modern Mathematics

Francine Blanchet-Sadri, Algorithmic Combinatorics on Partial Words

Miklós Bóna, Combinatorics of Permutations, Second Edition

Miklós Bóna, Handbook of Enumerative Combinatorics

Miklós Bóna, Introduction to Enumerative and Analytic Combinatorics, Second Edition

Anthony Bonato and Pralat Pawel, Graph Searching Games and Probabilistic Methods

Jason I. Brown, Discrete Structures and Their Interactions

Richard A. Brualdi and Dragoš Cvetković, A Combinatorial Approach to Matrix Theory and Its Applications

Kun-Mao Chao and Bang Ye Wu, Spanning Trees and Optimization Problems

Charalambos A. Charalambides, Enumerative Combinatorics

Gary Chartrand and Ping Zhang, Chromatic Graph Theory

Henri Cohen, Gerhard Frey, et al., Handbook of Elliptic and Hyperelliptic Curve Cryptography

Charles J. Colbourn and Jeffrey H. Dinitz, Handbook of Combinatorial Designs, Second Edition

Titles *(continued)*

Abhijit Das, Computational Number Theory

Matthias Dehmer and Frank Emmert-Streib, Quantitative Graph Theory:
Mathematical Foundations and Applications

Martin Erickson, Pearls of Discrete Mathematics

Martin Erickson and Anthony Vazzana, Introduction to Number Theory

Steven Furino, Ying Miao, and Jianxing Yin, Frames and Resolvable Designs: Uses,
Constructions, and Existence

Mark S. Gockenbach, Finite-Dimensional Linear Algebra

Randy Goldberg and Lance Riek, A Practical Handbook of Speech Coders

Jacob E. Goodman and Joseph O'Rourke, Handbook of Discrete and Computational Geometry,
Third Edition

Jonathan L. Gross, Combinatorial Methods with Computer Applications

Jonathan L. Gross and Jay Yellen, Graph Theory and Its Applications, Second Edition

Jonathan L. Gross, Jay Yellen, and Ping Zhang Handbook of Graph Theory, Second Edition

David S. Gunderson, Handbook of Mathematical Induction: Theory and Applications

Richard Hammack, Wilfried Imrich, and Sandi Klavžar, Handbook of Product Graphs,
Second Edition

Darrel R. Hankerson, Greg A. Harris, and Peter D. Johnson, Introduction to Information Theory
and Data Compression, Second Edition

Darel W. Hardy, Fred Richman, and Carol L. Walker, Applied Algebra: Codes, Ciphers, and
Discrete Algorithms, Second Edition

Daryl D. Harms, Miroslav Kraetzl, Charles J. Colbourn, and John S. Devitt, Network Reliability:
Experiments with a Symbolic Algebra Environment

Silvia Heubach and Toufik Mansour, Combinatorics of Compositions and Words

Leslie Hogben, Handbook of Linear Algebra, Second Edition

Derek F. Holt with Bettina Eick and Eamonn A. O'Brien, Handbook of Computational Group Theory

David M. Jackson and Terry I. Visentin, An Atlas of Smaller Maps in Orientable and
Nonorientable Surfaces

Richard E. Klima, Neil P. Sigmon, and Ernest L. Stitzinger, Applications of Abstract Algebra
with Maple™ and MATLAB®, Second Edition

Richard E. Klima and Neil P. Sigmon, Cryptology: Classical and Modern with Maplets

Patrick Knupp and Kambiz Salari, Verification of Computer Codes in Computational Science
and Engineering

William L. Kocay and Donald L. Kreher, Graphs, Algorithms, and Optimization, Second Edition

Donald L. Kreher and Douglas R. Stinson, Combinatorial Algorithms: Generation Enumeration
and Search

Graph Searching Games and Probabilistic Methods

Anthony Bonato

Pralat Pawel

CRC Press
Taylor & Francis Group
Boca Raton London New York

CRC Press is an imprint of the
Taylor & Francis Group, an **informa** business

A CHAPMAN & HALL BOOK

CRC Press
Taylor & Francis Group
6000 Broken Sound Parkway NW, Suite 300
Boca Raton, FL 33487-2742

First issued in paperback 2022

© 2018 by Taylor & Francis Group, LLC
CRC Press is an imprint of Taylor & Francis Group, an Informa business

No claim to original U.S. Government works

Version Date: 20170925

ISBN 13: 978-1-03-247641-4 (pbk)
ISBN 13: 978-1-138-62716-1 (hbk)

DOI: 10.1201/9781315212135

This book contains information obtained from authentic and highly regarded sources. Reasonable efforts have been made to publish reliable data and information, but the author and publisher cannot assume responsibility for the validity of all materials or the consequences of their use. The authors and publishers have attempted to trace the copyright holders of all material reproduced in this publication and apologize to copyright holders if permission to publish in this form has not been obtained. If any copyright material has not been acknowledged please write and let us know so we may rectify in any future reprint.

Except as permitted under U.S. Copyright Law, no part of this book may be reprinted, reproduced, transmitted, or utilized in any form by any electronic, mechanical, or other means, now known or hereafter invented, including photocopying, microfilming, and recording, or in any information storage or retrieval system, without written permission from the publishers.

For permission to photocopy or use material electronically from this work, please access www.copyright.com (http://www.copyright.com/) or contact the Copyright Clearance Center, Inc. (CCC), 222 Rosewood Drive, Danvers, MA 01923, 978-750-8400. CCC is a not-for-profit organization that provides licenses and registration for a variety of users. For organizations that have been granted a photocopy license by the CCC, a separate system of payment has been arranged.

Trademark Notice: Product or corporate names may be trademarks or registered trademarks, and are used only for identification and explanation without intent to infringe.

Publisher's Note
The publisher has gone to great lengths to ensure the quality of this reprint but points out that some imperfections in the original copies may be apparent.

Visit the Taylor & Francis Web site at
http://www.taylorandfrancis.com

and the CRC Press Web site at
http://www.crcpress.com

Contents

List of Figures

List of Tables

Preface

The intersection of graph searching and probabilistic methods is a new topic within graph theory, with applications to graph searching problems such as the game of Cops and Robbers and its many variants, graph cleaning, Firefighting, and acquaintance time. Research on this topic emerged only over the last few years, and as such, it represents a rapidly evolving and dynamic area. Before we give a definition of this topic, we give some background on three of its key constituents: the probabilistic method, random graphs, and graph searching.

The *probabilistic method* is a powerful nonconstructive tool in mathematics. While it has found tremendous success in combinatorics and graph theory, it has been successfully applied to many other areas of mathematics (such as number theory, algebra, and analysis) as well as theoretical computer science (for example, randomized algorithms). As one of its goals, the method may prove the existence of an object with given properties without actually finding it. A *random graph* is a graph that is generated by some random process. Although technically a topic within the probabilistic method, random graphs are an important topic in their own right. The theory of random graphs lies at the intersection between graph theory and probability theory, and studies the properties of typical random graphs. Random graphs have also found a natural home in the study of real-world complex networks such as the web graph and on-line social networks.

Graph searching deals with the analysis of games and graph processes that model some form of intrusion in a network, and efforts to eliminate or contain that intrusion. For example, in Cops and Robbers, a robber is loose on the network, and a set of cops attempts to capture the robber. How the players move and the rules of capture depend on which variant is studied. In graph cleaning, the network begins as contaminated, and brushes move between vertices and along edges to clean them. There are many variants of graph searching studied in the literature, which are either motivated by problems in practice, or are inspired by foundational issues in computer science, discrete mathematics, and artificial intelligence, such as robotics, counterterrorism, and network security. In the past few years, a number of problems have emerged from applications related to the structure of real-world networks that are expected to be large-scale and dynamic, and where agents can be probabilistic, decentralized, and even selfish or antagonistic. This is one of the reasons why the field of graph searching is nowadays rapidly expanding. Several new models, problems, or approaches have appeared, relating it to diverse fields such as random walks, game theory, logic, probabilistic analysis, complex networks, mobile robotics, and distributed computing.

Graph searching games and probabilistic methods take two separate, but intertwined approaches: the study of graph searching games on random graphs and processes, and the use of the probabilistic method to prove results about deterministic graphs. We will see both approaches many times throughout the book. One of goals of this monograph is to bring the intersection of probabilistic methods and graph searching games into a place more readily visible to researchers. While we do not claim to make an exhaustive account, the material presented here is a survey of some main results in this new field. Our intended audience is broad, including both mathematicians and computer scientists. Since our approach is to be self-contained wherever possible, much of the material is accessible to students (mainly graduate but also advanced undergraduates) with some background in graph theory and probability.

We present eleven chapters that can be read in order, or the first three chapters read first, with the reader then moving to whichever chapter captures their interest. Chapter 1 supplies the required background and notation in graph theory, asymptotics, and discrete probability used throughout the remaining chapters. Chapter 2 focuses on one of the most popular graph searching games, Cops and Robbers. We discuss there the cop number of random graphs, properties of almost all cop-win and k-cop-win graphs. Variants of Cops and Robbers, where for example, the players play on edges or the robber can move at infinite speed, are considered in Chapter 3. We devote Chapter 4 to the new vertex pursuit game of Zombies and Survivors, where the zombies (the cops) appear randomly and always move along shortest paths to the survivor (the robber). In Chapter 5, we discuss one of the most important conjectures in the area, Meyniel's conjecture on the cop number. We summarize there some recent work on the proof of the conjecture for random graphs. Chapter 6 focuses on graph cleaning, and Chapter 7 discusses acquaintance time. Chapter 8 focuses the Firefighter graph process, and Chapter 9 focuses on acquisition number. In Chapter 10, we present topics on temporal parameters such as capture time. The final chapter presents a number of miscellaneous topics, ranging from Revolutionaries and Spies, robot crawler, and toppling number, to the game of Seepage played on acyclic directed graphs.

We would like to give a collective thanks to our co-authors, post-doctoral fellows, and students, who inspire us every day. Indeed, many of their results are highlighted throughout. Thanks to Deepak Bal, William Kinnersley, and Dieter Mitsche for their careful reading of early drafts of the book. Thank you to Doug West for supporting the book through his role as editor at CRC Press, and we highly valued the assistance received there from Bob Ross and Jose Soto. Last but certainly not least we would like to thank our families for their constant love and support: Douglas, Anna Maria, and Lisa and Anna, Piotr, Adam, and Julia.

Chapter 1

Introduction

The discussion in this first chapter will give us a common reference to present the results on the intersection of probabilistic methods and graph searching games. As the name suggests, this is a book on graphs and probability (we will deal with the searching part more explicitly in the following chapters). With a combinatorial audience in mind, we devote a few brief pages to summarize some notation from graph theory, and spend more time covering a few elementary but key theorems in discrete probability. An advanced reader may safely skim these pages and move directly to Chapter 2; this chapter may be used, nevertheless, as a quick reference for statements of key facts (like the Chernoff bounds) that we freely use later. More involved tools, such as martingales or the differential equations method, will be introduced in later chapters as needed.

Some basic notation comes first. The set of *natural numbers* (excluding 0 for notation simplicity, although this notation often includes 0) is written $\mathbb{N}$ while the *rationals* and *reals* are denoted by $\mathbb{Q}$ and $\mathbb{R}$, respectively. If n is a natural number, then define

$$[n] = \{1, 2, \ldots n\}.$$

The *Cartesian product* of two sets A and B is written $A \times B$. The difference of two sets A and B is written $A \backslash B$. We use the notation $\log n$ for the logarithm in the natural base.

1.1   Graphs

Graphs are our main objects of study. For further background in graph theory, the reader is directed to any of the texts [34, 76, 180].

A *graph* $G = (V, E)$ is a pair consisting of a *vertex set* $V = V(G)$, an *edge set* $E = E(G)$ consisting of pairs of vertices. Note that E is taken as a multiset, as its elements may occur more than once. We write uv if u and v form an edge, and say that u and v are *adjacent* or *joined*. For consistency, we will use the former term only. We refer to u and v as *endpoints* of the edge uv. The *order* of a graph is $|V(G)|$, and its *size* is $|E(G)|$. Graphs are often depicted by their drawings; see Figure 1.2.

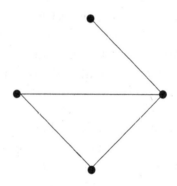

FIGURE 1.1: An example of a graph of order and size 4.

A *loop* is an edge whose endpoints are equal. *Multiple edges* are edges having the same pair of endpoints. If u and v are the endpoints of an edge, then we say that they are *neighbors*. The *neighborhood* $N(v) = N_G(v)$ of a vertex v is the set of all neighbors of v. We usually restrict our attention to *simple graphs*; that is, graphs without loops and multiple edges. Further, we only consider finite graphs.

The *degree* of a vertex v in G, written $\deg_G(v)$, is the number of neighbors of v in G; that is, $\deg_G(v) = |N(v)|$. We will drop the subscript G if the graph is clear from context. The number $\delta(G) = \min_{v \in V(G)} \deg(v)$ is the *minimum degree* of G, and the number $\Delta(G) = \max_{v \in V(G)} \deg(v)$ is the *maximum degree* of G. A graph is *k-regular* if each vertex has degree k.

The *complement* $\overline{G}$ of a graph G is the graph with vertex set $V(\overline{G}) = V(G)$ and edge set $E(\overline{G})$ defined by $uv \in E(\overline{G})$ if and only if $uv \notin E(G)$. See Figure 1.2. A *clique* (sometimes called a *complete graph*) is a set of pairwise-adjacent vertices. The clique of order n is denoted by K_n. An *independent set* (sometimes called an *empty graph*) is a set of pairwise-nonadjacent vertices. Note that an independent set is the complement of a clique.

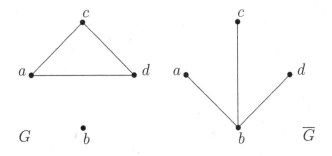

FIGURE 1.2: The graph G and its complement $\overline{G}$.

A graph G is *bipartite* if $V(G) = X \cup Y$, where $X \cap Y = \emptyset$, and every edge is of the form xy, where $x \in X$ and $y \in Y$; here X and Y are called *partite sets*. The *complete bipartite graph* $K_{m,n}$ is the graph with partite sets X, Y with $|X| = m$, $|Y| = n$, and edge set

$$E = \{xy : x \in X, y \in Y\}.$$

A graph $G' = (V', E')$ is a *subgraph* of $G = (V, E)$ if $V' \subseteq V$ and $E' \subseteq E$. We say that G' is a *spanning subgraph* if $V' = V$. If $V' \subseteq V$, then

$$G[V'] = (V', \{uv \in E : u, v \in V'\})$$

is the subgraph of G *induced* by V'. Similarly, if $E' \subseteq E$, then $G[E'] = (V', E')$ where

$$V' = \{v \in V : \text{there exists } e \in E' \text{ such that } v \in e\}$$

is an *induced subgraph of G by E'*. Given a graph $G = (V, E)$ and a vertex $v \in V$, we define $G - v = G[V \setminus \{v\}]$. For an edge e, $G - e$ is the subgraph formed by deleting e.

An *isomorphism* from a graph G to a graph H is a bijection $f : V(G) \to V(H)$ such that $uv \in E(G)$ if and only if $f(u)f(v) \in E(H)$. G is *isomorphic* to H, written $G \cong H$, if there is an isomorphism from G to H. See Figure 1.3 for two isomorphic graphs.

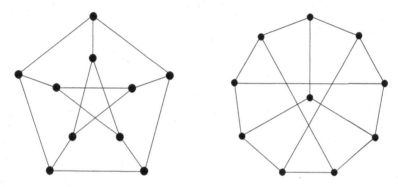

FIGURE 1.3: Two graphs isomorphic to the Petersen graph.

A *walk* in a graph $G = (V, E)$ from vertex u to vertex v is a sequence $W = (u = v_0, v_1, \ldots, v_l = v)$ if $v_i v_{i+1} \in E$ for $0 \le i < l$. The *length* $l(W)$ of a walk W is the number of vertices in W *minus 1* (that is, the number of edges). A walk is *closed* if $v_0 = v_l$. A *path* is a walk in which the internal vertices are distinct. The path of order n is denoted by P_n. A *cycle* is a closed path of length at least 3. We use the notation C_n for a cycle of order n. A graph G is *connected* if there is a walk (equivalently, a path) between every pair of vertices; otherwise, G is *disconnected*. See Figure 1.4. A *connected component* (or just *component*) of a graph G is a maximal connected subgraph. A connected component consisting of a single vertex is called an *isolated vertex*. A vertex adjacent to all other vertices is called *universal*.

A *forest* is a graph with no cycle. A *tree* is a connected forest; hence, every component of a forest is a tree. Each tree on n vertices has size $n-1$. An *end-vertex* is a vertex of degree 1; note that every *nontrivial* tree (that is, a tree of order at least 2) has at least two end-vertices. A *spanning tree* is a spanning subgraph that is a tree. The graph P_n and an n-vertex *star* $K_{1,n-1}$ are trees. A *hypercube* of dimension n, written Q_n, has vertices elements of $\{0,1\}^n$, with two vertices adjacent if they differ in exactly one coordinate. In particular, Q_n has order 2^n and size $n2^{n-1}$.

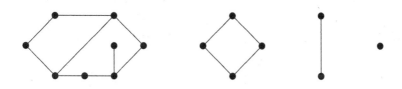

FIGURE 1.4: A disconnected graph with 4 components.

For distinct vertices u and v, the *distance* between u and v, written $d_G(u,v)$ (or just $d(u,v)$) is the length of a shortest path connecting u and v if such a path exists, and ∞, otherwise. We take the distance between a vertex and itself to be 0. The *diameter* of a connected graph G, written $\operatorname{diam}(G)$, is the maximum of all distances between vertices. If the graph is disconnected, then $\operatorname{diam}(G)$ is ∞. For a nonnegative integer r and vertex u in G, define $N_r(u)$ to be set of those vertices of distance r from u in G. Note that $N_0(u) = \{u\}$ and $N_1(u) = N(u)$.

The *breadth-first search* (BFS) process is a graph search algorithm that begins at the root vertex v and explores all the neighboring vertices. Then for each of those neighboring vertices, it explores their unexplored neighbors, and so on, until it explores the whole connected component containing vertex v. Formally, the algorithm starts by putting vertex v into a *FIFO queue*; that is, *First In, First Out*. In each round, one vertex is taken from the queue and all neighbors that have not yet been discovered are added to the queue. The process continues until the queue is empty. It may be shown that the BFS process naturally yields the breadth-first search tree.

A *dominating set* of a graph $G = (V, E)$ is a set $U \subseteq V$ such that every vertex $v \in V \setminus U$ has at least one neighbor in U. The *domination number* of G, written $\gamma(G)$, is the minimum cardinality of a dominating set in G. Note that the vertex set V is a dominating set. However, it is usually possible to find a much smaller dominating set (for example, consider a graph with a universal vertex).

A *matching* in a graph G is a 1-regular subgraph. A matching is *maximal* if it cannot be extended by adding an edge. A matching is *maximum* if it contains the largest possible number of edges. A *perfect matching* in a graph G is a matching in G that is a spanning subgraph of G.

The *line graph* of a graph G, written $L(G)$, is the graph whose vertices are the edges of G, with $ef \in E(L(G))$ when $e = uv$ and $f = vw$ are both in $E(G)$. See Figure 1.5 for an example. For

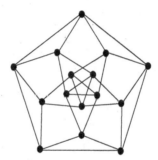

FIGURE 1.5: The line graph of the Petersen graph (see Figure 1.3).

graphs G and H, define the *Cartesian product* of G and H, written $G \square H$, to have vertices $V(G) \times V(H)$, and vertices (a, b) and (c, d) are adjacent if $a = c$ and $bd \in E(H)$ or $ac \in \mathbb{E}(G)$ and $b = d$.

We can assign a direction to each edge of a graph G. A simple *directed graph* (or *digraph*) $G = (V, E)$ is a pair consisting of a vertex set $V = V(G)$ and an edge set $E = E(G) \subseteq \{(x, y) : x, y \in V(G), x \neq y\}$. See Figure 1.6; we use the arrow notation to depict an edge pointing from vertex to vertex. The *in-degree* of a vertex

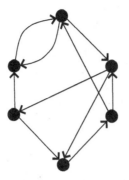

FIGURE 1.6: An example of a digraph.

v, written $\deg^-(v)$, is the number of in-neighbors of v; that is,

$$\deg^-(v) = |\{u \in V : (u, v) \in E\}|.$$

Similarly, the *out-degree* of a vertex v, written $\deg^+(v)$, is the number of out-neighbors of v; that is,

$$\deg^+(v) = |\{u \in V : (v, u) \in E\}|.$$

1.2 Probability

We next introduce some basic definitions and theorems from discrete probability theory. For more details and any proofs not provided here, see, for example, [8, 104].

Definitions

The set of possible outcomes of an experiment is called the *sample space* and is denoted by Ω. An *elementary event* is an event that contains only a single outcome in the sample space. For example, a coin is tossed. There are two possible outcomes: heads (H) and tails (T), so $\Omega = \{H, T\}$. We might be interested in the following events:

(i) the outcome is H,

(ii) the outcome is H or T,

(iii) the outcome is not H, and so on.

Note that we think of *events* as subsets of Ω.

A collection $\mathcal{F}$ of subsets of Ω is called a *σ-field* if it satisfies the following conditions:

(i) $\emptyset \in \mathcal{F}$,

(ii) if $A_1, A_2, \ldots \in \mathcal{F}$, then $\bigcup_i A_i \in \mathcal{F}$, and

(iii) if $A \in \mathcal{F}$, then $A^c \in \mathcal{F}$.

The smallest σ-field associated with Ω is the collection $\mathcal{F} = \{\emptyset, \Omega\}$. If A is any subset of Ω, then $\mathcal{F} = \{\emptyset, A, A^c, \Omega\}$ is a σ-field. The *power set* of Ω, which contains all subsets of Ω, is obviously a σ-field. For reasons beyond the scope of this book, when Ω is infinite, its power set is too large for probabilities to be assigned reasonably to all its members. Fortunately, we are going to deal with finite graphs and related structures only, and so from now on it will always be assumed that a σ-field is the power set of Ω.

A *probability measure* $\mathbb{P}$ on $(\Omega, \mathcal{F})$, where $\mathcal{F}$ is a σ-field, is a function $\mathbb{P} : \mathcal{F} \to [0,1]$ such that

(i) $\mathbb{P}(\Omega) = 1$, and

(ii) if $A_1, A_2, \ldots$ is a sequence of pairwise disjoint events, then

$$\mathbb{P}\left(\bigcup_i A_i\right) = \sum_i \mathbb{P}(A_i).$$

The triple $(\Omega, \mathcal{F}, \mathbb{P})$ is called a *probability space*.

Basic Properties

Below we present a few elementary properties of a probability space.

Theorem 1.2.1. *If $(\Omega, \mathcal{F}, \mathbb{P})$ is a probability space, then for any $A, B \in \mathcal{F}$ we have that* *(i)* $\mathbb{P}(\emptyset) = 0,$

(ii) $\mathbb{P}(A^c) = 1 - \mathbb{P}(A),$

(iii) if $A \subseteq B$, then $\mathbb{P}(A) \leq \mathbb{P}(B),$ *and*

(iv) $\mathbb{P}(A \cup B) = \mathbb{P}(A) + \mathbb{P}(B) - \mathbb{P}(A \cap B).$

The last equality can be generalized to the *Inclusion-Exclusion Principle*. Let $A_1, A_2, \ldots, A_n$ be events, where $n \geq 2$.

$$\mathbb{P}\left(\bigcup_{i=1}^{n} A_i\right) = \sum_i \mathbb{P}(A_i) - \sum_{i<j} \mathbb{P}(A_i \cap A_j)$$
$$+ \cdots + (-1)^{n+1}\mathbb{P}(A_1 \cap A_2 \cap \cdots \cap A_n).$$

If $\mathbb{P}(B) > 0$, then the *conditional probability* that A occurs given that B occurs is defined to be

$$\mathbb{P}(A|B) = \frac{\mathbb{P}(A \cap B)}{\mathbb{P}(B)}.$$

We state a number of facts about conditional probabilities that will be used (sometimes implicitly) in later discussions.

Theorem 1.2.2 (Law of total probabilities). *If $A \in \mathcal{F}$ is an event with $P(A) > 0$, and $\{A_i\}_{i=1}^{n}$ is a partition of A, then we have that*

$$\mathbb{P}(B|A) = \frac{1}{\mathbb{P}(A)} \sum_{i=1}^{n} \mathbb{P}(B|A_i)\mathbb{P}(A_i).$$

In particular, taking $A = \Omega$, we derive the following corollary.

Corollary 1.2.3. *If $\{A_i\}_{i=1}^{n}$ is a partition of the sample space Ω, then we have that*

$$\mathbb{P}(B) = \sum_{i=1}^{n} \mathbb{P}(B|A_i)\mathbb{P}(A_i).$$

We also have the following.

Theorem 1.2.4 (Chain law). *If $A_1, A_2, \ldots, A_n \in \mathcal{F}$, then for any event $B \in \mathcal{F}$ such that $\mathbb{P}\left(\left(\bigcap_{i=1}^{n-1} A_i\right) \cap B\right) > 0$, we have that*

$$\mathbb{P}\left(\bigcap_{i=1}^{n} A_i \Big| B\right) = \mathbb{P}(A_1|B) \cdot \mathbb{P}(A_2|A_1 \cap B)$$

$$\cdots \mathbb{P}\left(A_n \Big| \left(\bigcap_{i=1}^{n-1} A_i\right) \cap B\right).$$

In particular, taking $B = \Omega$, we obtain the following corollary.

Corollary 1.2.5 (Principle of deferred decision). *If we have $A_1, A_2, \ldots, A_n \in \mathcal{F}$ such that $\mathbb{P}\left(\bigcap_{i=1}^{n-1} A_i\right) > 0$, then*

$$\mathbb{P}\left(\bigcap_{i=1}^{n} A_i\right) = \mathbb{P}(A_1)\mathbb{P}(A_2|A_1) \cdots \mathbb{P}\left(A_n \Big| \bigcap_{i=1}^{n-1} A_i\right).$$

We will make use of the following equalities.

Theorem 1.2.6 (Bayes' law). *If $\{A_i\}_{i=1}^{n}$ is a partition of the sample space Ω and $B \in \mathcal{F}$, then for any $j \in [n]$*

$$\mathbb{P}(A_j|B) = \frac{\mathbb{P}(B|A_j)\mathbb{P}(A_j)}{\sum_{i=1}^{n} \mathbb{P}(B|A_i)\mathbb{P}(A_i)}.$$

Proof. Note that

$$\mathbb{P}(A_j|B) = \frac{\mathbb{P}(A_j \cap B)}{\mathbb{P}(B)} = \frac{\mathbb{P}(B|A_j)\mathbb{P}(A_j)}{\sum_{i=1}^{n} \mathbb{P}(B|A_i)\mathbb{P}(A_i)}. \qquad \square$$

The following elementary fact, also known as the *union bound*, proves useful.

Lemma 1.2.7 (Boole's inequality). *If $A_1, A_2, \ldots, A_n$ are events, then*

$$\mathbb{P}\left(\bigcup_{i=1}^{n} A_i\right) \leq \sum_{i=1}^{n} \mathbb{P}(A_i).$$

We also mention a generalization of Boole's inequality, known as Bonferroni's inequality (see Lemma 1.2.8). This inequality can be used to find upper and lower bounds on the probability of a finite union of events. Boole's inequality is recovered by setting $k = 1$. If $k = n$, then equality holds, and the resulting identity is the inclusion-exclusion principle.

Lemma 1.2.8 (Bonferroni inequalities). *Let*

$$B_k = \sum_{1 \le i_1 < i_2 < \cdots < i_k \le n} \mathbb{P}(A_{i_1} \cap A_{i_2} \cap \cdots \cap A_{i_k})$$

for all integers $k \in [n]$. Then for odd $k \in [n]$ we have that

$$\mathbb{P}\left(\bigcup_{i=1}^{n} A_i\right) \le \sum_{j=1}^{k} (-1)^{j-1} B_j,$$

and for even $k \in [n]$ we have that

$$\mathbb{P}\left(\bigcup_{i=1}^{n} A_i\right) \ge \sum_{j=1}^{k} (-1)^{j-1} B_j.$$

Useful Distributions

Here are some discrete probability distributions that we use throughout.

(i) *Bernoulli(p):* Fix $p \in (0, 1)$.

$$\mathbb{P}(X = x) = \begin{cases} p, & \text{if } x = 1; \\ 1 - p, & \text{if } x = 0. \end{cases}$$

Here p is the *success probability* and $1 - p$ is the *failure probability*.

(ii) *Binomial(n, p)* (we will use Bin(n, p) instead): Fix $p \in (0, 1)$ and $n \in \mathbb{N}$. Let X be the number of successes in n independent repetitions of the same Bernoulli(p) trial. Then we have that

$$\mathbb{P}(X = k) = \binom{n}{k} p^k (1 - p)^{n-k}, \qquad 0 \le k \le n.$$

(iii) *Hypergeometric*(N, K, n): Let $N, K, n, k \in \mathbb{N}$. The hypergeometric distribution describes the probability of k successes in n draws, without replacement, from a population of size N that contains exactly K successes, wherein each draw is either a success or a failure. A random variable X follows the hypergeometric distribution if

$$\mathbb{P}(X = k) = \frac{\binom{K}{k}\binom{N-K}{n-k}}{\binom{N}{n}}.$$

(iv) *Geometric*(p): Fix $p \in (0, 1)$. Let X be the *waiting time* (that is, the number of trials) for the first success in independent repetitions of the same Bernoulli(p) trial. Then it follows that

$$\mathbb{P}(X = k) = (1 - p)^{k-1}p, \qquad k = 1, 2, \ldots.$$

Note that $\mathbb{P}(X > k) = (1-p)^k$. Note also that the geometric distribution is "memory-less"; that is, for $r > k$,

$$\begin{aligned}
\mathbb{P}(X > r \mid X > k) &= \frac{\mathbb{P}(X > r)}{\mathbb{P}(X > k)} = \frac{(1-p)^r}{(1-p)^k} \\
&= (1-p)^{r-k} = \mathbb{P}(X > r - k).
\end{aligned}$$

(v) *Poisson*(λ) (we will use Po(λ) instead): Fix $\lambda > 0$.

$$\mathbb{P}(X = k) = \frac{\lambda^k}{k!}e^{-\lambda}, \qquad k = 0, 1, 2, \ldots.$$

1.3   Asymptotic Notation and Useful Inequalities

Since many of our results will be asymptotic, in this section we recall some asymptotic notation and some inequalities that we frequently use.

Asymptotic Notation

Let $f(n)$ and $g(n)$ be two functions whose domain is some fixed subset of $\mathbb{R}$, and assume that $g(n) > 0$ for all n. We say that *f is of order at most g*, written $f(n) = O(g(n))$, if there exist constants $A > 0$ and $N > 0$ such that for all $n > N$, we have that

$$|f(n)| \leq A|g(n)|.$$

Observe that $f(n)$ could be negative or even oscillate between negative and positive values (for example, $2\sin(3n)n^2 = O(n^2)$). We say that *f is of order at least g*, written $f(n) = \Omega(g(n))$, if there exist constants $A > 0$ and $N > 0$ such that for all $n > N$,

$$f(n) \geq Ag(n).$$

Finally, we say that *f is of order g*, written $f(n) = \Theta(g(n))$, if $f(n) = O(g(n))$ and $f(n) = \Omega(g(n))$.

Note that for a polynomial $p(n)$ of degree k, $p(n) = \Theta(n^k)$. Here are some useful properties of the O- and Ω- notation, which are by no means exhaustive.

Theorem 1.3.1. *For positive functions $f(n)$ and $g(n)$ we have the following.*

(i) $O(f(n)) + O(g(n)) = O(f(n) + g(n))$.

(ii) $f(n)O(g(n)) = O(f(n)g(n))$ *and*
$f(n)\Omega(g(n)) = \Omega(f(n)g(n))$.

(iii) *If $f(n) = O(1)$, then $f(n)$ is bounded by a constant.*

(iv) $n^r = O(n^s)$ *for any real numbers r, s with $r \leq s$.*

(v) $n^r = O(a^n)$ *for any real numbers r, a with $a > 1$.*

(vi) $\log n = O(n^r)$ *for any real number $r > 0$ (note that this could be the logarithm to any base, since $\log_a x = \frac{\log_b x}{\log_b a} = \Theta(\log_b x)$ for any $a, b > 1$).*

(vii) $\log \log n = O(\log n)$ *and* $\log \log \log n = O(\log \log n)$.

We say that *f is of order smaller than g*, written $f(n) = o(g(n))$ or $f(n) \ll g(n)$, if

$$\lim_{n \to \infty} \frac{f(n)}{g(n)} = 0.$$

Note that we do not control the sign of function $f(n)$ while in the next definition we do. The function f is *of order larger than g,* written $f(n) = \omega(g(n))$ or $f(n) \gg g(n)$, if

$$\lim_{n \to \infty} \frac{f(n)}{g(n)} = \infty.$$

Finally, f is *asymptotically equal* to g, written $f(n) \sim g(n)$ or $f(n) = (1 + o(1))g(n)$, if

$$\lim_{n \to \infty} \frac{f(n)}{g(n)} = 1.$$

For more details about asymptotic notation see, for example, [177].

Useful Inequalities

We collect a few inequalities that will be used throughout. For all $x \in \mathbb{R}$, $e^x \geq 1 + x$. Hence, for all $x \in \mathbb{R}$ that are positive, we have that $\log(1 + x) \leq x$. For the factorial function $n!$, we have

$$\left(\frac{n}{e}\right)^n \leq n! \leq en \left(\frac{n}{e}\right)^n.$$

Stirling's formula, as stated in the following lemma, is often useful for our estimates.

Lemma 1.3.2.
$$n! \sim \sqrt{2\pi n} \left(\frac{n}{e}\right)^n.$$

More precisely,

$$n! = \sqrt{2\pi n} \left(\frac{n}{e}\right)^n e^{\lambda_n}, \text{ with } \frac{1}{12n + 1} < \lambda_n < \frac{1}{12n},$$

so

$$n! = \sqrt{2\pi n} \left(\frac{n}{e}\right)^n \left(1 + \frac{1}{12n} + \frac{1}{288n^2} + O(n^{-3})\right).$$

For the binomial coefficient $\binom{n}{k}$, we have the inequalities

$$\left(\frac{n}{k}\right)^k \leq \binom{n}{k} \leq \left(\frac{en}{k}\right)^k.$$

For the middle binomial coefficient $\binom{2m}{m}$, we have the better estimate

$$\frac{2^{2m}}{2\sqrt{m}} \leq \binom{2m}{m} \leq \frac{2^{2m}}{\sqrt{2m}},$$

and from Stirling's formula the asymptotic behavior may be obtained as follows:

$$\binom{2m}{m} = \frac{(2m)!}{(m!)^2} \sim \frac{\sqrt{4\pi m}(2m/e)^{2m}}{\left(\sqrt{2\pi m}(m/e)^m\right)^2} = \frac{2^{2m}}{\sqrt{\pi m}}.$$

1.4 Random Graphs

In this section, we supply five examples of probability spaces. These notions of *random graphs*, especially the second and to a lesser extent the third and fourth, will be central to our discussion.

Let Ω be the family of all graphs with n vertices and exactly M edges, $0 \leq M \leq \binom{n}{2}$. To every graph $G \in \Omega$ we assign a uniform probability; that is,

$$\mathbb{P}(\{G\}) = \binom{\binom{n}{2}}{M}^{-1}.$$

We denote this associated probability space by $\mathbb{G}(n, M)$.

Now, let $0 \leq p \leq 1$ and let Ω be the family of all graphs on n vertices. To every graph $G \in \Omega$ we assign a probability

$$\mathbb{P}(\{G\}) = p^{|E(G)|}(1 - p)^{\binom{n}{2}-|E(G)|}.$$

Note that this indeed is a probability measure, since

$$\sum_G \mathbb{P}(\{G\}) = \sum_{m=0}^{\binom{n}{2}} \binom{\binom{n}{2}}{m} p^m(1 - p)^{\binom{n}{2}-m} = (p + (1 - p))^{\binom{n}{2}} = 1.$$

We denote this probability space by $\mathbb{G}(n, p)$. The space $\mathbb{G}(n, p)$ is often referred to as the *binomial random graph* or *Erdős-Rényi random graph.* Note also that this probability space can informally be viewed as a result of $\binom{n}{2}$ independent coin flips, one for each pair of vertices u, v, where the probability of success (that is, adding an edge uv) is equal to p. Let us also note that if $p = 1/2$, then $\mathbb{P}(\{G\}) = 2^{-\binom{n}{2}}$ for any graph G on n vertices. We obtain a uniform probability space.

Next, let Ω be the family of all d-regular graphs on n vertices, where $0 \leq d \leq n - 1$ and dn is even. (Note that the condition dn is needed; otherwise, $\Omega = \emptyset$.) To every graph $G \in \Omega$ we assign a uniform probability; that is,

$$\mathbb{P}(\{G\}) = \frac{1}{|\Omega|} \, .$$

We refer to this space as the *random regular graph of degree d*, and write $\mathbb{G}_{n,d}$.

As typical in random graph theory, we shall consider only asymptotic properties of $\mathbb{G}(n, M)$ and $\mathbb{G}(n, p)$ as $n \to \infty$, where $M = M(n)$ and, respectively, $p = p(n)$ may and usually do depend on n. For $\mathbb{G}_{n,d}$ we typically concentrate on d being a constant but it is also interesting to consider $d = d(n)$ tending to infinity with n. We say that an event in a probability space holds *asymptotically almost surely* (*a.a.s.*) if its probability tends to one as n goes to infinity. For mode details see, for example, the two classic books [35, 115] or more recent monograph [94]. Random d-regular graphs are also discussed in the survey [184].

Finally, we introduce the *random geometric graph* $\mathbb{G}(\mathcal{X}_n, r_n)$, where (i) $\mathcal{X}_n$ is a set of n points located independently uniformly at random in $[0, \sqrt{n}]^2$, (ii) $(r_n)_{n \geq 1}$ is a sequence of positive real integers, and (iii) for $\mathcal{X} \subseteq \mathbb{R}^2$ and $r > 0$, the graph $\mathbb{G}(\mathcal{X}, r)$ is defined to have vertex set $\mathcal{X}$, with two vertices connected by an edge if and only if their spatial locations are at Euclidean distance at most r from each other. As before, we shall consider only asymptotic properties of $\mathbb{G}(\mathcal{X}_n, r_n)$ as $n \to \infty$. We will therefore write $r = r_n$, we will identify vertices with their spatial locations, and we will define $\mathbb{G}(n, r)$ as the graph with vertex set $[n]$ corresponding to n locations chosen independently uniformly at random in

$[0, \sqrt{n}]^2$ and a pair of vertices within Euclidean distance r appear as an edge.

We will also consider the *percolated random geometric graph* $\mathbb{G}(n, r, p)$, which is defined as a random graph with vertex set $[n]$ corresponding to n locations chosen independently uniformly at random in $[0, \sqrt{n}]^2$, and for each pair of vertices within Euclidean distance at most r we flip a biased coin with success probability p to determine whether there is an edge (independently for each such a pair, and pairs at distance bigger than r never share an edge). In particular, for $p = 1$ we simply have the random geometric graph $\mathbb{G}(n, r)$. Percolated random geometric graphs were recently studied by Penrose [154] under the name *soft random geometric graphs*.

Alternatively, we can scale the space and define the model in $[0, 1]^2$. Of course, results from one model can be translated to the other one and we will make sure that it is always clear which model we have in mind. For more details, see, for example, the monograph [155].

We note that we use the notations $\mathbb{G}(n, p)$, $\mathbb{G}(n, M)$, and $\mathbb{G}(n, r)$, but this will not cause confusion as it will be clear from the context which of the corresponding three models we are discussing.

1.5 Tools: First and Second Moment Methods

Events A, B are *independent* if

$$\mathbb{P}(A \cap B) = \mathbb{P}(A)\mathbb{P}(B).$$

In general, events $A_1, A_2, \ldots, A_n$ are *independent* if for any $I \subseteq [n]$,

$$\mathbb{P}\left(\bigcap_{i \in I} A_i\right) = \prod_{i \in I} \mathbb{P}(A_i).$$

Intuitively, the property of independence means that the knowledge of whether some of the events $A_1, A_2, \ldots, A_n$ occurred does not affect the probability that the remaining events occur.

A *random variable* on a probability space $(\Omega, \mathcal{F}, \mathbb{P})$ is a function $X : \Omega \to \mathbb{R}$ that is $\mathcal{F}$-measurable; that is, for any $x \in \mathbb{R}$,

$$\{\omega \in \Omega : X(\omega) \le x\} \in \mathcal{F}.$$

Random variables X, Y are *independent* if for every pair of events $\{X \in A\}$, $\{Y \in B\}$, where $A, B \subseteq \mathbb{R}$, we have that

$$\mathbb{P}(\{X \in A\} \cap \{Y \in B\}) = \mathbb{P}(\{X \in A\})\mathbb{P}(\{Y \in B\}).$$

Thus, two random variables are independent if and only if the events related to those random variables are independent events.

The First Moment Method

Let $(\Omega, \mathcal{F}, \mathbb{P})$ be a finite probability space. The *expectation* of a random variable X is defined as

$$\mathbb{E}[X] = \sum_{\omega \in \Omega} \mathbb{P}(\omega) X(\omega).$$

A simple but useful property of expectation is the following.

Lemma 1.5.1 (Linearity of expectation). *For any two random variables X, Y and $a, b \in \mathbb{R}$, we have that*

$$\mathbb{E}[aX + bY] = a\mathbb{E}[X] + b\mathbb{E}[Y].$$

Linearity of expectation implies that for all random variables $X_1, X_2, \ldots, X_n$ and all $c_1, c_2, \ldots, c_n \in \mathbb{R}$,

$$\mathbb{E}\left[\sum_{i=1}^{n} c_i X_i\right] = \sum_{i=1}^{n} c_i \mathbb{E}[X_i].$$

This is a simple observation, but a powerful one. It is important to point out that it holds for both dependent and independent random variables.

For an event $A \in \mathcal{F}$, we define the *indicator random variable* as follows:

$$I_A(\omega) = \begin{cases} 1, & \text{if } \omega \in A, \\ 0, & \text{otherwise.} \end{cases}$$

It is evident that

$$\mathbb{E}[I_A] = \sum_{\omega \in \Omega} \mathbb{P}(\omega) I_A(\omega) = \sum_{\omega \in A} \mathbb{P}(\omega) = \mathbb{P}(A).$$

It is common that a random variable can be expressed as a sum of indicators. In such case, the expected value can also be expressed as a sum of expectations of corresponding indicators.

The first moment method that we are about to introduce is a standard tool used in investigating random graphs and the probabilistic method. It is a useful tool to bound the probability that a random variable X satisfies $X \geq 1$.

Theorem 1.5.2 (Markov's inequality). *If $(\Omega, \mathcal{F}, \mathbb{P})$ is a probability space, and X is a nonnegative random variable, then for all $\varepsilon > 0$,*

$$\mathbb{P}(X \geq \varepsilon) \leq \frac{\mathbb{E}[X]}{\varepsilon}.$$

Proof. Note that

$$X = X I_{X \geq \varepsilon} + X I_{X < \varepsilon} \geq X I_{X \geq \varepsilon} \geq \varepsilon I_{X \geq \varepsilon}.$$

Hence, by linearity of expectation, we have that

$$\mathbb{E}[X] \geq \varepsilon \mathbb{E}[I_{X \geq \varepsilon}] = \varepsilon \mathbb{P}(X \geq \varepsilon),$$

and the theorem holds. □

Markov's inequality has a simple corollary, proved by setting $\varepsilon = 1$.

Corollary 1.5.3 (The first moment method). *If X is a nonnegative integer-valued random variable, then*

$$\mathbb{P}(X > 0) \leq \mathbb{E}[X].$$

The Second Moment Method

Next, we will use the variance to bound the probability that a random variable X satisfies $X = 0$. The second moment method that we are about to introduce is another standard tool used in investigating random graphs. Let X be a random variable. The *variance* of X is defined as

$$\mathrm{Var}[X] = \mathbb{E}[(X - \mathbb{E}X)^2] = \mathbb{E}[X^2] - (\mathbb{E}[X])^2,$$

that is,

$$\mathrm{Var}[X] = \sum_{i=1}^{\infty} (x_i - \mathbb{E}[X])^2 \mathbb{P}(X = x_i).$$

We collect some elementary properties of the variance. For random variables X and Y, the *covariance* $\mathbb{C}\mathrm{ov}(X, Y)$ is defined as

$$\mathbb{C}\mathrm{ov}(X, Y) = \mathbb{E}[XY] - (\mathbb{E}[X])(\mathbb{E}[Y]).$$

Theorem 1.5.4. *Let X and Y be random variables, and let a and b be real numbers. The variance operator has the following properties.*

(i) $\mathrm{Var}[X] \geq 0$.

(ii) $\mathrm{Var}[aX + b] = a^2 \mathrm{Var}[X]$.

(iii)

$$
\begin{aligned}
\mathrm{Var}[X + Y] &= \mathbb{C}\mathrm{ov}(X, X) + \mathbb{C}\mathrm{ov}(Y, Y) + 2\mathbb{C}\mathrm{ov}(X, Y) \\
&= \mathrm{Var}[X] + \mathrm{Var}[Y] + 2\mathbb{C}\mathrm{ov}(X, Y).
\end{aligned}
$$

(iv) *If X, Y are independent, then*

$$\mathrm{Var}[X + Y] = \mathrm{Var}[X] + \mathrm{Var}[Y].$$

The following theorem is a key tool to achieve our goal.

Theorem 1.5.5 (Chebyshev's inequality). *Let $(\Omega, \mathcal{F}, \mathbb{P})$ be a probability space. If X is a random variable, then for any $\varepsilon > 0$,*

$$\mathbb{P}\left(|X - \mathbb{E}[X]| \geq \varepsilon \right) \leq \frac{\mathrm{Var}[X]}{\varepsilon^2}.$$

Proof. By Markov's inequality

$$\mathbb{P}(|Y| \geq \varepsilon) = \mathbb{P}(Y^2 \geq \varepsilon^2) \leq \frac{\mathbb{E}[Y^2]}{\varepsilon^2} .$$

Setting $Y = X - \mathbb{E}[X]$ we derive the desired assertion. $\square$

In combinatorial applications of probability, the following consequence of Chebyshev's inequality plays an important role, as we will develop in later chapters.

Theorem 1.5.6 (The second moment method). *If X is a nonnegative integer-valued random variable, then*

$$\mathbb{P}(X = 0) \leq \frac{\mathrm{Var}[X]}{(\mathbb{E}[X])^2} = \frac{\mathbb{E}[X^2]}{(\mathbb{E}[X])^2} - 1 .$$

The second moment method can easily be strengthened using the Cauchy-Schwarz inequality.

Theorem 1.5.7 (The Cauchy-Schwarz inequality). *For all vectors* $\mathbf{x} = (x_1, x_2, \ldots, x_n)$ *and* $\mathbf{y} = (y_1, y_2, \ldots, y_n)$ *in* $\mathbb{R}^n$, *we have that*

$$\left(\sum_{i=1}^{n} x_i y_i \right)^2 \leq \left(\sum_{i=1}^{n} x_i^2 \right) \left(\sum_{i=1}^{n} y_i^2 \right) .$$

Theorem 1.5.8 (The strong second moment method). *If X is a nonnegative integer-valued random variable, then*

$$\mathbb{P}(X = 0) \leq \frac{\mathrm{Var}[X]}{\mathbb{E}[X^2]} = 1 - \frac{(\mathbb{E}[X])^2}{\mathbb{E}[X^2]} .$$

Proof. Note that $X = X \cdot I_{X>0}$. But

$$\begin{aligned}
(\mathbb{E}[X])^2 &= \left(\mathbb{E}[X \cdot I_{X>0}] \right)^2 \leq \mathbb{E}[X^2] \cdot \mathbb{E}[I_{X>0}^2] \\
&= \mathbb{E}[X^2] \cdot \mathbb{E}[I_{X>0}] = \mathbb{E}[X^2] \cdot \mathbb{P}(X > 0)
\end{aligned}$$

yields

$$\mathbb{P}(X = 0) = 1 - \mathbb{P}(X > 0) \leq 1 - \frac{(\mathbb{E}[X])^2}{\mathbb{E}[X^2]} = \frac{\mathrm{Var}[X]}{\mathbb{E}[X^2]} .$$

The last equality follows immediately from the definition of the variance. $\square$

The bound in Theorem 1.5.8 is better than the bound in Theorem 1.5.6, since $\mathbb{E}[X^2] \geq (\mathbb{E}[X])^2$. For many applications, however, these bounds are equally powerful.

Examples of Both Methods

We will be using both the first and second moment methods many times in this book, especially the former. For illustrative purposes, we conclude this section with simple examples of both methods. In particular, we will prove that $p = \log n/n$ is the threshold for the disappearance of isolated vertices in $\mathbb{G}(n, p)$.

First, we will show that a.a.s. there is no isolated vertex in a random graph $\mathbb{G}(n, p)$ for

$$p = p(n) = \frac{\ln n + \omega(n)}{n},$$

where $\omega(n)$ is any function tending to infinity. (The case when $\omega(n)$ tends to a constant $c \in \mathbb{R}$ will be discussed later.) Let I_i denote the indicator random variable for the event when the vertex i is isolated, where $i \in [n]$. The number of isolated vertices in $\mathbb{G}(n, p)$ is $X = \sum_{i=1}^{n} I_i$. Since for every $i \in [n]$

$$\mathbb{P}(I_i = 1) = (1 - p)^{n-1} \sim \exp\left(-\ln n - \omega(n)\right) = \frac{e^{-\omega(n)}}{n},$$

$\mathbb{E}[X] \sim e^{-\omega(n)} \to 0$ as $n \to \infty$, and the claim holds by the first moment method.

Now, we will show that a.a.s. there is at least one isolated vertex in a random graph $\mathbb{G}(n, p)$ for

$$p = p(n) = \frac{\ln n - \omega(n)}{n}.$$

In this case, $\mathbb{E}[X] \sim e^{\omega(n)} \to \infty$ as $n \to \infty$, and

$$
\begin{aligned}
\mathrm{Var}[X] &= \mathrm{Var}\left[\sum_{i=1}^{n} I_i\right] = \sum_{1 \leq i,j \leq n} \mathrm{Cov}(I_i, I_j) \\
&= \sum_{1 \leq i,j \leq n} (\mathbb{E}[I_i I_j] - (\mathbb{E}[I_i])(\mathbb{E}[I_j])).
\end{aligned}
$$

Hence,

$$\mathbb{Var}[X] = \sum_{1 \le i, j \le n, i \ne j} \left(\mathbb{P}(I_i = 1, I_j = 1) - (\mathbb{P}(I_i = 1))^2 \right)$$
$$+ \sum_{i=1}^{n} \left(\mathbb{P}(I_i = 1) - (\mathbb{P}(I_i = 1))^2 \right).$$

The second term in the last sum can be dropped to derive the following bound:

$$\mathbb{Var}[X] \le \sum_{1 \le i, j \le n, i \ne j} \left((1-p)^{2n-3} - (1-p)^{2n-2} \right) + \mathbb{E}[X]$$
$$= \sum_{1 \le i, j \le n, i \ne j} (1-p)^{2n-3}(1 - (1-p)) + \mathbb{E}[X]$$
$$\sim \sum_{1 \le i, j \le n, i \ne j} \frac{e^{2\omega(n)}}{n^2} p + \mathbb{E}[X] \sim e^{2\omega(n)} p + \mathbb{E}[X].$$

Hence, from the second moment method we derive that

$$\mathbb{P}(X = 0) \le \frac{\mathbb{Var}[X]}{(\mathbb{E}[X])^2} \le (1 + o(1))p + \frac{1}{\mathbb{E}[X]} = o(1).$$

1.6 Tools: Chernoff Bounds

Suppose that S is a random variable and $t > 0$. We would like to find the upper and lower tails of the distribution; that is, bounds for $\mathbb{P}(S \ge \mathbb{E}[S] + t)$ and $\mathbb{P}(S \le \mathbb{E}[S] - t)$. Let $u \ge 0$. Then

$$\mathbb{P}(S \ge \mathbb{E}[S] + t) = \mathbb{P}(e^{uS} \ge e^{u(\mathbb{E}[S]+t)}) \le e^{-u(\mathbb{E}[S]+t)} \mathbb{E}[e^{uS}],$$

by Markov's inequality. Similarly, for $u \le 0$,

$$\mathbb{P}(S \le \mathbb{E}[S] - t) \le e^{-u(\mathbb{E}[S]-t)} \mathbb{E}[e^{uS}].$$

Combining these inequalities, we obtain a bound for $\mathbb{P}(|S - \mathbb{E}[S]| \ge t)$.

Now, let $S_n = \sum_{i=1}^n X_i$, where $X_i, i \in [n]$ are independent random variables. Then for $u \geq 0$, we have that

$$\mathbb{P}(S_n \geq \mathbb{E}[S_n] + t) \leq e^{-u(\mathbb{E}[S_n]+t)} \prod_{i=1}^n \mathbb{E}[e^{uX_i}],$$

whereas for $u \leq 0$, we have that

$$\mathbb{P}(S_n \leq \mathbb{E}[S_n] - t) \leq e^{-u(\mathbb{E}[S_n]-t)} \prod_{i=1}^n \mathbb{E}[e^{uX_i}].$$

After calculating $\mathbb{E}[e^{uX_i}]$ and finding the value of u that minimizes the right side, we derive the desired bound.

To illustrate this general approach, we focus on Bernoulli(p) random variables. Then S_n has a $\mathrm{Bin}(n, p)$ distribution with expectation $\mu = \mathbb{E}[S_n] = np$. For $u \geq 0$, we have that

$$\mathbb{P}(S_n \geq \mu + t) \leq e^{-u(\mu+t)}\left(pe^u + (1-p)\right)^n.$$

To minimize the right side, we take

$$e^u = \frac{(\mu + t)(1-p)}{(n - \mu - t)p}.$$

Hence, assuming that $\mu + t < n$,

$$\mathbb{P}(S_n \geq \mu + t) \leq \left(\frac{\mu}{\mu + t}\right)^{\mu+t}\left(\frac{n - \mu}{n - \mu - t}\right)^{n-\mu-t},$$

whereas for $\mu + t > n$ this probability is zero.

Now, let

$$\varphi(x) = \begin{cases} (1+x)\log(1+x) - x, & x > -1, \\ \infty, & \text{otherwise.} \end{cases}$$

For $0 \leq t < n - \mu$, we have that

$$\begin{aligned} \mathbb{P}(S_n \geq \mu + t) &\leq \exp\left(-\mu\varphi\left(\frac{t}{\mu}\right) - (n - \mu)\varphi\left(\frac{-t}{n - \mu}\right)\right) \\ &\leq e^{-\mu\varphi(t/\mu)}, \end{aligned}$$

since $\varphi(x) \geq 0$ for every x. By a similar argument, for $0 \leq t < \mu$ we obtain that

$$\mathbb{P}(S_n \leq \mu - t) \leq \exp\left(-\mu\varphi\left(\frac{-t}{\mu}\right) - (n-\mu)\varphi\left(\frac{t}{n-\mu}\right)\right)$$
$$\leq e^{-\mu\varphi(-t/\mu)}.$$

Now, observe that $\varphi(0) = 0$ and

$$\varphi'(x) = \log(1+x) \leq x = \left(x^2/2\right)'.$$

Thus, $\varphi(x) \geq x^2/2$ for $-1 \leq x \leq 0$. Further, $\varphi'(0) = 0$ and

$$\varphi''(x) = \frac{1}{1+x} \geq \frac{1}{(1+x/3)^3} = \left(\frac{x^2}{2(1+x/3)}\right)'',$$

so for $x \geq 0$

$$\varphi(x) \geq \frac{x^2}{2(1+x/3)}.$$

The functions $\varphi(x)$ (brown), $\frac{x^2}{2}$ (in green), and $\frac{x^2}{2(1+x/3)}$ (in red) are presented in Figure 1.7.

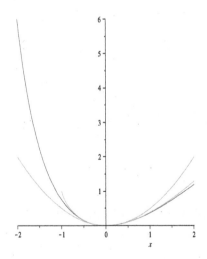

FIGURE 1.7: The functions: $\varphi(x)$, $\frac{x^2}{2}$, and $\frac{x^2}{2(1+x/3)}$.

Therefore, we arrive at the following result.

Theorem 1.6.1 (Chernoff bounds; see, for example, [115]). *If S_n is a random variable with the binomial distribution* $\mathrm{Bin}(n, p)$ *and* $\mu = \mathbb{E}[S_n] = np$, *then for* $t \geq 0$ *we have that*

$$\mathbb{P}(S_n \geq \mu + t) \;\leq\; \exp\left(-\frac{t^2}{2(\mu + t/3)}\right) \quad and$$

$$\mathbb{P}(S_n \leq \mu - t) \;\leq\; \exp\left(-\frac{t^2}{2\mu}\right).$$

The following corollary is sometimes more convenient.

Corollary 1.6.2 ([115]). *If S_n is a random variable with the binomial distribution* $\mathrm{Bin}(n, p)$ *and* $\mu = \mathbb{E}[S_n] = np$, *then for* $\varepsilon \leq 3/2$ *we have that*

$$\mathbb{P}(|S_n - \mu| \geq \varepsilon\mu) \leq 2\exp\left(-\frac{\varepsilon^2}{3}\mu\right).$$

We mention another, sometimes useful version of the Chernoff bounds.

Theorem 1.6.3 ([8]). *If S_n is a random variable with the binomial distribution* $\mathrm{Bin}(n, p)$ *and* $\mu = \mathbb{E}[S_n] = np$, *then for* $a > 0$ *we have that*

$$\mathbb{P}(S_n > \mu + a) < e^{-2a^2/n} \quad and \quad \mathbb{P}(S_n < \mu - a) < e^{-2a^2/n}.$$

In addition, all of the above bounds hold for the general case in which $X_i \in \mathrm{Bernoulli}(p_i)$ with (possibly) different p_i. Indeed, we can repeat all calculations with the only difference being that now

$$\prod_{i=1}^{n} \mathbb{E}[e^{uX_i}] = \prod_{i=1}^{n} \left(p_i e^u + (1 - p_i)\right).$$

We need the following inequality known as the *arithmetic-geometric mean inequality*.

Lemma 1.6.4. *For all sequences of nonnegative numbers* $(a_1, a_2, \ldots, a_n)$ *we have that*

$$\frac{1}{n}\sum_{i=1}^{n} a_i \geq \left(\prod_{i=1}^{n} a_i\right)^{1/n}.$$

Using the arithmetic-geometric mean inequality we derive that

$$\prod_{i=1}^{n} \mathbb{E}[e^{uX_i}] \leq \left(\frac{1}{n}\sum_{i=1}^{n}\left(p_i e^u + (1-p_i)\right)\right)^n = \left(pe^u + (1-p)\right)^n,$$

where $p = \sum_{i=1}^{n} p_i/n$. This is exactly the same expression as we had before with p taken as the arithmetic mean of the p_i's. The rest of the proof is not affected.

Finally, recall that the hypergeometric distribution describes the probability of k successes in n draws, without replacement, from a population of size N that contains exactly K successes, wherein each draw is either a success or a failure. Note that drawing with replacement would yield a binomial random variable. It seems reasonable that drawing without replacement tends to produce smaller random fluctuations, and indeed the bounds obtained above (Theorem 1.6.1 and Corollary 1.6.2) still hold for *Hypergeometric*(N, K, n) with $\mu = nK/N$. For more details, see, for example [115].

Lower Bound

Until now, we have focused on bounding from above the probability that a random variable is far away from the expectation. However, sometimes we are more interested in bounding the probability of this rare event from below. The well-known Central Limit Theorem suggests that the distribution of the sum of many independent random variables is approximately normal, and so the bounds we obtained earlier should not be far from the truth. This is actually the case under general circumstances.

Theorem 1.6.5. *If S_n is a random variable with the binomial distribution* Bin$(n, 1/2)$ *and* $\mathbb{E}[S_n] = n/2 = \mu$, *then for any integer* $t \in [0, n/8]$ *we have that*

$$\mathbb{P}(S_n \geq \mu + t) \geq \frac{1}{15}e^{-16t^2/n}.$$

Such general and precise bounds can be found in [85]. We use some elementary calculations to prove Theorem 1.6.5.

Proof of Theorem 1.6.5. We note that

$$\mathbb{P}(S_n \geq n/2 + t) = \sum_{j=t}^{n/2} \mathbb{P}(S_n = n/2 + j)$$

$$= 2^{-n} \sum_{j=t}^{n/2} \binom{n}{n/2+j} \geq 2^{-n} \sum_{j=t}^{2t-1} \binom{n}{n/2+j}$$

$$= 2^{-n} \sum_{j=t}^{2t-1} \binom{n}{n/2} \frac{n/2}{n/2+j} \cdot \frac{n/2-1}{n/2+j-1}$$

$$\cdots \frac{n/2-j+1}{n/2+1}.$$

As $\binom{n}{n/2} \geq 2^n/2\sqrt{n/2} = 2^n/\sqrt{2n}$ (see Lemma 1.3.2), we derive that

$$\mathbb{P}(S_n \geq n/2 + t) \geq \frac{1}{\sqrt{2n}} \sum_{j=t}^{2t-1} \prod_{i=1}^{j} \left(1 - \frac{j}{n/2+i}\right)$$

$$\geq \frac{t}{\sqrt{2n}} \left(1 - \frac{2t}{n/2}\right)^{2t}$$

$$\geq \frac{t}{\sqrt{2n}} \exp\left(-\frac{16t^2}{n}\right),$$

since $1 - x \geq e^{-2x}$ for $0 \leq x \leq 1/2$. For $t \geq \frac{1}{4\sqrt{2}}\sqrt{n}$, the probability is at least $\frac{1}{8}e^{-16t^2/n}$. For $0 \leq t < \frac{1}{4\sqrt{2}}\sqrt{n}$, we have that

$$\mathbb{P}(S_n \geq n/2 + t) \geq \mathbb{P}\left(S_n \geq n/2 + \frac{1}{4\sqrt{2}}\sqrt{n}\right)$$

$$\geq \frac{1}{8}e^{-1/2} \geq \frac{1}{15} \geq \frac{1}{15}e^{-16t^2/n},$$

and so the claimed bound also holds for this range of t. $\square$

Chapter 2

The game of Cops and Robbers

We focus now on our first and arguably most famous graph searching game, Cops and Robbers. As such, this chapter and the following one play a prominent role in our discussion. Although the
game is over 30 years old, it still presents some deep questions for
mathematicians and theoretical computer scientists. Over the last
few years, we have seen a wealth of probabilistic arguments arise in
the study of the game. As we discussed in the preface, research in
this direction either analyzes Cops and Robbers on random struc-
tures (such as $\mathbb{G}(n, p)$ or $\mathbb{G}_{n,d}$), or uses the probabilistic method
to derive properties in the deterministic case. We will see both
applications in this chapter.

We begin with a brief presentation of some background on Cops
and Robbers. We then turn our attention to the cop number of
dense random graphs $\mathbb{G}(n, p)$ in Section 2.1 (a more detailed discussion will be left to Chapter 5). Random graphs with constant
cop number are discussed in Section 2.2. In the following section,
we prove that almost all cop-win graphs have a universal vertex;
an extension of this is also shown in Section 2.4, by showing that
almost all k-cop-win graphs have a dominating set of order k. The
two final sections, Sections 2.5 and 2.6, are devoted to analyzing
Cops and Robbers games in random geometric graphs and percolated random geometric graphs, respectively.

Background

In Cops and Robbers, a two-player game of perfect information, a set of cops tries to capture a robber by moving at unit
speed from vertex to vertex. More precisely, *Cops and Robbers* is
a game played on a reflexive graph (that is, there is a loop at each
vertex). Graphs that we study in this book (such as $\mathbb{G}(n, p)$) are

not reflexive, so we will always add loops to make them reflexive. There are two players consisting of a set of *cops* and a single *robber*. The game is played over a sequence of discrete time-steps or *turns*, with the cops going first on turn 0 and then playing on alternate time-steps. A *round* of the game is a cop move together with the subsequent robber move. The cops and robber occupy vertices; for simplicity, the player is identified with the vertices they occupy. The set of cops is referred to as C and the robber as R. If a player is ready to move in a round, then they must move to a neighboring vertex. Because of the loops, players can *pass*, or remain on their own vertex. Observe that any subset of C may move in a given round. The cops win if after some finite number of rounds, one of them can occupy the same vertex as the robber (in a reflexive graph, this is equivalent to the cop landing on the robber). This is called a *capture*. The robber wins if they can evade capture indefinitely. A *winning strategy for the cops* is a set of rules that, if followed, results in a win for the cops. A *winning strategy for the robber* is analogously defined.

If a cop is placed at each vertex, then the cops are guaranteed to win. Therefore, the minimum number of cops required to win in a graph G is a well-defined positive integer, named the *cop number* (or *copnumber*) of the graph G. The notation $c(G)$ is used for the cop number of a graph G. If $c(G) \leq k$, then G is k-*cop-win*. In the special case $k = 1$, G is *cop-win* (or *copwin*). For example, graphs with universal vertices are cop-win (in particular, cliques), as are trees. Cycles of length at least 4 are 2-cop-win.

The game of Cops and Robbers was first considered by Quilliot [169, 170] in his doctoral thesis, and was independently considered by Nowakowski and Winkler [152]. Both [152, 169] refer only to one cop. The introduction of the cop number came in 1984 with Aigner and Fromme [1] (see also the early paper [23]). Many papers have now been written on cop number since these three early works; see the surveys [9, 105] and the book of Bonato and Nowakowski [52].

The *closed neighbor set* of a vertex x, written $N[x]$, is the set of vertices adjacent to x (including x itself); that is, $N[x] = N(x) \cup$

$\{x\}$. A vertex u is a *corner* if there is some vertex v such that $N[u] \subseteq N[v]$. We say that u is a *child* with *parent* v.

A graph is *dismantlable* if some sequence of deleting corners results in the graph K_1. For example, each tree is dismantlable (delete end-vertices), as are chordal graphs (delete *simplicial* vertices; that is, ones whose neighbor sets are cliques). For application of dismantlable graphs, see [57]. The following result characterizes cop-win graphs, and can be appropriately cast as the "First Theorem on Cops and Robbers."

Theorem 2.0.1 ([152]). *A graph is cop-win if and only if it is dismantlable.*

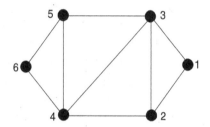

FIGURE 2.1: A cop-win ordering of a cop-win graph.

Cop-win (or dismantlable) graphs have a recursive structure, made explicit in the following sense. Observe that a graph is dismantlable if the vertices can be labeled by positive integers from $[n]$ in such a way that for each $i < n$, the vertex i is a corner in the subgraph induced by $\{i, i+1, \ldots, n\}$. This ordering of $V(G)$ is called a *cop-win ordering*. See Figure 2.1 for a graph with vertices labeled by a cop-win ordering.

A generalization of Theorem 2.0.1 exists for graphs with higher cop number as found in [69] and [106]. Given a graph G, define the *kth strong power* of G, written $G_{\boxtimes}^k$, to have vertices the ordered k-tuples from G, with two distinct tuples adjacent if they are equal or adjacent in each coordinate. See Figure 2.2 for an example. The positions of k-many cops in G are naturally identified with a single vertex in $G_{\boxtimes}^k$. The definition of the strong power allows us

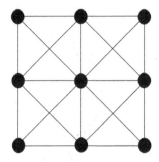

FIGURE 2.2: The strong power $(P_3)^2_\boxtimes$.

to simulate movements of the cops in G by movements of a single cop in $G^k_\boxtimes$.

Let $P = G^k_\boxtimes$. For $i \in \mathbb{N} \cup \{0\}$, the relation $\leq_i$ on $V(G) \times V(P)$ is defined as follows by induction on i. For $x \in V(G)$ and $p \in V(P)$, $x \leq_0 p$ if in position p, at least one of the k cops is occupying x. For $i > 0$, $x \leq_i p$ if and only if for each $u \in N[x]$ there exists a $v \in N[p]$ such that $u \leq_j v$ for some $j < i$. The relations $\leq_i$ are nondecreasing sets in i, and hence, there is a nonnegative M such that $\leq_M = \leq_{M+1}$ and define $\preceq = \leq_M$.

Theorem 2.0.2 ([69]). *A graph G is k-cop-win if and only if there exists $p \in V(P)$ such that $x \preceq p$ for every $x \in V(G)$.*

There are not too many useful, known bounds on the cop number in terms of other graph parameters. One elementary and sometimes useful upper bound for the cop number is

$$c(G) \leq \gamma(G), \tag{2.1}$$

where $\gamma(G)$ is the domination number of G. However, the inequality (2.1) is far from tight (consider a path, for example, for which $\gamma(P_n) = \lceil n/3 \rceil$ and $c(P_n) = 1$).

We reference another bound for the cop number in terms of treewidth. In a tree decomposition, each vertex of the graph is represented by a subtree, such that vertices are adjacent only when the corresponding subtrees intersect. Formally, given a graph

$G = (V, E)$, a *tree decomposition* is a pair (X, T), where $X = \{X_1, X_2, \ldots, X_n\}$ is a family of subsets of V called *bags*, and T is a tree whose vertices are the subsets X_i, satisfying the following three properties.

(i) $V = \bigcup_{i=1}^{n} X_i$. That is, each graph vertex is associated with at least one tree vertex.

(ii) For every edge vw in the graph, there is a subset X_i that contains both v and w. That is, vertices are adjacent in G only when the corresponding subtrees have a vertex in common.

(iii) If X_i, X_j and X_k are bags, and X_k is on the path from X_i to X_j, then $X_i \cap X_j \subseteq X_k$.

The *width* of a tree decomposition is the cardinality of its largest set X_i minus one. The *treewidth* of a graph G, written $tw(G)$, is the minimum width among all possible tree decompositions of G. For more on treewidth, see [76]. The following upper bound was given in [116]:

$$c(G) \leq tw(G)/2 + 1. \tag{2.2}$$

The bound (2.2) is tight if the graph has small treewidth (in particular, up to treewidth 5). Further, it gives a simple proof that outerplanar graphs have cop number at most 2 (first proved in [67]). For many families of graphs, however, the bound (2.2) is far from tight; for example, for a positive integer n, the clique K_n has treewidth $n - 1$, but is cop-win. Similarly, $n \times n$ Cartesian grids have cop number 2, but have treewidth n. Better bounds in this case using tree decompositions are discussed in [43].

The question of how large the cop number of a connected graph can be as a function of its order turns is highly nontrivial. For a positive integer n, let $c(n)$ be the maximum value of $c(G)$, where G is a connected graph of order n. *Meyniel's conjecture* states that

$$c(n) = O(\sqrt{n}).$$

The conjecture was mentioned in Frankl's paper [91] as a personal communication to him by Henri Meyniel in 1985 (see page 301 of [91] and reference [8] in that paper). Meyniel's conjecture stands out as one of the deepest problems on the cop number, and is the subject of Chapter 5.

We finish this section with a useful theorem of Aigner and Fromme [1] that supplies a good lower bound on the cop number in certain cases. The *girth* of a graph is the length of a shortest cycle.

Theorem 2.0.3 ([1]). *If G has girth at least 5, then $c(G) \geq \delta(G)$.*

2.1   Binomial Random Graphs

The cop number of $\mathbb{G}(n, p)$, where $p = p(n)$ is a function of n such that $(2\log n)/\sqrt{n} \leq p = p(n) \leq 1 - \varepsilon$ for some $\varepsilon > 0$, was first considered in [53]. The results of this section summarize those results. We will give a much more detailed discussion of the cop number of random graphs in Chapter 5.

For constant $p \in (0, 1)$ or $p = p(n)$ tending to 0 as n tends to ∞, define $\mathbb{L}n = \log_{\frac{1}{1-p}} n$.

Theorem 2.1.1 ([53]). *Suppose that $(2\log n)/\sqrt{n} \leq p = p(n) \leq 1 - \varepsilon$ for some $\varepsilon > 0$. If $\omega(n)$ is any function tending to infinity arbitrarily slowly, then a.a.s. $G \in \mathbb{G}(n, p)$ satisfies*

$$\mathbb{L}n - \mathbb{L}\big((p^{-1}\mathbb{L}n)(\log n)\big) \leq c(G) \leq \mathbb{L}n + \mathbb{L}(\omega(n)).$$

From Theorem 2.1.1, we have immediately that if $p = n^{-o(1)}$ and $p \leq 1 - \varepsilon$ for some $\varepsilon > 0$, then a.a.s. $G \in \mathbb{G}(n, p)$ satisfies

$$c(G) \sim \mathbb{L}n. \tag{2.3}$$

Indeed, the lower and upper bounds in (2.3) are straightforward to verify when p is a constant. If $p = o(1)$, then we have that

$$\mathbb{L}n = \frac{\log n}{-\log(1-p)} \sim \frac{\log n}{p}, \tag{2.4}$$

and for $p = n^{-o(1)}$ tending to zero with n, the lower bound in Theorem 2.1.1 is

$$
\begin{aligned}
\mathbb{L}n - \mathbb{L}\big((p^{-1}\mathbb{L}n)(\log n)\big) &= \mathbb{L}n - 2\mathbb{L}\big((1+o(1))p^{-1}\log n\big) \\
&= \mathbb{L}n - 2\mathbb{L}\big(n^{o(1)}\big) \\
&\sim \mathbb{L}n.
\end{aligned}
$$

Note also that for $p = n^{-a(1+o(1))}$ $(0 < a < 1/2)$ we do not have concentration for $c(G)$, but the following bounds hold a.a.s.:

$$(1 + o(1))(1 - 2a)\mathbb{L}n \le c(G) \le (1 + o(1))\mathbb{L}n.$$

We may improve the upper bound to $(1+o(1))(1 - a)\mathbb{L}n$ by using the fact that

$$\gamma(G) \le n\frac{1 + \log(\delta + 1)}{\delta + 1}, \tag{2.5}$$

which for $\mathbb{G}(n, p)$ it is a.a.s. asymptotic to $\log(pn)/p \sim (1 - a)\mathbb{L}n$. Note that (2.5) may be obtained by the probabilistic method; see, for example, Theorem 1.2.2 in [8]. Finally, we note that for denser graphs, it is actually known that a.a.s. the domination number takes one of two consecutive integer values [101, 181].

Proof of Theorem 2.1.1. The upper bound follows from the fact that $c(G) \le \gamma(G)$ and that a.a.s.

$$\gamma(G) \le k = \lceil \mathbb{L}n + \mathbb{L}(\omega(n)) \rceil.$$

Indeed, the probability that the domination number of a random graph is at most k is bounded from below by the probability that any fixed set of k vertices is a dominating set. But the latter probability is equal to

$$
\begin{aligned}
\big(1 - (1 - p)^k\big)^{n-k} &\ge 1 - (n - k)(1 - p)^k \\
&\ge 1 - n(1 - p)^k \\
&\ge 1 - n(1 - p)^{\mathbb{L}n + \mathbb{L}(\omega(n))} \\
&= 1 - \frac{1}{\omega(n)} \sim 1.
\end{aligned}
$$

For the lower bound, we employ the following adjacency property. For a fixed $k \in \mathbb{N}$, we say that G is $(1, k)$-*existentially closed* (or $(1, k)$-*e.c.*) if G has order at least $k + 2$, and for each k-set S of vertices of G (that is, each $S \subseteq V(G)$ with $|S| = k$) and every vertex $u \notin S$, there is a vertex $z \notin S \cup \{u\}$ adjacent to u but not adjacent to any vertex in S. See Figure 2.3.

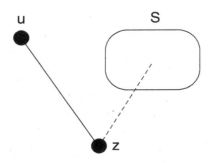

FIGURE 2.3: The $(1, k)$-e.c. property.

If G is $(1, k)$-*e.c.*, then notice that $c(G) \geq k + 1$. To see that k cops cannot win the game, suppose that the cops occupy a set of at most k vertices at the start of the game. By enlarging this set to a k-set S and finding a vertex u outside of S, there exists a vertex z adjacent to u and not S. The robber, at the beginning of the game, moves to z and is safe in the next round. In a similar fashion, on each turn the robber may use the property to escape to a vertex not adjacent to any vertex occupied by a cop. The lower bound will follow once we prove that a.a.s. $G \in \mathbb{G}(n, p)$ is $(1, k)$-e.c. for

$$k = \left\lfloor \mathbb{L}n - \mathbb{L}\big((p^{-1}\mathbb{L}n)(\log n)\big) \right\rfloor.$$

Assume that $(2 \log n)/\sqrt{n} \leq p = o(1)$. Note that we do not use the lower bound for p in the proof of the theorem. The condition is introduced in order to obtain a nontrivial result; the value of k is negative, otherwise. Then using (2.4) we obtain that

$$
\begin{aligned}
k &= \mathbb{L}n - \mathbb{L}\left((1 + o(1))\frac{\log^2 n}{p^2}\right) \\
&= \mathbb{L}n - 2\mathbb{L}\left((1 + o(1))\frac{\log n}{p}\right) \\
&= \mathbb{L}n - 2(1 + o(1))\frac{\log \log n + \log(1/p)}{p}.
\end{aligned}
\tag{2.6}
$$

Fix a k-set S and a vertex u not in S. For a vertex $x \in V(G)\backslash(S \cup \{u\})$, the probability that x is adjacent to u and to no vertex of S is $p(1-p)^k$. As edges are chosen independently, we find that the probability that no suitable vertex can be found for this particular S and u is $(1 - p(1-p)^k)^{n-k-1}$.

Let X be the random variable counting the number of S and u for which no suitable x can be found. We then have that

$$
\begin{aligned}
\mathbb{E}[X] &= \binom{n}{k}(n-k)\left(1-p(1-p)^k\right)^{n-k-1} \\
&\leq n^{k+1}\left(1 - \frac{(\mathbb{L}n)(\log n)}{n}\right)^{n(1-(\mathbb{L}n)/n)},
\end{aligned}
$$

since $(1-p)^k \geq (1-p)^{\mathbb{L}n - \mathbb{L}\left((p^{-1}\mathbb{L}n)(\log n)\right)} = (n^{-1}p^{-1}\mathbb{L}n)(\log n)$. Hence,

$$
\begin{aligned}
\mathbb{E}[X] &\leq n^{k+1}\exp\left(-(\mathbb{L}n)(\log n)(1-(\mathbb{L}n)/n)\right) \\
&= n^{k+1}\exp\left(-(\mathbb{L}n - (\mathbb{L}n)^2/n)(\log n)\right) \\
&\leq n^{k+1}\exp\left(-\left(k + \frac{2\log\log n + 2\log(1/p)}{p} - \frac{\log^2 n}{p^2 n}\right)\right. \\
&\quad\left.\cdot(\log n)(1 + o(1))\right) \\
&= o(1),
\end{aligned}
$$

where the second inequality follows from (2.4) and (2.6). It is also straightforward to show that the same argument holds for constant p. The theorem now follows by the first moment method. $\qquad\square$

2.2 Graphs with a Given Cop Number

Results presented earlier in this chapter may be used to show that for a given $k \in \mathbb{N}$, there exists a connected graph G such that $c(G) = k$. For instance, take $p = p(n)$ such that Theorem 2.2.2 implies that the probability that a random graph $G \in \mathbb{G}(n,p)$ satisfies $c(G) = k$ is at least, say, $1/2$ for n large enough. This shows that there exists a graph with the desired property (via a nonconstructive argument). This is a useful observation since it is not so easy to explicitly construct a graph with a given cop number. But what if we want to check if there is a graph on n vertices with cop number equal to, say, $\lfloor n^{1/3} \rfloor$? The results for random graphs cannot be used.

Constant Cop Number

We consider when the random graph has a constant cop number, which was first studied in [166]. We first need the following result, proved in [166], on the domination number of random graphs. We omit the proof, which uses the convergence of moments method known also as Brun's sieve (see Chapter 6, Section 6.1). For more details about this method, see, for example, Section 8.3 in [8].

Theorem 2.2.1 ([166]). *Let $k \in \mathbb{N}$, and for a given sequence $(a_n)_{n \in \mathbb{N}}$ let*

$$p = p(n) = 1 - \left(\frac{k \log n + a_n}{n} \right)^{\frac{1}{k}}.$$

If $G \in \mathbb{G}(n, p)$, then we have that

$$\lim_{n \to \infty} \mathbb{P}(\gamma(G) \leq k) = \begin{cases} 0 & \text{if } a_n \to \infty, \\ 1 - e^{-e^{-a}/k!} & \text{if } a_n \to a \in \mathbb{R}, \\ 1 & \text{if } a_n \to -\infty. \end{cases}$$

Additionally, the above inequality can be replaced by an equality provided that, say, $|a_n| = n^{o(1)}$.

The main theorem of this section is the following.

Theorem 2.2.2 ([166]). *Let $k \in \mathbb{N}$, and for a given sequence $(a_n)_{n \in \mathbb{N}}$ let*

$$p = p(n) = 1 - \left(\frac{k \log n + a_n}{n} \right)^{\frac{1}{k}}.$$

If $G \in \mathbb{G}(n, p)$, then the following statements hold.

(i) *If $a_n \to -\infty$, then a.a.s. $c(G) \leq k$.*

(ii) *If $a_n \to a \in \mathbb{R}$, then*

(a) $c(G) = k$ *with probability tending to* $1 - e^{-e^{-a}/k!}$, *and*

(b) $c(G) = k + 1$ *with probability tending to* $e^{-e^{-a}/k!}$.

(iii) *If $a_n \to \infty$, then a.a.s. $c(G) \geq k + 1$.*

If we add an additional condition that the sequence $(|a_n|)_{n \in \mathbb{N}}$ does not grow too fast with n (say, $|a_n| = n^{o(1)}$), then the above inequalities can be replaced by equalities. Note that to obtain a constant cop number, $p(n)$ has to tend to 1 as n tends to ∞. However, it is possible that $p(n)$ tends to 1 and the cop number grows together with n. In this case, $p(n) = 1 - n^{-o(1)}$ and a.a.s. $c(G) = o(\log n)$, $G \in \mathbb{G}(n, p)$.

The first step we take toward proving Theorem 2.2.2 is the following lemma, which shows that the cop number is at least $k+1$ provided that we are below the threshold we want to establish. The proof uses a similar argument as in the proof of Theorem 2.1.1; that is, if G is $(1, k)$-e.c., then $c(G) \geq k + 1$.

Lemma 2.2.3 ([166]). *Fix $k \in \mathbb{N}$ and let $G \in \mathbb{G}(n, p)$. If*

$$p = p(n) < 1 - \left(\frac{(k+1) \log n + \log \log n}{n} \right)^{\frac{1}{k}},$$

then a.a.s. $c(G) > k$.

Proof. Fix $k \in \mathbb{N}$. As we already mentioned, we may assume that $p(n)$ tends to 1 as n goes to infinity; otherwise, a.a.s. the cop number grows with n.

Fix a k-subset S of vertices and a vertex u not in S. For a vertex $x \in [n] \setminus (S \cup \{u\})$, the probability that the vertex x is adjacent to u and to no vertex of S is $p(1 - p)^k$. Since the edges are chosen independently, the probability that no suitable vertex can be found for this particular S and u is $(1 - p(1 - p)^k)^{n-k-1}$.

Let X be the random variable counting the number of ordered pairs of S and u for which no suitable x can be found. We then have that

$$
\begin{aligned}
\mathbb{E}[X] &= \binom{n}{k}(n-k)\left(1 - p(1-p)^k\right)^{n-k-1} \\
&= O(n^{k+1})\left(1 - \frac{(k+1)\log n + 0.5 \log \log n}{n}\right)^{n-k-1} \\
&= O\left(\frac{1}{\sqrt{\log n}}\right) = o(1).
\end{aligned}
$$

Observe that the second equality follows from the fact that

$$
\begin{aligned}
p(1-p)^k &> p\frac{(k+1)\log n + \log \log n}{n} \\
&> \frac{(k+1)\log n + 0.5 \log \log n}{n}
\end{aligned}
$$

for $p = 1 - o(\log \log n / \log n)$. The same inequality holds for smaller values of p, since then $p(1-p)^k \sim (1-p)^k \gg \log n / n$. The proof now follows by the first moment method.   □

Finally, we are ready to prove the main theorem of this section.

Proof of Theorem 2.2.2. Fix $k \in \mathbb{N}$. Given the threshold for the appearance of a dominating set of cardinality k (see Theorem 2.2.1 and Lemma 2.2.3), it is enough to show that the cop number is greater than k if

$$
\left(\frac{k\log n - \log \log n}{n}\right)^{\frac{1}{k}} \leq 1 - p
$$

$$
\leq \left(\frac{(k+1)\log n + \log \log n}{n}\right)^{\frac{1}{k}},
$$

and there is no dominating set of cardinality k.

For a vertex u, define $N^c(u) = V(G) \setminus N[u]$. A vector $(r, c_1, c_2, \ldots, c_k)$ is *dangerous* if $\bigcap_{i=1}^{k} N^c(c_i) \subseteq N^c(r)$. A vertex r is *dangerous* if there is a vector $(c_1, c_2, \ldots, c_k)$ such that $(r, c_1, c_2, \ldots, c_k)$ is dangerous. It is evident that a graph with cop number at most k must have at least one dangerous vector. If the game can be won by the cops, then a position at the end of the game yields such a vector. Regardless of the moves of the robber, at least one cop can capture them (in other words, the robber should try to escape to $\bigcap_{i=1}^{k} N^c(c_i)$, but those vertices cannot be reached by them).

The expected number of dangerous vectors that satisfy $|\bigcap_{i=1}^{k} N^c(c_i)| = \ell$ is at most

$$O(n^{k+\ell+1})(1-p)^{(k+1)\ell} \left(1 - (1-p)^k\right)^{n-k-\ell-1}$$

$$= O(n^{k+\ell+1}) \left(\frac{\log n}{n}\right)^{\frac{(k+1)\ell}{k}}$$

$$\cdot \left(1 - \frac{k \log n - \log \log n}{n}\right)^{n-O(1)}$$

$$\leq n^{1-\frac{\ell}{k}+o(1)},$$

provided ℓ is a constant. For $\ell = \ell(n)$ tending to infinity with n, we derive that the expectation is at most

$$O(n^{k+\ell+1})(1-p)^{(k+1)\ell}$$

$$= O(n^{k+\ell+1}) \left(\frac{(1+o(1))(k+1)\log n}{n}\right)^{\frac{(k+1)\ell}{k}}$$

$$= n^{(-\ell/k)(1+o(1))} \left((1+o(1))(k+1)\log n\right)^{\ell(1+1/k)}$$

$$= n^{(-\ell/k)(1+o(1))}.$$

Thus, a.a.s. there is no dangerous vector with $\ell > k$ by Markov's inequality. Further, we may assume that $\ell \geq 1$ since we will be interested in the case where there is no dominating set of cardinality k. Thus, Markov's inequality can be used to show that a.a.s. the number of dangerous vertices (which can be bounded from above by the number of dangerous vectors) is at most $n^{1-1/k+o(1)}$.

We will show now that the robber can avoid dangerous configurations. Since a.a.s. $G \in \mathbb{G}(n,p)$ is connected, without loss of generality, we may assume that at the beginning of the game all cops are placed at the same vertex w of degree at most $n - \Omega(\log n)$.

Such a vertex must exist since there must be a vertex of degree at most the average. Indeed, if there is a winning strategy for the cops that starts with an initial configuration X, then the cops can always move from w to X and continue from there. The robber starts on any vertex in $N^c(w)$, and a.a.s. the corresponding vector is not dangerous (since a.a.s. there is no dangerous vector with $|\bigcap_{i=1}^{k} N^c(c_i)| = \Omega(\log n)$). Suppose the robber occupies some vertex R, which is not dangerous. To force the robber to go to a dangerous vertex, the cops must occupy vertices $c_1, c_2, \ldots, c_k$ such that $(\bigcap_{i=1}^{k} N^c(c_i)) \setminus N^c(r)$ contains only s vertices, all dangerous; that is, the robber has to escape to $\bigcap_{i=1}^{k} N^c(c_i)$ to be safe, but the only reachable vertices are dangerous. Observe that $s \geq 1$ since $(r, c_1, c_2, \ldots, c_k)$ is not dangerous. Using exactly the same argument as before, we derive that

$$ t = \left| \left(\bigcap_{i=1}^{k} N^c(c_i) \right) \cap N^c(r) \right| \leq k. $$

The expected number of sets of vertices $\{c_1, c_2, \ldots, c_k\}$ with the property that $\bigcap_{i=1}^{k} N^c(c_i)$ contains s dangerous vertices ($s \geq 1$) and

$$ \left| \bigcap_{i=1}^{k} N^c(c_i) \right| = s + t $$

(where $0 \leq t \leq k$) is at most

$$ \sum_{t=0}^{k} \sum_{s \geq 1} n^{k+t} \left(n^{1 - \frac{1}{k} + o(1)} \right)^s (1-p)^{k(s+t)} (1 - (1-p)^k)^{n-k-s-t}. $$

The latter expression equals

$$ \sum_{t=0}^{k} n^{-1/k + o(1)} = o(1). $$

Therefore, a.a.s. there is no way to force the robber to move to a dangerous vertex. □

Large Cop Number

Consider a connected graph $\hat{G}$ on n vertices with $k = c(\hat{G})$, some large value. We will show in Section 5.4 that there are connected graphs on n vertices with cop number at least $D\sqrt{n}$, for some universal constant D. A simple but key observation is that for every graph G and any edge $e \in E(G)$ we have that

$$c(G) \leq c(G - e) + 1.$$

Indeed, suppose that we have a winning strategy for $c(G - e)$ cops on $G - e$. These cops may then use this strategy to play on G and one additional cop guards edge e. The game on G is played as if it were played on $G - e$ and so the cops win, proving the claimed inequality.

Now, we can start with any spanning tree T of $\hat{G}$ and keep adding one missing edge from $\hat{G}$ at the time until we obtain the final graph $\hat{G}$. Since $c(T) = 1$, $c(\hat{G}) = k$, and each time the cop number goes up by at most 1, we derive that for each value of j, where $1 \leq j \leq k$, there exists a connected graph on n vertices with the cop number equal to j.

2.3 Properties of Almost All Cop-Win Graphs

In this section, our goal is to investigate the structure of *random cop-win* graphs; the random graph model we use is $\mathbb{G}(n, 1/2)$. Recall that $\mathbb{G}(n, 1/2)$ is a uniform probability space over all labeled graphs on n vertices. We use this interpretation of $\mathbb{G}(n, 1/2)$ in the proofs of this and the next section. Random cop-win graphs were first considered in [47] and the results of this section are taken from that work.

Let **cop-win** be the event that the random graph $G \in \mathbb{G}(n, 1/2)$ is cop-win and let **universal** be the event that there is a universal vertex. If a graph has a universal vertex w, then it is cop-win. Adding a universal vertex clearly makes a graph cop-win; see Figure 2.4. Hence, for $n \in \mathbb{N}$, there are at least as many nonisomorphic cop-win graphs of a given order $n + 1$ as there are graphs of order n.

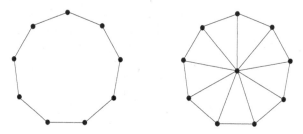

FIGURE 2.4: A graph becomes cop-win after a universal vertex is added.

We make an elementary but important observation. The probability that a random graph is cop-win can be estimated as follows.

$$\mathbb{P}\left(\textbf{cop-win}\right) \geq \mathbb{P}\left(\textbf{universal}\right) \;=\; n2^{-n+1} - O(n^2 2^{-2n+3})$$
$$\sim \; n2^{-n+1}. \tag{2.7}$$

Surprisingly, (2.7) is the correct asymptotic value for $\mathbb{P}\left(\textbf{cop-win}\right)$. Our main result in this section is the following theorem.

Theorem 2.3.1 ([47]). *In $G \in \mathbb{G}(n, 1/2)$, we have that*

$$\mathbb{P}\left(\textit{cop-win}\right) \sim n2^{-n+1}.$$

Using Theorem 2.3.1, we derive the asymptotic number of labeled cop-win graphs.

Corollary 2.3.2 ([47]). *The number of cop-win graphs on n labeled vertices is*

$$(1 + o(1))2^{\binom{n}{2}} n2^{-n+1} \sim n2^{n^2/2 - 3n/2 + 1}.$$

We therefore, have the following surprising fact that almost all cop-win graphs contain a universal vertex.

Corollary 2.3.3 ([47]).

$$\mathbb{P}\left(\textit{universal}|\textit{cop-win}\right) = 1 - o(1).$$

To prove Theorem 2.3.1, we use the approach from [47] of bounding the probability of **cop-win** for graphs of maximum degree at most $n - 2$. Since the proof for $\Delta \leq n - 3$ is more detailed, we omit the proof of part (a) here. We include the proof of part (b) to illustrate the methods involved (in particular, using the dismantling property of cop-win graphs stated in Theorem 2.0.1).

Theorem 2.3.4 ([47]).

(a) *For some $\varepsilon > 0$ we have that*

$$\mathbb{P}\left(\textbf{\textit{cop-win}} \text{ and } \Delta \leq n - 3\right) \leq 2^{-(1+\varepsilon)n}.$$

(b) $\mathbb{P}\left(\textbf{\textit{cop-win}} \text{ and } \Delta = n - 2\right) \leq 2^{-(3 - \log_2 3)n + o(n)}.$

Observe that Theorem 2.3.1 follows immediately from Theorem 2.3.4 and (2.7).

Proof of Theorem 2.3.4(b). In the proof, we make heavy use of Theorem 2.0.1. Let D be the event that a graph is **cop-win** and has $\Delta \leq n - 3$. First, we deduce from Theorem 2.3.4(a) a rough upper bound for the probability that the graph is cop-win: there exists $\varepsilon > 0$ such that

$$
\begin{aligned}
\mathbb{P}\left(\textbf{cop-win}\right) &\leq \mathbb{P}\left(D\right) + \mathbb{P}\left(\Delta \geq n - 2\right) \\
&\leq 2^{-(1+\varepsilon)n} + n^2 2^{-n+1} \\
&\leq 2^{-n+o(n)}.
\end{aligned}
\tag{2.8}
$$

The vertex set of every cop-win graph G with $\Delta = n - 2$ can be partitioned as follows: it must have a vertex w of degree $n - 2$, a (unique) vertex v which is not adjacent to w, a set B of vertices adjacent to v (and also to w), and a set A of vertices that are not adjacent to v (but are adjacent to w). See Figure 2.5. We claim that the graph induced by B is cop-win.

Since G is cop-win, by Theorem 2.0.1 we can dismantle all vertices in A (using w as a parent), leaving us with the cop-win subgraph H induced by v, w, and B. If B contains one vertex only, then the graph induced by B is clearly cop-win. Otherwise, either B has a universal vertex in H (and so the graph induced by B

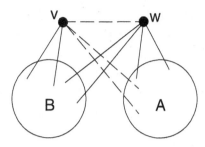

FIGURE 2.5: A partition of $V(G)$, where the subgraph induced by B is claimed to be cop-win.

is cop-win and we can dismantle all remaining vertices of B), or B must have a corner (since if there is no universal vertex in B, we cannot dismantle either v or w but H is cop-win). In either case, we can dismantle some vertex x in B, such that the following properties hold.

(i) $H - x$ is a cop-win subgraph induced by v, w, and $B \setminus \{x\}$, and

(ii) v and w are adjacent to all the vertices of $B \setminus \{x\}$.

Hence, by (1) and (2) we may use induction to dismantle B starting from the subgraph H. In addition, the same sequence of vertices can be used to dismantle the graph induced by the set B, since all the parents were in B. Therefore, B is cop-win.

Finally, we estimate the number of labeled cop-win graphs with $\Delta = n - 2$. There are n choices for w, $n - 1$ choices for v, and $2^{n-2} = \sum_{i=0}^{n-2} \binom{n-2}{i}$ choices for A. The probability that w and v have the correct neighborhoods is $2^{-n+1}2^{-n+2}$. If $|A| = i$, then the probability that the graph induced by B is cop-win is at most $2^{-n+2+i+o(n)}$ using (2.8) (note that $|B| = n - 2 - i$ and that there are no other restrictions on B except that the subgraph it induces is cop-win). Thus,

$\Pr(\textbf{cop-win and } \Delta = n - 2)$

$$\leq n^2 \sum_{i=0}^{n-2} \binom{n-2}{i} 2^{-n+1} 2^{-n+2} 2^{-n+2+i+o(n)}$$

$$= 2^{-3n+o(n)} \sum_{i=0}^{n-2} \binom{n-2}{i} 2^i$$

$$= 2^{-3n+o(n)} (1+2)^{n-2}$$

$$= 2^{-(3-\log_2 3)n+o(n)},$$

where the third equality follows by the binomial theorem.   □

2.4   Properties of Almost All k-Cop-Win Graphs

After the discussion in Section 2.3, we may think that analogous results hold for k-cop-win graphs. This intuition is indeed correct, as was shown in [159].

Let k-**cop-win** be the event that the random graph $G \in \mathbb{G}(n, 1/2)$ is k-cop-win, and let k-**dom** be the event that there is a dominating set of cardinality k. If a graph has a dominating set of cardinality k, then it is k-cop-win. See Figure 2.6; note that the Petersen graph has cop number exactly 3.

The probability that a random graph is k-cop-win can be estimated as follows:

$$\mathbb{P}\left(k\text{-}\textbf{cop-win}\right) \geq \mathbb{P}\left(k\text{-}\textbf{dom}\right) \sim \left(1 - 2^{-k}\right)^{-k} \binom{n}{k} \left(1 - 2^{-k}\right)^n.$$

$$(2.9)$$

For any $S \subseteq V = [n]$ of cardinality k, let A_S denote the event that S is a dominating set. By the union bound,

$$\mathbb{P}\left(k\text{-}\textbf{dom}\right) = \mathbb{P}\left(\bigcup_S A_S\right) \leq \sum_S \mathbb{P}\left(A_S\right) = \binom{n}{k} \left(1 - 2^{-k}\right)^{n-k}.$$

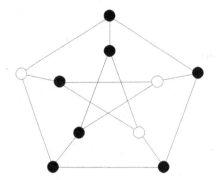

FIGURE 2.6: The white vertices dominate the Petersen graph.

Note that it follows from Bonferroni's inequalities (see Lemma 1.2.8) that

$$\mathbb{P}\left(k\text{-}\mathbf{dom}\right) = \mathbb{P}\left(\bigcup_{S} A_S\right) \geq \sum_{S} \mathbb{P}\left(A_S\right) - \sum_{S \neq T} \mathbb{P}\left(A_S \cap A_T\right).$$

Let $S, T \subseteq V$ be such that $|S| = |T| = k$ and $|S \cap T| = \ell$ for some $0 \leq \ell < k$. Let $x = 1 - 2^{-(k-\ell)}$. The probability that a vertex $v \in V \setminus (S \cup T)$ is dominated by both S and T is equal to

$$\begin{aligned}
\left(1 - 2^{-\ell}\right) + x^2 2^{-\ell} &= 1 - 2^{-k} - 2^{-k}x \\
&\leq 1 - \frac{3}{2} \cdot 2^{-k} < 1 - 2^{-k}.
\end{aligned}$$

Hence,

$$\sum_{S \neq T} \mathbb{P}\left(A_S \cap A_T\right) \leq n^{2k}\left(1 - \frac{3}{2} \cdot 2^{-k}\right)^{n-2k} = o\left(\sum_{S} \mathbb{P}\left(A_S\right)\right)$$

and so the lower bound (2.9) holds.

The main result in this section is the following theorem.

Theorem 2.4.1 ([159]). *Let $k \in \mathbb{N}$. In $\mathbb{G}(n, 1/2)$, we have that*

$$\begin{aligned}
\mathbb{P}\left(k\text{-}\boldsymbol{cop\text{-}win}\right) &\sim \left(1 - 2^{-k}\right)^{-k}\binom{n}{k}\left(1 - 2^{-k}\right)^{n} \\
&= \Theta\left(n^k\left(1 - 2^{-k}\right)^{n}\right).
\end{aligned}$$

From Theorem 2.4.1, we derive the asymptotic number of labeled k-cop-win graphs.

Corollary 2.4.2 ([159]). *Let $k \in \mathbb{N}$. The number of k-cop-win graphs on n labeled vertices is*

$$\mathbb{P}\left(k\text{-}\boldsymbol{cop\text{-}win}\right) 2^{\binom{n}{2}} \sim (1 - 2^{-k})^{-k} \binom{n}{k} 2^{n^2/2 - n/2 + n \log_2(1 - 2^{-k})}.$$

As an additional corollary, we have that almost all k-cop-win graphs contain a dominating set of cardinality k, a fact not obvious a priori.

Corollary 2.4.3 ([159]). *Let $k \in \mathbb{N}$. In $\mathbb{G}(n, 1/2)$, we have that*

$$\mathbb{P}\left(k\text{-}\boldsymbol{dom}|k\text{-}\boldsymbol{cop\text{-}win}\right) = 1 - o(1).$$

Recall that we use $N^c(v) = V(G) \setminus (N(v) \cup \{v\})$ for the set of nonneighbors of v, and for every set $S \subseteq V(G)$, let $N^c(S) = \bigcap_{u \in S} N^c(u)$. For a given $k \in \mathbb{N}$, let

$$\delta_k = \min_{S \subseteq V, |S| = k} |N^c(S)|.$$

Note that $\delta_k = 0$ if and only if there exists a dominating set of cardinality k. To prove Theorem 2.4.1 we bound the probability of k-**cop-win** for graphs with $\delta_k \geq 1$. We provide here only the proof of the first part (a).

Theorem 2.4.4 ([159]). *Let $k \in \mathbb{N}$. There exist $\xi > 0$ and $\varepsilon > 0$ such that the following hold. In $\mathbb{G}(n, 1/2)$, we have that*

(a) $\mathbb{P}\left(k\text{-}\boldsymbol{cop\text{-}win}$ and $1 \leq \delta_k \leq \xi n\right) \leq 2^{(\log_2(1 - 2^{-k}) - \varepsilon)n}$, and

(b) $\mathbb{P}\left(k\text{-}\boldsymbol{cop\text{-}win}$ and $\delta_k > \xi n\right) \leq 2^{(\log_2(1 - 2^{-k}) - \varepsilon)n}$.

Theorem 2.4.1 follows immediately from Theorem 2.4.4 and (2.9).

Let G be a random graph drawn from the $\mathbb{G}(n, 1/2)$ distribution. Our goal is to investigate the probability that $\delta_k \geq 1$ and that $c(G) \leq k$. We show that this event holds *with extremely small probability* (w.e.s.p.), which means that the probability it holds is at most $2^{(\log_2(1 - 2^{-k}) - \varepsilon)n}$ for some $\varepsilon > 0$. Observe that if we can show that each of a polynomial number of events holds w.e.s.p., then the same is true for the union of these events.

Proof of Theorem 2.4.4(a)

Let $S \subseteq V(G)$ be a set of vertices that dominates all but $\delta_k \geq 1$ vertices; let $v \in N^c(S)$ be some vertex not dominated by S. To estimate the probability from above that $c(G) \leq k$, we introduce a strategy for the robber and show that w.e.s.p. k cops can win (against this given strategy). To motivate the desired structure of a graph G, we start with the outline of the strategy for the robber. However, at this point, it is not clear that this strategy can be successfully applied; this will be shown next. The outline is described via the following items.

(a) If cops move such that they are located in the whole set S, then the robber goes to v.

(b) If cops move to a proper subset T of S and no cop occupies v, then the robber goes to a vertex of $N^c(T) \cap N(v)$ that is adjacent to no cop.

(b') In particular, if no cop is located in $S \cup \{v\}$ (corresponding to $T = \emptyset$ above), then the robber goes to a vertex of $N(v)$ that is adjacent to no cop.

(c) If cops move to a proper subset T of S and some cop occupies v, then the robber goes to a vertex of $(N^c(T) \cap N^c(v)) \setminus S$ that is adjacent to no cop.

(c') In particular, if no cop is located in S but some cop occupies v (corresponding to $T = \emptyset$ above), then the robber goes to a vertex of $N^c(v) \setminus S$ that is adjacent to no cop.

This outlines the strategy for the robber when the game is underway, and the same applies to the beginning of the game. For example, if cops start the game such that they are located in the whole set S, then the robber starts by occupying v, and so on. If at some point in the game the robber has more than one vertex to choose from, then they can make an arbitrary choice. This strategy is a greedy one that guarantees that the robber stays alive for at least one more round but does not think about the future. If all the cops pause, then we may assume that the robber also pauses. Hence, we may assume that in each round at least one cop moves.

We need the following lemma. To simplify the notation, we consider the random graph $\mathbb{G}(n+k+1,1/2)$ instead of $\mathbb{G}(n,1/2)$.

Lemma 2.4.5 ([159]). *Let* $S \subseteq V = \{1,2,\ldots,n+k+1\}$ *with* $|S| = k$, *and let* $v \notin S$. *Let* $T \subset S$ *with* $0 \leq |T| = \ell < k$. *Let* $x = x(n)$, $y = y(n)$, $z = z(n)$ *be any deterministic functions such that* $xn, yn, zn \in \mathbb{Z}$, $0 \leq y, z \leq 1$, *and* $0 \leq x \leq y$. *Consider the following properties:*

(p1) $v \in N^c(S)$,

(p2) yn *vertices of* $W = V \setminus (S \cup \{v\})$ *are adjacent to* v,

(p3) S *dominates all but* zn *vertices of* W,

(p4) xn *vertices of* $N(v)$ *that are dominated by* S *are not adjacent to any vertex of* T,

(p5) *the following is false: for every* $U \subseteq W$ *with* $|U| = k - \ell$ *and every* $w \in W \setminus U$, *there exists a vertex* $w' \in N(v)$ *adjacent to* w *but not adjacent to any vertex of* $T \cup U$.

Let $G = (V, E) \in \mathbb{G}(n+k+1,1/2)$. *Note that all the variables are chosen in advance before a random graph* G *is generated. Then there exists* $\xi \in (0,1/2]$ *such that if* $z \leq \xi$, *then properties (p1)-(p5) hold simultaneously w.e.s.p. Note that the statement holds when in both (p4) and (p5),* $N(v)$ *is replaced by* $N^c(v)$.

Before we prove the lemma, we show how the lemma implies Theorem 2.4.4(a). Let $G \in \mathbb{G}(n+k+1,1/2)$. We want to estimate $\mathbb{P}(k\text{-}\mathbf{cop\text{-}win}$ and $\delta_k = \delta)$ for some $\delta \geq 1$. Let $S \subseteq V(G)$ be a set of vertices that dominates all but δ_k vertices, and let $v \in N^c(S)$ be a vertex not dominated by S.

The robber will try to follow the strategy we outlined above. The goal is to show that if this strategy fails (that is, k cops have a winning strategy), then G satisfies properties (p1)-(p5) in Lemma 2.4.5 for some specific choice of $S, v, T, x(n), y(n)$, and $z(n) = 1 + \delta_k$. Since the number of possible choices for $S, v, T, x(n), y(n), z(n)$ to consider is at most $O(n^{k+1+3}) = n^{O(1)}$ (recall that $k = O(1)$) and for each fixed choice we obtain a statement that holds w.e.s.p., we derive the desired upper bound for $\mathbb{P}(k\text{-}\textbf{cop-win}$ and $1 \le \delta_k \le \xi n)$, where $\xi > 0$ is a constant implied and guaranteed by Lemma 2.4.5.

Suppose that the cops start the game by going to the whole set S. The robber starts at v and the desired position is achieved (see rule (a)). Suppose then that the cops start the game by going to set T, a proper subset of S (possibly the empty set), and possibly some vertices of $U \subseteq W = V(G) \setminus (S \cup \{v\})$. Let $w \in W \setminus U$ be any vertex. The robber can start at $w' \in N(v)$ that is not adjacent to any cop, and the desired position is achieved (see rule (b)) unless property (p5) in Lemma 2.4.5 holds, and so (p1)-(p5) hold for some specific choice of parameters. (Note that w' is adjacent to w but this is not important at this point; w was used to show the existence of w' only.) Finally, suppose that the cops start the game by going to $T \subset S$ (perhaps $T = \emptyset$), to some $U \subseteq W$, and at least one cop starts at v. Again, we take an arbitrary vertex $w \in W \setminus U$ and use it to show that there exists a vertex $w' \in N^c(v)$ that is adjacent to no cop, the desired position in rule (c), unless (p1)-(p5) hold when in both (p4) and (p5), $N(v)$ is replaced by $N^c(v)$.

Now, suppose that the robber is in their desired position and it is the cops turn to move. We will show that regardless of what they do, the robber will be able to move to another desired position. If cops move so that they occupy the whole S, the robber wants to move to v (see rule (a) of the robber's strategy). Since there is no edge between v and S, no cop occupied v in the previous round. Hence, the robber must be currently in $N(v)$ (see rule (b)) and can easily move to v.

Suppose then that cops move to a proper subset T of S but not to v; that is, some cops perhaps go to $U \subseteq W$. The robber is at v (if cops were located in the whole S in the previous round) or some vertex of W, and it wants to move to a vertex of $N^c(T) \cap N(v)$ that is adjacent to no cop (see rule (b)). If for every $T \subset S$, $U \subseteq W$ such that $|T \cup U| = k$, and $w \in W \setminus U$, there exists a vertex $w' \in N(v)$ adjacent to w but not adjacent to any vertex of $T \cup U$, the robber can move from w to w' and survive for at least one more round, reaching another desired position. Hence, if the robber's strategy fails at this point, then G satisfies properties (p1)-(p5) in Lemma 2.4.5 for some specific choice of $S, v, T, x(n), y(n), z(n)$. A similar argument can be used to analyze the case when cops move to $T \subset S$ and to v (see rule (c)) to derive that if the robber's strategy fails because of this event, then G must satisfy properties (p1)-(p5), where $N(v)$ is replaced by $N^c(v)$.

It remains to prove Lemma 2.4.5.

Proof of Lemma 2.4.5. Since edges of a random graph are generated independently, the probability that properties (p2) and (p3) hold simultaneously is equal to

$$\binom{n}{yn} 2^{-n} \binom{n}{zn} \left(2^{-k}\right)^{zn} \left(1 - 2^{-k}\right)^{(1-z)n}$$

$$\leq \binom{n}{yn} 2^{-n} \binom{n}{\xi n} \left(1 - 2^{-k}\right)^{n}, \qquad (2.10)$$

since $0 < \xi \leq 1/2$. Set

$$\alpha = -y \log y - (1-y) \log(1-y) - \xi \log \xi - (1-\xi) \log(1-\xi) - \log 2.$$

Using Stirling's formula (see Lemma 1.3.2) and taking the exponential part we obtain an upper bound of

$$\exp\left(\alpha n\right) \left(1 - 2^{-k}\right)^{n}.$$

It is straightforward to see that

$$f(t) = -t \log t - (1-t) \log(1-t)$$

tends to zero as $t \to 0$, and that $f(t)$ is maximized at $t = 1/2$ giving $f(1/2) = \log 2$. If $y \leq 1/2 - \varepsilon_1$ or $y \geq 1/2 + \varepsilon_1$ for some $\varepsilon_1 > 0$, then we derive that $f(y) \leq \log 2 - \varepsilon_2$ for some $\varepsilon_2 = \varepsilon_2(\varepsilon_1) > 0$, and so properties (p2) and (p2) hold w.e.s.p. after taking ξ sufficiently small such that, for example, $f(\xi) \leq \varepsilon_2/2$ (and so it is also the case that w.e.s.p. (p1)-(p5) hold). Indeed, for such choice of y and ξ we obtain a bound of

$$\exp\left(\left(\log 2 - \varepsilon_2 + \varepsilon_2/2 - \log 2\right)n\right)\left(1 - 2^{-k}\right)^n = 2^{(\log_2(1-2^{-k})-\varepsilon_2/2)n}.$$

Hence, we may assume that $1/2 - \varepsilon_1 \leq y \leq 1/2 + \varepsilon_1$ for some $\varepsilon_1 > 0$. (The constant ε_1 can be made arbitrarily small by assuming that ξ is small enough.) For this range of y, we lose a negligible term by using a weaker bound (see (2.10)). The probability that properties (p2) and (p3) hold simultaneously is at most

$$\binom{n}{yn} 2^{-n} \binom{n}{zn} \left(2^{-k}\right)^{zn} \left(1 - 2^{-k}\right)^{(1-z)n} \leq \binom{n}{\xi n} \left(1 - 2^{-k}\right)^n.$$

We need to consider (on top of that) property (p4) to obtain the desired bound for the event that (p1)-(p5) hold simultaneously. We first expose edges from v to the vertices of W. For a vertex $u \in N(v)$, let $A(u)$ be the event that u is dominated by S and let $B(u)$ be the event that u is nonadjacent to T. We can perform a "double exposure" for each vertex. First, we determine whether u is dominated by S; $\mathbb{P}\left(A(u)\right) = (1 - 2^{-k})$. If this is the case, then we determine whether u is also nonadjacent to T. This time,

$$\eta = \mathbb{P}\left(B(u)|A(u)\right) = \frac{\mathbb{P}\left(B(u) \wedge A(u)\right)}{\mathbb{P}\left(A(u)\right)} = \frac{2^{-\ell}(1 - 2^{-(k-\ell)})}{1 - 2^{-k}}.$$

(Note that, as expected, $\eta = 1$ if $T = \emptyset$; in any case, $\eta > 0$, since T is a proper subset of S.) Since the number of vertices not dominated by S is $zn \leq \xi n$ and $|N(v)| = yn \geq (1/2 - \varepsilon_1)n$, the expected number of neighbors of v that are dominated by S but not adjacent to T is at least

$$\left(\frac{1}{2} - \varepsilon_1 - \xi\right) n \cdot \eta > \frac{\eta n}{3},$$

provided ξ (and so ε_1 as well) are small enough. The events associated with two distinct vertices u_1 and u_2 are independent. It follows from the Chernoff bounds that the probability that (p4) holds for $x \leq \eta/6$ (conditioned on (p2) and (p3) holding) is at most

$$\mathbb{P}\left(\mathrm{Bin}(n/3, \eta) \leq \eta n/6\right) \leq \exp\left(-\frac{(1/2)^2(\eta n/3)}{3}\right)$$

$$= \exp\left(-\frac{\eta}{36}n\right).$$

Hence, if $x \leq \eta/6$, then properties (p2)-(p4) hold simultaneously w.e.s.p., after taking ξ sufficiently small such that, for example, $f(\xi) \leq \eta/40$. This implies that w.e.s.p. (p1)-(p5) hold.

We now may assume, in addition to assuming that $1/2 - \varepsilon_1 \leq y \leq 1/2 + \varepsilon_1$ for some $\varepsilon_1 > 0$, that $x \geq \eta/6$. (As before, the constant ε_1 can be made arbitrarily small by assuming that ξ is small enough but recall that η is not a function of ξ and depends only on k and ℓ.) As before, we note that the probability that properties (p2) and (p3) hold is at most $\binom{n}{\xi n}\left(1 - 2^{-k}\right)^n$, but this time we need to consider property (p5) to obtain the desired bound. To accomplish this, we first expose edges from $S \cup \{v\}$ to W (to estimate the probability that (p2) and (p3) hold) but do not yet expose any edge between vertices of W. Hence, we may estimate the probability that (p5) holds by exposing the edges of the subgraph induced by W, and all events are independent. The number of choices of U and w is $n^{O(1)}$. For a particular choice of U and w, the probability that no suitable v' can be found can be estimated by

$$\left(1 - 2^{-1}2^{-(k-\ell)}\right)^{xn-(k-\ell)-1} \leq \left(1 - 2^{-(k-\ell)-1}\right)^{\eta n/7}$$

$$= \exp\left(-\varepsilon_3 n\right),$$

where $\varepsilon_3 = \varepsilon_3(k, \ell) > 0$ and does not depend on ξ. Indeed, with probability 2^{-1} a given candidate vertex v' (neighbor of v, dominated by S but not adjacent to T) is adjacent to w, and with probability $2^{-(k-\ell)}$ it is not adjacent to any vertex of U. The bound holds, since we have at least $xn - (k - \ell) - 1 \geq \eta n/7$ candidates to test and the corresponding events are independent. The properties (p2), (p3), and (p5) hold w.e.s.p. after taking ξ sufficiently small such that, for example, $f(\xi) \leq \varepsilon_3/2$. Hence, w.e.s.p. (p1)-(p5) hold. $\qquad \square$

2.5   Random Geometric Graphs

In this section, we present the following result and its proof from [6] for the random geometric graph $\mathbb{G}(n, r)$ defined on the unit square $[0, 1]^2$.

Theorem 2.5.1 ([6]). *There exists an absolute constant $c_2 > 0$ such that if $r^5 > c_2 \frac{\log n}{n}$, then a.a.s. $c(\mathbb{G}(n, r)) = 1$.*

The same result was obtained earlier and independently in [26] but the proof from [6] described here gives a tight $O(1/r^2)$ bound for the number of rounds required to capture the robber, and can be generalized to higher dimensions. In the proof, we describe a strategy for the cop that is a winning one a.a.s. In [26], the known necessary and sufficient condition for a graph to be cop-win (see Theorem 2.0.1 for more details) is used; that is, it is shown that the random geometric graph is dismantlable a.a.s. In all dimensions, the proof gives that a.a.s. the cop can win in $O(1/r^2)$ moves and, as we mention below, this is tight; namely, a.a.s. the robber can ensure not to be caught in less moves. Therefore, the capture time for this range of parameters is $\Theta(1/r^2)$ a.a.s.

To prove Theorem 2.5.1, we first describe the proof of the corresponding result for the continuous infinite graph $\mathbb{G}_2(r)$ whose vertices are all of the points of $[0, 1]^2$, where two of them are adjacent if and only if their distance is at most r. This is a natural variant of the well-known problem of *the Lion and the Christian* in which (perhaps surprisingly) the Christian (counterpart of the robber in our game) has a winning strategy; see, for example, [31] for more details. In our game, the cop (counterpart of the lion) has a winning strategy. This essentially is a known result [127, 175], but the proof described may be modified to yield a proof of Theorem 2.5.1.

Theorem 2.5.2 ([6]). *For all $r > 0$, we have that $c(\mathbb{G}_2(r)) = 1$.*

Proof. We show that $c(\mathbb{G}_2(r)) = 1$ for any $r > 0$, by describing a winning strategy for the cop. In the first round, the cop places

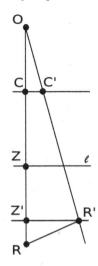

FIGURE 2.7: Catching the robber on $\mathbb{G}_2(r)$.

themselves at the center O of $[0, 1]^2$. After each move of the robber, when they are located at a point R, the cop catches them if possible (that is, if the distance between them and the robber is at most r); otherwise, they move to a point C that lies on the segment OR, ensuring that their distance from the robber is at least, say, $r^2/100$. In addition, we will show that the cop can do it and also ensure that in each step the square of the distance between the location of the cop and O increases by at least $r^2/5$. As this square distance cannot be more than $1/2$, this implies that the cop catches the robber in at most $O(1/r^2)$ moves.

We show that the cop may achieve the above in each round of the game. Suppose that the cop is located at C and the robber at R, where C lies on OR (and the distance between C and R is at least $r^2/100$). The cop can ensure this will be the case after their first move (unless the robber gives up prematurely without fighting and starts the game too close to the cop). A round, now, consists of a move of the robber from R to R' followed by a move of the cop from C to C'. Let Z denote the midpoint of CR and let ℓ be the line through Z perpendicular to OZ—see Figure 2.7. Without loss of generality, choose a coordinate system such that OR is a vertical line (and hence, ℓ is a horizontal one), and assume R (as well as C and Z) are below O. Note that R'—the new position of the robber—may be assumed to be below the line ℓ, since otherwise, the distance between C and R' is at most the distance between R and R'. This results in the cop capturing the robber. Hence, consider the case that R' is below ℓ and let C' be the intersection point of the horizontal line through C with the line OR'. Let Z' denote the point of intersection of the horizontal line through R' with the line containing OR. See Figure 2.7. Now it is straightforward to see that $RR' \geq R'Z' > CC'$, as the triangle $RR'Z'$ is a right-angle triangle and the two triangles OCC' and $OZ'R'$ are similar. Hence, the cop may move to C' if they decide to do so, as $CC' < RR' \leq r$. There are two cases.

Case 1. $CC' > r/2$. If the cop moves to C', then their square distance to O increases by $CC'^2 > r^2/4$. If C' is too close to R', then we shift them toward O (that is, by less than $r^2/100$) to ensure the distance between C' and R' is at least $r^2/100$. Note that such a shift decreases the square distance from O by less than $2r^2/100 = r^2/50$; hence, the square distance still increases by at least $r^2/4 - r^2/50 > r^2/5$. Thus, in this case the cop can make a step as required.

Case 2. $|CC'| \leq r/2$. In this case, the cop can move to C' and then walk along the line OR' at least distance $r/2$ toward R' (without passing it, since otherwise, the game ends and the cop wins). Since $OC' \geq OC$, in this case, the cop increases their distance from O by more than $r/2$. Hence, the cop increases their square distance by more than $r^2/4$. As before, it may be the case that they get too close to R' and then they back up by less than $r^2/100$, which is still fine.

This shows that in $\mathbb{G}_2(r)$ the cop can indeed increase their square distance from O by at least $r^2/5$ in each step (which is not a winning step ending the game), staying on the segment connecting the center and the robber (and being closer to the center than the robber). This discussion implies that the game ends with a cop capturing the robber in $O(1/r^2)$ moves. $\qquad\square$

In $\mathbb{G}(n, r)$, the cop follows an analogous strategy, but will always place themselves at a vertex of the graph that is sufficiently close to where they want to be in the continuous game. More precisely, for each point X of the unit square whose distance from the center O is at least $r/2$ (to ensure that the triangle $T(x)$ defined below will indeed be well-defined; in our argument this will always be the case) and whose distance from the boundary is at least $r^2/10^3$, we define an isosceles triangle $T(X)$ as follows. One vertex of the triangle is X, and the segment of length $r^2/100$ on the line OX starting at X and going toward O is the height of $T(X)$. The base is orthogonal to it and of length $r^3/10^5$. Despite the fact that there are infinitely many triangles, it is not difficult to show that if the area of such a triangle is large enough, then a.a.s. $\mathbb{G}_2(n,r)$ contains a vertex inside each such triangle.

Lemma 2.5.3 ([6]). *There exists an absolute constant $d > 0$ such that a.a.s. every triangle $T(X)$ contains a vertex of $\mathbb{G}(n,r)$, provided $r^5 > d\frac{\log n}{n}$.*

Proof. Consider a fixed collection F of $O((1/r)^6)$ rectangles, each of area $\Omega(r^5)$, such that every triangle $T(X)$ contains at least one of these rectangles. To do so, for each point Y in an $10^6 r^3$ by $10^6 r^3$ grid in the unit square, take the rectangle of width $r^3/10^6$ and height $r^2/10^6$ in which Y is the midpoint of the edge of length $10^6 r^3$ and the other edge is in direction YO. It is evident that every $T(X)$ under consideration (X not too close to O nor to the boundary) fully contains at least one such a rectangle.

To complete the proof it is enough to show that a.a.s. each rectangle in F contains at least one vertex of $\mathbb{G}(n,r)$. The area of each such rectangle is $r^5/10^{12}$, and hence, the probability it contains no vertex is

$$\left(1 - \frac{r^5}{10^{12}}\right)^n \le e^{-d\log n/10^{12}}.$$

Since there are $O((1/r)^6) = O(n^2)$ rectangles, the desired result follows by the union bound by setting $d = 10^{13}$. $\qquad\square$

Now, let us come back to the main result of this section, since we have all the necessary ingredients.

Proof of Theorem 2.5.1. Since we aim for a statement that holds a.a.s., it follows from Lemma 2.5.3 that we may assume that every triangle $T(X)$ contains at least one vertex. The cop plays the continuous strategy, but whenever they want to place themselves at a point X, they choose an arbitrary vertex $x \in V$ of $T(X)$ to go to. The line $R'x$ is now not necessarily identical to the line $R'O$, but the angle between them is sufficiently small to ensure that in the computations above for the continuous case we do not lose much. That was the reason we ensured that R' and X are never too close in the continuous algorithm, and as the triangle $T(X)$ is thin, the angle between these two lines is smaller than $r/10^3$.   □

Finally, note that a.a.s. the robber may remain uncaptured for $\Omega(1/r^2)$ rounds. We focus on the proof in two dimensions (although the proof for higher dimensions is analogous). As before, we start with the continuous variant of the game. In the first round, the robber places themselves at distance bigger than r from the cop ensuring they are not too far from the center O of the square. At each step, when the robber located at R has to move, they move distance r in the direction perpendicular to RC, where the choice of the direction (among the two options), is such that their square distance from O increases by at most r^2 (that is, the angle ORR' is at most $\pi/2$). This suffices for the continuous case, as it is evident that the distance from the cop will exceed r after each such step. In the discrete case, the robber chooses a nearby point, making sure their distance from the cop is at least what it would have been in the continuous case.

2.6   Percolated Random Geometric Graphs

In the final section of this chapter, we consider the cop number of percolated random geometric graphs $\mathbb{G}(n, r, p)$ (introduced in Chapter 1). These results were presented first in [131]. The principal result here is the following.

Theorem 2.6.1 ([131]). *For every $\varepsilon > 0$, and functions $p = p(n)$*

and $r = r(n)$ *such that* $p^2 r^2 \geq n^{-1+\varepsilon}$ *and* $p \leq 1 - \varepsilon$, *we have that a.a.s.*

$$c(\mathbb{G}(n, r, p)) = \Theta\left(\frac{\log n}{p}\right).$$

Observe that the asymptotics of the cop number for a large range of the parameters does not depend on r but only on p. It was conjectured in [131] that, under the conditions of our theorem, a.a.s. the cop number is $(1 + o(1)) \log_{1/(1-p)} n$.

As before, for $0 < p \leq 1 - \varepsilon$ for some $\varepsilon > 0$, it is convenient to define

$$\mathbb{L} = \mathbb{L}(n) = \log_{1/(1-p)} n,$$

and to state our intermediate results in terms of $\mathbb{L}$. For the proof of the lower bound, we use the property that was used for dense binomial random graphs; see Section 2.1. Writing $k = \lfloor \varepsilon \mathbb{L}/2 \rfloor$, where $\varepsilon > 0$ is as provided by the conditions of Theorem 2.6.1. We have that a.a.s., $\mathbb{G}(n, r, p)$ is $(1, k)$-e.c. In particular, a.a.s. $c(G(n, r, p)) > k$. We omit the proof as it is similar to the one for binomial random graphs.

We concentrate here on the upper bound. We show that, a.a.s., $(21,000)\mathbb{L}$ cops suffice to capture the robber. Before presenting a winning strategy of the cops, we give some preparatory lemmas. We denote by

$$B(x, s) = \{y \in \mathbb{R}^2 : \|x - y\| \leq s\}$$

the ball of radius s around x.

Lemma 2.6.2 ([131]). *A.a.s. for every* $v, w \in V$ *with* $\|v - w\| \leq (0.99)r$, *there is a subset* $A \subseteq N(v)$ *with* $|A| \leq 1000\mathbb{L}$ *that dominates* $\{w\} \cup N(w)$.

Proof. We will consider the number of "bad" (ordered) pairs $(v, w) \in V^2$ such that $\|v - w\| \leq 0.99 \cdot r$, yet no set A as required by the lemma exists. We will compute the probability that (X_1, X_2) form such a bad pair. To do this, we reveal the graph in three stages. In the first stage, we reveal V, which is the positions of the points. In the second stage, we reveal all edges (or coin flips) that have X_1 as an endpoint. In the third stage, we reveal all other edges.

We condition on the event that $\|X_1 - X_2\| \le 0.99 \cdot r$. Note that this does not affect the locations of the other points or the status of any of the coin flips. We now define, for $i, j \in \{-1, +1\}$:

$$B_{i,j} = B(X_2 + i(r/10^{10})e_1 + j(r/10^{10})e_2, r/10^{10}),$$
$$U_{i,j} = N(X_1) \cap B_{i,j}.$$

Here, of course, $e_1 = (1, 0)$ and $e_2 = (0, 1)$. See Figure 2.8 for a depiction. The $B_{i,j}$ have been chosen such that, no matter where

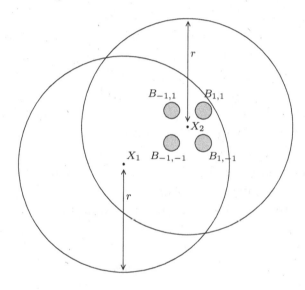

FIGURE 2.8: The definition of the $U_{i,j}$, not drawn to scale.

in the unit square X_2 falls, for every $z \in B(X_2, r) \cap [0, 1]^2$ there is at least one pair $(i, j) \in \{-1, 1\}^2$ such that $B_{i,j} \subseteq B(z, r) \cap [0, 1]^2$.

Observe that conditioning on the event that the position of X_2 is such that $B_{i,j} \subseteq [0, 1]^2$, we have that $|U_{i,j}|$ is the random variable $\mathrm{Bin}(n-2, p\pi(r/10^{10})^2)$. In particular, $\mathbb{E}|U_{i,j}| = \Omega(pnr^2) = \Omega(n^\varepsilon/p) \gg \mathbb{L}$. Using the Chernoff bounds, it follows that

$$\mathbb{P}(|U_{i,j}| < \mathbb{E}|U_{i,j}|/2) \le \exp[-\Omega(n^\varepsilon)].$$

Note that, to find the $U_{i,j}$ we have to reveal the first two stages, but we do not need to reveal the coin flips corresponding to potential edges not involving X_1. Assuming that in the first two stages we managed to find $U_{i,j}$'s of size at least half of the expected size, we can now fix, for each $i, j \in \{-1, 1\}$ with $B_{i,j} \subseteq [0,1]^2$, an arbitrary subset $A_{i,j} \subseteq U_{i,j}$ with $|A_{i,j}| = 250\mathbb{L}$. We let A be the union of these $A_{i,j}$'s. Since each $z \in B(X_2, r) \cap [0,1]^2$ satisfies $A_{i,j} \subseteq B(z,r)$ for at least one pair $(i,j) \in \{-1,1\}^2$, the probability that there is a vertex $X_j \in N(X_2) \cup \{X_2\}$ not connected by an edge to any vertex of A is at most $n(1-p)^{250\mathbb{L}} = n^{-249}$. It follows that

$$\mathbb{P}((X_1, X_2) \text{ is a bad pair}) \leq 4e^{-\Omega(n^\varepsilon)} + n^{-249} \leq 2n^{-249},$$

the last inequality holding for sufficiently large n. This shows that the expected number of bad pairs is at most $\binom{n}{2} 2n^{-249} = o(1)$. The lemma now follows by Markov's inequality. $\qquad\square$

Lemma 2.6.3 ([131]). *A.a.s. for every $v \in V$ and every $z \in B(v,r) \cap [0,1]^2$ there is a vertex $w \in N(v) \cap B(z, r/1000)$.*

Proof. We dissect $[0,1]^2$ into squares of side $s = 1/\lceil \frac{10^{10}}{r} \rceil$ (note $s \leq r/10^{10}$ and $s = \Theta(r)$). Observe that if $v, z \in [0,1]^2$ with $\|v - z\| \leq r$ then there is at least one square of our dissection contained in $B(v,r) \cap B(z, r/1000)$. It suffices to count the number Z of "bad pairs" consisting of a vertex v and a square S of the dissection contained in $B(v,r)$ such that $N(v) \cap S = \emptyset$, and to show this number is zero a.a.s. Note that the number of squares is $O(1/r^2) = O(n)$. Hence, we have that

$$\begin{aligned}
\mathbb{E}Z &= O(n^2) \cdot (1 - ps^2)^{n-1} = O(n^2) \cdot \exp[-\Omega(pnr^2)] \\
&= \exp[O(\log n) - \Omega(n^\varepsilon)] = o(1),
\end{aligned}$$

and the proof of the lemma follows by applying Markov's inequality. $\qquad\square$

Lemma 2.6.4 ([131]). *A.a.s. for every $v, w \in V$ with $\|v - w\| \leq 1.99r$ there is a vertex u such that $uv, uw \in E$ and $\|u - (v + w)/2\| \leq r/1000$.*

Proof. We use the same dissection into small squares of side $s = 1/\lceil \frac{10^{10}}{r} \rceil$ as in the proof of Lemma 2.6.3. Note that if $v, w \in [0, 1]^2$ then $B((v + w)/2, r/1000)$ contains at least one square of the dissection. It is sufficient to count the number Z of "bad triples" consisting of two vertices $v \neq w$ at distance at most $1.99r$ and one square S of the dissection that is contained in $B((v+w)/2, r/1000)$, such that $N(v) \cap N(w) \cap S = \emptyset$. We have

$$
\begin{aligned}
\mathbb{E}Z &\leq O(n^3) \cdot (1 - p^2 s^2)^{n-2} = O(n^3) \cdot \exp[-\Omega(p^2 nr^2)] \\
&= \exp[O(\log n) - \Omega(n^\varepsilon)] = o(1),
\end{aligned}
$$

proving the lemma. □

The proof of the following lemma is elementary and so is omitted.

Lemma 2.6.5 ([131]). *If $x_1, x_2, y_1, y_2 \in \mathbb{R}^2$ are such that $\|x_1 - x_2\|, \|y_1 - y_2\| \leq r$ and the line segments $[x_1, x_2], [y_1, y_2]$ cross, then $\|x_i - y_j\| \leq r/\sqrt{2}$ for at least one pair $(i, j) \in \{1, 2\}^2$.*

We say that a cop C *guards* a path P in a graph G if whenever the robber moves onto P, then they either step onto C or is caught by C on their responding move. The terminology "shortest path" here refers to the graph distance. Aigner and Fromme in [1] proved the following useful result.

Lemma 2.6.6 ([1]). *Let G be any graph, $u, v \in V(G)$, $u \neq v$ and $P = \{u = v_0, v_1, \ldots v_s = v\}$ a shortest path between u and v. A single cop C can guard P after at most $\operatorname{diam}(G) + s$ moves.*

Proof of the upper bound in Theorem 2.6.1. Since we aim for a statement that holds a.a.s., we assume that we are given a realization of $\mathbb{G}(n, p, r)$ that is connected (which is true a.a.s. for our choice of parameters as, for instance, follows from [154]) and for which the conclusions of Lemmas 2.6.2, 2.6.3, and 2.6.4 hold. We will show that under these conditions, a team of $21000\mathbb{L}$ cops is able to capture the robber.

Our strategy is an adaptation of the strategy of Aigner and Fromme showing $c(G) \leq 3$ for connected planar graphs. We will have three teams T_1, T_2, T_3 of cops, each consisting of $(7,000)\mathbb{L}$ cops that are each charged with guarding a particular shortest path. In more detail, a team T_i that patrols a shortest path $P = v_0 v_1 \ldots v_m$ is divided into 7 subteams $T_{i,-3}, T_{i,-2}, T_{i,-1}, T_{i,0}, T_{i,1}, T_{i,2}, T_{i,3}$ of $1000\mathbb{L}$ cops each. These subteams will move in unison (that is, the cops in a particular subteam will always be on the same vertex of P). The team $T_{i,0}$ moves exactly according to the strategy given by Lemma 2.6.6. That is, after an initial period, the $T_{i,0}$-cops are able to move along P in such a way that, whenever the robber moves onto a vertex $v_k \in P$ then either the entire team $T_{i,0}$ is already on v_k or they are on v_{k-1} or v_{k+1}. Team $T_{i,j}$ will be j places along $T_{i,0}$ (that is, if $T_{i,0}$ is on v_k then $T_{i,j}$ is on v_{k+j}). If this is not possible because $T_{i,0}$ is too close to the respective endpoint of P then $T_{i,j}$ just stays on that endpoint (that is, if $T_{i,0}$ is on v_k and $k + j > m$ then $T_{i,j}$ is on v_m and if $k + j < 0$ then $T_{i,j}$ stays on v_0). We now claim that the robber cannot cross (in the sense that the edge they use crosses an edge of P when both are viewed as line segments) the path P without getting caught by the cops of team T_i.

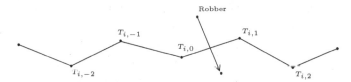

FIGURE 2.9: The robber tries to cross a path guarded by team T_i.

To see this, we first observe that if the robber moves along an edge that crosses some edge of P, then either their position before the move or their position right after the move is within distance at most $r/\sqrt{2}$ of some vertex of P by Lemma 2.6.5. Next, we remark that whenever the robber moves onto a vertex u within distance $0.99 \cdot r$ of some vertex $v_k \in P$, then the cops can capture them in at most two further moves. This is because from u, the robber could move to v_k in at most two moves (by Lemma 2.6.4). As the cops of subteam $T_{i,0}$ follow the strategy prescribed by Lemma 2.6.6, they are guaranteed to be on one of $v_{k-3}, v_{k-2}, v_{k-1}, v_k, v_{k+1}, v_{k+2}, v_{k+3}$

when the robber arrives on u. But then there must be some team $T_{i,j}$ that inhabits the vertex v_k at the very moment when the robber arrived on u. This team now acts as follows: at the time the robber arrives on u, the subteam occupies the set A provided by Lemma 2.6.2 (this one time the subteam members do not all stay on the same vertex; instead they spread following the strategy implied by the lemma) and in the next move the cops are able to capture the robber, since they now dominate the closed neighborhood of the vertex they inhabit. Thus, each of our three teams can (after an initial number of rounds) prevent the robber from crossing a chosen path. What is more, the robber can never get to within distance $0.99r$ of any vertex of such a path.

The remainder of the proof focuses on proving that we may restrict the movements of the robber to a subgraph that decreases their order. Hence, we shrink *the robber territory* of the graph. This approach was used first by Aigner and Fromme [1] in their proof that planar graphs have cop number at most three (see also [52]).

Consider district vertices u, v. We let P_1 be the shortest uv-path, and we let P_2 be the shortest uv-path in the graph with all internal vertices of P_1, and all edges that cross P_1 removed. (Using Lemmas 2.6.3 and 2.6.4 it is easily seen that at least one such path exists.) Note that $P_1 \cup P_2$ constitutes a Jordan curve and hence, $\mathbb{R}^2 \setminus (P_1 \cup P_2)$ consists of two connected regions: the interior and the exterior. Once the game starts, we send T_1 to guard P_1 and T_2 to guard P_2. After an initial phase, the robber will either be trapped in the interior region or the exterior region of $\mathbb{R}^2 \setminus (P_1 \cup P_2)$. Let us denote the region they are trapped on by R. If it happens that every vertex inside R is within distance $0.99r$ of some vertex of $P_1 \cup P_2$, then we are done by the previous argument. We assume this property fails.

We next remove all vertices not on $P_1 \cup P_2$ or inside R, and we remove all edges that cross P_1 or P_2. There can be vertices that lie inside R, but with an edge between them that passes through $P_1 \cup P_2$. We let P_3 be a uv-path in the remaining graph that is shortest among all uv-paths that are distinct from P_1, P_2. To see that at least one such path exists, we first find a vertex $u \in R$

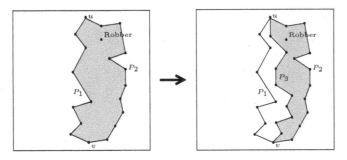

FIGURE 2.10: The adapted Aigner and Fromme strategy.

that has distance at least $(0.99)r$ to every vertex of $P_1 \cup P_2$. Then we use Lemmas 2.6.3 and 2.6.4 to construct vertex-disjoint paths between u and two distinct vertices of $P_1 \cup P_2$. See Figure 2.10 for a depiction.

Note that P_3 does not cross P_1 or P_2, but it may share some edges with them. In particular, $R \setminus P_3$ consists of two or more connected parts, each of which is either bounded by parts of P_1 and P_3 or by parts of P_2 and P_3. We now send T_3 to guard P_3. After an initial phase, the robber will be caught in one of the connected parts R' of $R \setminus P_3$. Without loss of generality, R' is bounded by P_2, P_3. Discarding unneeded parts of P_2, P_3 (namely, those that do not bound R') and relabeling we can also assume that P_2, P_3 only meet in their endpoints u, v. If every vertex inside R' is within distance $(0.99)r$ of a vertex of P_2, P_3, then we are done. Otherwise, the team T_1 abandons guarding path P_1, we remove all vertices not on P_2, P_3 or inside R' and all edges that cross P_2 or P_3, we find a uv-path P_4 in the remaining graph, shortest among all uv-paths different from P_2, P_3, and we let T_1 guard P_4. Now, P_4 will dissect R' into two or more connected paths. The process repeats resulting in the robber being captured or being restricted to a region of smaller size.

It is evident that in each iteration of this process, at least one edge is removed from the subgraph under consideration. Hence, the process must stop eventually and the robber loses. $\square$

Chapter 3

Variations of Cops and Robbers

The game of Cops and Robbers is lucky to possess many beautiful and natural variations. For example, we could speed up the robber, give the cops greater reach to capture the robber, or even have the players move only on edges. We focus on only a few variations in this chapter, with no claim of being exhaustive. We focus especially on those variations where the settings of the probabilistic method have been applied.

In Section 3.1, we consider a variant of Cops and Robbers played on edges. In Section 3.2, we provide bounds on the cop number where the robber is allowed to move with unbounded speed. The next section considers Lazy Cops and Robbers, where only one cop may move in a given round. In Section 3.4, we consider a variant of the game on a hypercube that restricts the initial positions and the allowed moves. Another variant, the containment game, is discussed in Section 3.5.

3.1 Playing on Edges

Consider the game of Cops and Robbers where the players start the game on edges and use vertices to move from edges to edges. The other rules of the game remain unchanged. We define the *edge cop number*, $\bar{c}(G)$, to be the counterpart of the cop number for the classic game. It is evident that playing the new game on a graph G is equivalent to playing the classic game on $L(G)$, the line graph of G; that is, $\bar{c}(G) = c(L(G))$. However, it seems that it is not beneficial to look at the problem this way, especially if one is interested in discovering properties of the new graph parameter and trying to relate it to some other parameters, including the classical cop number. We note that this variant of the game was recently introduced in [79], and the results of this section come from that work.

We first state the following result, which will be needed later on. We direct the reader to [79] for the proof and more details.

Theorem 3.1.1 ([79]). *If G is a connected graph, then*

$$\left\lceil \frac{c(G)}{2} \right\rceil \leq \bar{c}(G) \leq c(G) + 1.$$

Since there is only a multiplicative constant factor difference between the two cop numbers, all known bounds for random graphs also hold for the edge cop number (up to a constant factor). However, the asymptotic value of the classic cop number is known only for dense random graphs; that is, when $p = n^{-o(1)}$. We will show that in this case, the edge cop number is essentially two times smaller than its classic counterpart. Recall that for $n \geq 1$, $\mathbb{L}n = \log_{1/(1-p)} n$.

Theorem 3.1.2 ([79]). *If $p = p(n)$ is such that $p = n^{-o(1)}$ and $p < 1 - \varepsilon$ for some $\varepsilon > 0$, and $G \in \mathbb{G}(n, p)$, then a.a.s.*

$$\bar{c}(G) \sim \frac{1}{2} \mathbb{L}n.$$

An elementary bound is $c(G) \leq \gamma(G)$. We would like an analogous bound for the edge cop number. An *edge cover* of a vertex set S is a set of edges X in a graph induced by S such that each vertex in S is incident with at least one edge in X. See Figure 3.1 for an example. A *minimum edge cover* of S is an edge cover of smallest possible cardinality. The *edge cover number* of a set S that induces a graph with no isolated vertices is the cardinality of a minimum edge cover of S. In this section, we use $\rho(S)$ for the edge cover number of S. Finally, for a graph G with no isolated vertices, let

$$\xi(G) = \min\{\rho(S) : S \in \mathcal{S}\},$$

where $\mathcal{S}$ is the family of dominating sets that induce a graph with

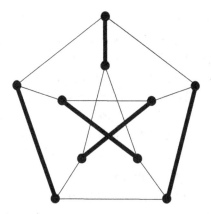

FIGURE 3.1: A minimum edge cover of the Petersen graph.

no isolated vertices. (The reason G is assumed to have no isolated vertices is to make sure $\xi(G)$ is well defined.)

Now, we are ready to state a lemma (whose proof is omitted) that will be useful in investigating dense random graphs. Note that the result is sharp, since for any $n \geq 4$ we have that $\bar{c}(K_n) = 2$ and $\xi(K_n) = 1$.

Lemma 3.1.3 ([79]). *If $G = (V, E)$ is a connected graph, then*

$$\bar{c}(G) \leq \xi(G) + 1.$$

Now we come back to the proof of the main result of this section.

Proof of Theorem 3.1.2. It follows from Theorem 2.1.1 that a.a.s. $c(G) \sim \mathbb{L}n$. By Theorem 3.1.1 we have $\bar{c}(G) \geq c(G)/2$, and the lower bound immediately follows.

Let

$$k = \mathbb{L}n + 2\mathbb{L}\omega,$$

where $\omega = \omega(n) = p^{-1}\log n$. Note that $\omega p = \log n \to \infty$ as $n \to \infty$, but note also that $\mathbb{L}\omega = o(k)$ so $k \sim \mathbb{L}n$. Indeed, if $p = \Omega(1)$, then $\mathbb{L}\omega = \Theta(\log \log n) = o(\log n) = o(k)$. However, for $p = o(1)$ we have $\omega = n^{o(1)}$ and

$$\mathbb{L}\omega \sim \frac{\log \omega}{p} = o\left(\frac{\log n}{p}\right) = o(k).$$

For the upper bound we are going to use Lemma 3.1.3. Our goal is to find a.a.s. a set that is dominating and has small edge cover. Fix any set X of cardinality k. First of all, we will show that a.a.s. there exists a set $Y \subseteq X$ of cardinality $\mathbb{L}n + \mathbb{L}\omega = k - \mathbb{L}\omega$ that has a perfect matching. To accomplish this, we consider the following process. Choose an arbitrary vertex x_1 and expose all edges from x_1 to other vertices of X. If at least one edge is found, then select any neighbor x_2 of x_1 and pick another vertex $x_3 \in X \setminus \{x_1, x_2\}$. This time we expose all edges from x_3 to $X \setminus \{x_1, x_2, x_3\}$ with a hope that at least one edge is discovered and the process can be continued. We repeat the process until, in some round, no edge is found. The only reason for the process to terminate at a given round is when no edge is found. We will show that a.a.s. the process does not stop before all but $r = \mathbb{L}\omega$ vertices of X are matched.

Indeed, the probability that the process does not stop earlier is equal to

$$\left(1 - (1-p)^{k-1}\right)\left(1 - (1-p)^{k-3}\right) \cdot \ldots \cdot \left(1 - (1-p)^{r+1}\right)$$

$$= \prod_{i=r/2}^{k/2-1}\left(1 - (1-p)^{2i+1}\right)$$

$$= \exp\left(-(1+o(1))(1-p)\sum_{i=r/2}^{k/2-1}(1-p)^{2i}\right)$$

$$= \exp\left(-(1+o(1))(1-p)\frac{((1-p)^2)^{r/2} - ((1-p)^2)^{k/2}}{1 - (1-p)^2}\right)$$

$$= \exp\left(-(1+o(1))\frac{1-p}{1-(1-p)^2}\omega^{-1}\right)$$

$$= \exp\left(-O\left((p\omega)^{-1}\right)\right),$$

which tends to one, since $p\omega$ tends to infinity as $n \to \infty$.

Second, we will show that a.a.s. $Y \subseteq X$ dominates all vertices of $V \setminus X$. (Note that edges between Y and $V \setminus X$ are not exposed yet. However, edges between Y and $X \setminus Y$ are already exposed and so we cannot investigate if $X \setminus Y$ is dominated by Y or not.) Since $|Y| = \mathbb{L}n + \mathbb{L}\omega$ and $|X| = \mathbb{L}n + 2\mathbb{L}\omega$, the probability that this occurs is

$$\left(1 - (1-p)^{|Y|}\right)^{n-|X|} \geq 1 - (n - |X|)(1-p)^{|Y|}$$
$$\geq 1 - n(1-p)^{\mathbb{L}n+\mathbb{L}\omega}$$
$$= 1 - 1/\omega,$$

which tends to one as $n \to \infty$.

Third, we will show that a.a.s. the set $X \setminus Y$ induces a graph with no isolated vertex. (Note that edges within $X \setminus Y$ are not exposed yet.) The probability that a given vertex in $X \setminus Y$ is isolated is

$$(1-p)^{|X \setminus Y|-1} = O(\omega^{-1}).$$

Hence, the expected number of isolated vertices is

$$O(\omega^{-1}\mathbb{L}\omega) = O\left(\frac{\log \omega}{\log n}\right) = o(1),$$

and the result holds by the first moment method.

Putting the three claims together we will obtain the upper bound. It follows that a.a.s. X is a dominating set (since $Y \subseteq X$ dominates $V \setminus X$ a.a.s.). Further, a.a.s.

$$\rho(X) \leq \rho(Y) + \rho(X \setminus Y) \leq \frac{|Y|}{2} + |X \setminus Y| = \frac{1}{2}\mathbb{L}n + \frac{3}{2}\mathbb{L}\omega,$$

since a.a.s. Y has a perfect matching, and $X \setminus Y$ induces no isolated vertex so we can simply take any incident edge for each vertex of $X \setminus Y$. Hence, a.a.s.

$$\xi(G) \leq \rho(X) \leq (1/2 + o(1))\mathbb{L}n.$$

The proof of the upper bound follows, since $\bar{c}(G) \leq \xi(G) + 1$ by Lemma 3.1.3. □

3.2 Cops and Fast Robbers

We next consider a variant where the robber can move at essentially infinite speed. In this variant, which we call *Cops and Fast*

Robbers, the rules are identical as before, except for the movement of the robber: on the robber's turn, they may move to any vertex connected to their present position, so long as the path does not go through a vertex occupied by a cop. See Figure 7.1.

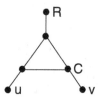

FIGURE 3.2: The robber can move to u but not to v.

We refer to the cop number for the game of Cops and Fast Robbers as c_∞. It is straightforward to see that for a graph G, $c(G) \leq c_\infty(G)$. Note that this bound is, however, far from tight: we have that $c(P_n \Box P_n) = 2$, while $c_\infty(P_n \Box P_n) = \Omega(n)$; see [4].

The parameter c_∞ was first studied in [90], and has since been investigated in [4, 95, 100]. Our focus here is to survey some results from [4] on c_∞ in random graphs. The cases in the following theorem essentially cover all $p = p(n)$, since if $np < 27$, then a.a.s. $G \in \mathbb{G}(n,p)$ has $\Omega(n)$ isolated vertices, and so has c_∞ equaling $\Theta(n)$. Recall that $\mathbb{L}(n) = \log_{1/(1-p)} n$.

Theorem 3.2.1 ([4]). *If $G \in \mathbb{G}(n,p)$, then the following statements hold a.a.s.*

(i) *If $27 \leq np = O(1)$, then there exist positive constants d_1, d_2 such that*

$$d_1 \log(np)/p \leq c_\infty(G) \leq d_2 \log(np)/p.$$

(ii) *If $np \gg 1$ and $p \leq 1 - \varepsilon$ for some $\varepsilon > 0$, then*

$$c_\infty(G) \sim \gamma(G) \sim \mathbb{L}(np).$$

We focus on the lower bound in Theorem 3.2.1 (1). The following elementary lemma (with proofs omitted) is useful.

Lemma 3.2.2 ([4]). *If $G \in \mathbb{G}(n,p)$, and $d = np \geq 27$, then the following hold a.a.s.*

(i) *For every two disjoint sets $A_1, A_2 \subseteq V(G)$ of cardinality at least $en \log d/d$, there exists an edge joining A_1 and A_2.*

(ii) *For every $A \subseteq V(G)$ of cardinality at least $3en \log d/d$, there is a component of $G[A]$ of order at least $en \log d/d$.*

Proof of Theorem 3.2.1 (1). Choose $\alpha \in (0,1)$ such that for sufficiently large n,

$$d^{-\alpha} - (3e + \alpha)\frac{\log d}{d} > 0,$$

and

$$-2\left(d^{-\alpha} - (3e + \alpha)\frac{\log d}{d}\right)^2$$
$$+ \alpha\frac{\log d}{d}(1 + \log d - \log \alpha - \log \log d) < 0.$$

Note that α exists for any d such that $d > 3e \log d$; the first integer-valued d that works is $d = 27$ (as appears in the statement of Theorem 3.2.1 (1)). By using the Chernoff bounds and a short argument (see [4]), it follows that a.a.s. for every set $X \subseteq V(G)$ of cardinality $\alpha n \log d/d$ we have that

$$|V(G) \setminus N(X)| > (3e + \alpha)n \log d/d.$$

Now if there are at most $\alpha n \log d/d$-many cops, then we show that the robber has a strategy to evade capture. Let G/S be the subgraph induced by $V(G) \setminus N[S]$. We note a few facts that hold a.a.s.

(i) By the discussion in the last paragraph, for all $S \subseteq V(G)$ of cardinality at most $\alpha n \log d/d$, G/S has order at least $3en \log d/d$.

(ii) By Lemma 3.2.2 (2) and item (1), for all $S \subseteq V(G)$ of cardinality at most $\alpha n \log d/d$, G/S has a component of order at least $en \log d/d$.

(iii) By Lemma 3.2.2 (1), for every two disjoint sets $A_1, A_2 \subseteq V(G)$ of cardinality at least $en \log d/p$, there exists an edge joining A_1 and A_2.

The strategy of the robber is to play so that if the cops are in a subset S, then the robber is in a component X of G/S with order at least $en \log d/d$. If the robber can maintain this condition, then the robber wins: the robber keeps moving to new such components outside of the reach of the cops.

Now assume that the condition holds at the end of round i. In round $i+1$, the cops move to some subset S' (note that $X \cap S'$ is empty). By item (2) above, let X' be a component of G/S' with order at least $en \log d/d$. If X and X' are not disjoint, then the condition holds trivially. If X and X' are disjoint, then by item (3), there is an edge joining them. Hence, in either case the robber may move to X' and evade capture.   □

3.3   Lazy Cops and Robbers

Our next version of Cops and Robbers slows the movement of the cops down, analogous to the game of chess, where at most one chess piece can move in a round. In the game of *Lazy Cops and Robbers*, the rules are almost exactly as in Cops and Robbers, with the exception that at most one cop moves in any round. The *lazy cop number* of the graph G, written $c_L(G)$, is the minimum number of cops needed to win the game of Lazy Cops and Robbers on G. This game and parameter were first considered by Offner and Ojakian [153]. In [17], binomial random graphs were investigated. In [153] it was proved for the hypercube Q_n that

$$2^{\lfloor \sqrt{n}/20 \rfloor} \leq c_L(Q_n) = O(2^n \log n/n^{3/2}). \qquad (3.1)$$

A recent result of [16] improves the lower bound in (3.1).

Theorem 3.3.1 ([16]). *For all $\varepsilon > 0$, we have that*

$$c_L(Q_n) = \Omega \left(\frac{2^n}{n^{5/2+\varepsilon}} \right).$$

Thus, the upper and lower bounds on $c_L(Q_n)$ differ by only a polynomial factor. The proof of Theorem 3.3.1 found in [16], presented below, relies on the probabilistic method coupled with a potential function argument.

Proof. We present a winning strategy for the robber provided that the number of cops is not too large. Let $\varepsilon \in (0,1)$ be fixed, and suppose there are $k = k(\varepsilon, n)$ cops (where k will be chosen later). We assume throughout that n is sufficiently large. We introduce a potential function that depends on each cop's distance to the robber. Let N_i represent the number of cops at distance i from the robber. With $\rho = \rho(n) = o(n)$, $\rho \to \infty$ as $n \to \infty$, a function to be determined later (but such that $n/2 - \rho$ is a positive integer), we let

$$P = \sum_{i=1}^{n} N_i w_i$$

where, for $1 \leq i \leq \frac{n}{2} - \rho$,

$$w_i = A \cdot \binom{n-2}{i}^{-1} \prod_{j=1}^{i} (1 + \varepsilon_j), \qquad A = \frac{n-2}{1+\varepsilon_1},$$

and

$$\varepsilon_i = \frac{2+\varepsilon}{n - 2i - 2} = o(1).$$

It will be desired that the sequence w_i is decreasing. Let $x = 1 + \frac{2+\varepsilon}{2\rho} + o(1/\rho)$. Since for $i = \frac{n}{2} - \rho$ we have

$$\frac{w_{i-1} - w_i}{w_i} = \frac{n-1-i}{i}(1+\varepsilon_i)^{-1} - 1$$

$$= \left(1 + \frac{2\rho}{n/2} + o(\rho/n)\right) x^{-1} - 1$$

$$= \left(1 + \frac{4\rho}{n} + o(\rho/n)\right) x - 1$$

$$= \frac{4\rho}{n} - \frac{2+\varepsilon}{2\rho} + o(\rho/n) + o(1/\rho),$$

the desired property holds for $1 \leq i \leq \frac{n}{2} - \rho$ provided that, say,

$\rho \geq \sqrt{n}$. For $\frac{n}{2} - \rho \leq i \leq n$, we let w_i decrease linearly from $w_{n/2-\rho}$ to $w_n = 0$. Formally, for such i, we have

$$w_i = (n - i) \cdot \frac{w_{n/2-\rho}}{\frac{n}{2} + \rho}.$$

We say that a cop at distance i from the robber has *weight* w_i; this represents that cop's individual contribution toward the potential. In particular, we have that $w_1 = 1$ and $w_2 \sim 2/n$. First, let us note that if the cops can capture the robber on their turn, then immediately before the cops' turn we must have $P \geq 1$, since some cop must be at distance 1 from the robber. Our goal is to show that the robber can always enforce that right before the cops' move

$$P \leq 1 - \frac{3}{n}, \tag{3.2}$$

from which it would follow that the robber can evade the cops indefinitely. Initially, we may assume that all cops start at the same vertex; the robber places themselves at the vertex at distance n from the cops. Therefore, $P = 0$, so (3.2) holds. Suppose that before the cops make their move, the potential function satisfies (3.2). There are three cases to consider: on the cops' turn, some cop moves to some vertex adjacent to the robber (creating a "deadly" neighbor for the robber), some cop moves to a vertex at distance 2 from the robber, or some cop moves to a vertex at distance 3 or higher from the robber.

Case 1. Suppose that on the cops' turn, a cop moves to some vertex adjacent to the robber. The robber's strategy is to move away from this deadly vertex, but to do so in a way that maintains the invariant (3.2). To show that this is possible, we compute the expected change in the potential function if the robber were to choose their next position at random from among all neighbors other than the deadly one.

Suppose that before the robber's move,

$$P_1 = \sum_{i=2}^{n/2-\rho-1} N_i w_i \quad \text{and} \quad P_2 = \sum_{n/2-\rho}^{n} N_i w_i.$$

Then by (3.2), we have that $P_1 + P_2 + w_2 \leq 1 - 3/n$, where the extra w_2 accounts for the weight of the cop who moved to the robber's neighborhood.

Consider a cop, C, at distance i from the robber, where $2 \leq i \leq n/2 - \rho - 1$. Before the robber's move, C has weight w_i. Let w_C represent the expected weight of C after the robber's move. If C's vertex and the deadly vertex differ on the deadly coordinate (that is, the coordinate in which the robber and their deadly neighbor differ), then $w_C = \frac{i-1}{n-1}w_{i-1} + \frac{n-i}{n-1}w_{i+1}$, whereas if they agree on this coordinate, then $w_C = \frac{i}{n-1}w_{i-1} + \frac{n-1-i}{n-1}w_{i+1}$. Since $w_{i-1} > w_{i+1}$, we may provide an upper bound on w_C as follows:

$$
\begin{aligned}
w_C &\leq \frac{i}{n-1}w_{i-1} + \frac{n-1-i}{n-1}w_{i+1} \\
&\leq \frac{i}{n-2}w_{i-1} + \frac{n-2-i}{n-2}w_{i+1} \\
&= \frac{i}{n-2} \cdot A \cdot \binom{n-2}{i-1}^{-1} \prod_{j=1}^{i-1}(1+\varepsilon_j) \\
&\quad + \frac{n-2-i}{n-2} \cdot A \cdot \binom{n-2}{i+1}^{-1} \prod_{j=1}^{i+1}(1+\varepsilon_j) \\
&= \left(\frac{i}{n-2}(1+\varepsilon_i)^{-1}\frac{(i-1)!(n-2-i+1)!}{(n-2)!} + \right. \\
&\quad \left. \frac{n-2-i}{n-2}(1+\varepsilon_{i+1})\frac{(i+1)!(n-2-i-1)!}{(n-2)!} \right) \\
&\quad \cdot A \cdot \prod_{j=1}^{i}(1+\varepsilon_j).
\end{aligned}
$$

Since

$$
(1+\varepsilon_i)^{-1} = 1 - \varepsilon_i + \varepsilon_i^2 - \varepsilon_i^3 + \ldots \leq 1 - \varepsilon_i + \varepsilon_i^2
$$

and

$$
1 + \varepsilon_{i+1} = 1 + \varepsilon_i\left(1 + \frac{2}{n-2i-4}\right) \leq 1 + \varepsilon_i + \varepsilon_i^2,
$$

we derive that

$$
w_C \leq w_i \left(\frac{n-i-1}{n-2}(1 - \varepsilon_i + \varepsilon_i^2) + \frac{i+1}{n-2}(1 + \varepsilon_i + \varepsilon_i^2) \right)
$$

$$
= w_i \left(1 + \frac{2}{n-2} - \varepsilon_i \left(\frac{n-i-1}{n-2} - \frac{i+1}{n-2} \right) \right.
$$

$$
\left. + \varepsilon_i^2 \left(\frac{n-i-1}{n-2} + \frac{i+1}{n-2} \right) \right)
$$

$$
= w_i \left(1 + \frac{2}{n-2} - \frac{2+\varepsilon}{n-2i-2} \cdot \frac{n-2i-2}{n-2} + \varepsilon_i^2{}^{(1+o(1))} \right)
$$

$$
\leq w_i \left(1 - \frac{\varepsilon/2}{n} \right).
$$

This last inequality holds as long as, say, $\varepsilon_i^2 \leq \frac{\varepsilon/4}{n}$. Since $i \leq \frac{n}{2} - \rho - 1$, we have $\varepsilon_i^2 \leq \left(\frac{2+\varepsilon}{2\rho} \right)^2$, and so we will take $\rho = \rho(\varepsilon, n)$ such that

$$
\rho^2 \geq \frac{4}{\varepsilon} \cdot \left(\frac{2+\varepsilon}{2} \right)^2 \cdot n = \frac{(2+\varepsilon)^2}{\varepsilon} \cdot n. \tag{3.3}
$$

Hence, after the robber's move, the expected sum of the weights of such cops has decreased by a multiplicative factor of at least $\left(1 - \frac{\varepsilon/2}{n} \right)$, making it at most

$$
P_1 \cdot \left(1 - \frac{\varepsilon/2}{n} \right). \tag{3.4}
$$

In addition, the cop that moved to the neighborhood of the robber would again be at distance 2, making their weight

$$
w_2 \sim \frac{2}{n}. \tag{3.5}
$$

Before dealing with cops at a distance of at least $\frac{n}{2} - \rho$, we estimate the weight of a single cop at distance $\frac{n}{2} - \rho$.

$$
w_{n/2-\rho} \sim n \cdot \binom{n-2}{n/2-\rho}^{-1} \prod_{i=1}^{n/2-\rho} \left(1 + \frac{2+\varepsilon}{n-2i-2} \right). \tag{3.6}
$$

We bound the product term in (3.6) by

$$\prod_{i=1}^{n/2-\rho}\left(1+\frac{2+\varepsilon}{n-2i-2}\right)\le\exp\left(\sum_{i=1}^{n/2-\rho}\frac{2+\varepsilon}{n-2i-2}\right)$$

$$=\exp\left(\frac{2+\varepsilon}{2}\sum_{i=\rho}^{n/2}\frac{1}{i}+O(1)\right)$$

$$=\exp\left(\frac{2+\varepsilon}{2}\left(\log(n/2)-\log\rho+O(1)\right)\right)$$

$$=O\left(\left(\frac{n}{\rho}\right)^{1+\varepsilon/2}\right).$$

In order to bound the binomial term, we note that $\binom{n-2}{n/2-\rho}=\Theta\left(\binom{n}{n/2-\rho}\right)$ and approximate:

$$\binom{n}{n/2-\rho}=\frac{n!}{(n/2-\rho)!(n/2+\rho)!}$$

$$\sim\frac{\sqrt{2\pi n}\left(\frac{n}{e}\right)^n}{\sqrt{2\pi(n/2-\rho)}\left(\frac{n/2-\rho}{e}\right)^{n/2-\rho}\sqrt{2\pi(n/2+\rho)}\left(\frac{n/2+\rho}{e}\right)^{n/2+\rho}}$$

$$=\Theta\left(\frac{2^n}{\sqrt{n}}\right)\cdot\left(1-\frac{2\rho}{n}\right)^{-\frac{n}{2}+\rho}\left(1+\frac{2\rho}{n}\right)^{-\frac{n}{2}-\rho}$$

$$=\Theta\left(\frac{2^n}{\sqrt{n}}\right)\cdot\exp\left(-(1+o(1))\frac{2\rho^2}{n}\right).$$

Now take $\rho(n)$ to be smallest such that $\rho\ge c_\varepsilon\sqrt{n}$ and $n/2-\rho$ is an integer, where, referring to (3.3), we set $c_\varepsilon=\frac{2+\varepsilon}{\sqrt{\varepsilon}}$. Then we have that

$$w_{n/2-\rho}=O\left(n\cdot\frac{\sqrt{n}}{2^n}\cdot n^{1/2+\varepsilon/4}\right)=O\left(\frac{n^{2+\varepsilon/4}}{2^n}\right).$$

Now let C be a cop at distance i from the robber, where $\frac{n}{2}-\rho\le i\le n$. Before the robber's move, C has weight w_i. Since the w_i are decreasing, we have that the change in weight of C is bounded above by $w_{i-1} - w_i$. For $i \geq \frac{n}{2} - \rho + 1$, this quantity is equal to $\frac{w_{n/2-\rho}}{\frac{n}{2}+\rho}$. The largest increase comes when $i = n/2 - \rho$. To bound this increase, we see that

$$
\begin{aligned}
\frac{w_{n/2-\rho-1}}{w_{n/2-\rho}} &= \frac{\binom{n-2}{n/2-\rho}}{\binom{n-2}{n/2-\rho-1}} \cdot \frac{1}{1 + \varepsilon_{n/2-\rho}} \\
&= \frac{n/2 + \rho - 1}{n/2 - \rho} \cdot \frac{1}{1 + \frac{2+\varepsilon}{2\rho-2}} \\
&\leq \left(1 + \frac{2\rho - 2}{n}\right) \cdot \left(1 + \frac{2\rho}{n} + O\left(\frac{\rho}{n}\right)^2\right) \\
&\quad \cdot \left(1 - \frac{2+\varepsilon}{2\rho-2} + \left(\frac{2+\varepsilon}{2\rho-2}\right)^2\right) \\
&\leq 1 + O\left(\frac{\rho}{n} + \frac{1}{\rho}\right).
\end{aligned}
$$

By our definition of ρ, this is $1 + O(1/\sqrt{n})$. Thus, we have that

$$
w_{n/2-\rho-1} - w_{n/2-\rho} = O\left(\frac{w_{n/2-\rho}}{\sqrt{n}}\right) = O\left(\frac{n^{3/2+\varepsilon/4}}{2^n}\right).
$$

Hence, if we let the total number of cops be $k = O(2^n/n^{5/2+\varepsilon})$, then we have that the total increase in weight of cops at distance at least $n/2 - \rho$ is at most

$$
O\left(\frac{2^n}{n^{5/2+\varepsilon}} \cdot \frac{n^{3/2+\varepsilon/4}}{2^n}\right) < \frac{\varepsilon/4}{n}.
$$

We then have that the total weight of such cops after the robber's move is at most

$$
P_2 + \frac{\varepsilon/4}{n}. \tag{3.7}
$$

Thus, after the robber's random move, combining estimates (3.4), (3.5) and (3.7), we can provide an upper bound on the total expected weight by

$$P_1 \cdot \left(1 - \frac{\varepsilon/2}{n}\right) + w_2 + P_2 + \frac{\varepsilon/4}{n}$$

$$\leq \left(1 - \frac{3}{n} - w_2 - P_2\right) \cdot \left(1 - \frac{\varepsilon/2}{n}\right) + w_2 + P_2 + \frac{\varepsilon/4}{n}$$

$$\leq 1 - \frac{3}{n} - \frac{\varepsilon/4}{n} + O\left(\frac{1}{n^2} + \frac{P_2}{n}\right)$$

$$\leq 1 - \frac{3}{n}.$$

To derive the last line, we used the fact that

$$P_2 = O(2^n/n^{5/2+\varepsilon} \cdot w_{n/2-\rho}) = O(n^{-1/2-3\varepsilon/4}) = o(1).$$

Some deterministic move produces a potential at least as low as the expectation, so the robber may maintain the invariant, as desired.

Case 2. Suppose now that on the cops' turn, some cop C^* moves to a vertex at distance 2 from the robber. The reader should note that at this point, there might be other cops at distance 2 from the robber; we only suppose that on the cops' turn, one particular cop has moved from distance 3 to distance 2. As in Case 1, we will see what happens if the robber moves away from C^*. In this case, there are two "deadly" coordinates for the robber. The robber will flip a coordinate randomly among the other $n - 2$ choices.

As before, suppose that before the robber moves, P_1 represents the total weight of all cops at distance i with $2 \leq i \leq \frac{n}{2} - \rho - 1$ other than the cop C^* who moved to distance 2. Let P_2 represent the total weight of all cops at distance at least $\frac{n}{2} - \rho$. Since C^* was at distance 3 before their move, we have that $P_1 + P_2 + w_3 \leq 1 - 3/n$. As in Case 1, for a cop $C \neq C^*$ at distance $2 \leq i \leq n/2 - \rho - 1$, we have that the expected weight after the robber's move satisfies $w_C \leq \frac{i}{n-2} w_{i-1} + \frac{n-2-i}{n-2} w_{i+1}$. So again we can provide an upper bound on the total expected weight of such cops by

$$P_1 \cdot \left(1 - \frac{\varepsilon/2}{n}\right).$$

The estimate for the change in P_2 remains the same, so we can provide an upper bound on the expected total weight after the robber's move by

$$P_1 \cdot \left(1 - \frac{\varepsilon/2}{n}\right) + w_3 + P_2 + \frac{\varepsilon/4}{n}$$

$$\leq \left(1 - \frac{3}{n} - w_3 - P_2\right) \cdot \left(1 - \frac{\varepsilon/2}{n}\right) + w_3 + P_2 + \frac{\varepsilon/4}{n}$$

$$\leq 1 - \frac{3}{n} - \frac{\varepsilon/4}{n} + O\left(\frac{1}{n^2} + \frac{w_3}{n} + \frac{P_2}{n}\right)$$

$$\leq 1 - \frac{3}{n}.$$

This time, in addition to our bound on P_2, we have used that $w_3 = o(1/n)$.

Case 3. Suppose now that some cop moves to a vertex at distance $i \geq 3$ from the robber. Keep in mind that, again, we allow for the possibility that other cops are at distance 2 from the robber. The resulting increase in the potential function is at most $w_3 = O(1/n^2)$, so the new potential function has value at most $1 - 3/n + o(1/n)$. Now, by the calculations from Case 1, the robber can move so that the total weight of all cops at distances 2 through $n/2 - \rho - 1$ decreases by a multiplicative factor of $(1 - \frac{\varepsilon/2}{n})$. Once again, define P_2 to be the weight of all cops at distance at least $n/2 - \rho$ before the robber moves. Then after the robber's move, the potential is at most

$$\left(1 - \frac{3}{n} - P_2 + o\left(\frac{1}{n}\right)\right) \cdot \left(1 - \frac{\varepsilon/2}{n}\right) + P_2 + \frac{\varepsilon/4}{n} \leq 1 - \frac{3}{n},$$

and the proof follows.   □

3.4   Cops and Falling Robbers

We consider a variant of the Cops and Robbers game on a hypercube, which we call *Cops and Falling Robbers*, introduced in the doctoral thesis of Alan Hill [110]. This variant restricts the initial positions and the allowed moves. Recall that the n-dimensional hypercube Q_n is the graph with vertex set $\{0, 1\}^n$ (the set of binary n-tuples) in which vertices are adjacent if and only if they differ in one coordinate. We view the vertices as subsets of $[n]$, and let the *kth level* consist of the k-sets; that is, the vertices whose size as subsets is k. We view $\varnothing$ as the bottom of the hypercube and $[n]$ as the top, and we say that S lies *below* T when $S \subseteq T$.

The robber starts at the full set $[n]$; the cops start at the empty set $\varnothing$. On the kth round, the cops all move from level $k - 1$ to level k, and then the robber moves from level $n + 1 - k$ to level $n - k$. If the cops capture the robber, then they do so on round $\lceil n/2 \rceil$ at level $\lceil n/2 \rceil$, when they move if n is odd, and by the robber moving onto them if n is even. Hence, we may view the game as played on Q_n as an ordered set, with the robber moving down a level each round, and the cops moving up. See Figure 3.3 for an example. Cops and Falling Robbers has some elements in common with Seepage, although it is distinct from that game; see Section 11.4.

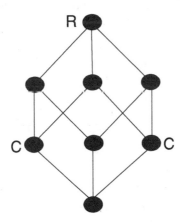

FIGURE 3.3: A winning first move for the cops on Q_3.

Let c_n denote the smallest number of cops that can guarantee winning the game in the hypercube. In [110], it was proven that there is a lower bound $2^{n/2}$ for even n and $\binom{n}{\lceil n/2 \rceil} 2^{-\lfloor n/2 \rfloor}$ for odd n; the former bound exceeds the latter by a factor of $\Theta(\sqrt{n})$. Note that here the cops have in some sense only one chance to capture the robber, on the middle level. When the players can move both up and down, the value is much smaller, with the cop number of the n-dimensional hypercube graph equaling $\lceil (n+1)/2 \rceil$ [137].

We begin with a proof of a lower bound and then show an upper bound within a factor of $O(\log n)$ of this lower bound.

Theorem 3.4.1 ([110]). $c_n \geq \begin{cases} 2^m, & n = 2m; \\ \binom{2m+1}{m+1} 2^{-m}, & n = 2m+1. \end{cases}$

Proof. After each move by the robber, some cops may no longer lie below the robber. Such cops are effectively eliminated from the game. We call them *evaded cops*; cops not yet evaded are *surviving cops*.

Consider the robber strategy that greedily evades as many cops as possible with each move. Deleting an element from the set at the robber's current position evades all cops whose set contains that element. On the kth round, the surviving cops occupy sets of size k, and the robber has $n - k + 1$ choices of an element to delete. Since each surviving cop can be evaded in k ways, the fraction of the surviving cops that the robber can evade on this move is at least $\frac{k}{n-k+1}$.

After the first m rounds, where $m = \lfloor n/2 \rfloor$, the fraction of the cops that survive is at most $\prod_{i=1}^{m} \left(1 - \frac{i}{n-i+1}\right)$. If $n = 2m$, then we compute that

$$\prod_{i=1}^{m} \left(1 - \frac{i}{2m - i + 1}\right) = \prod_{i=1}^{m} \frac{2m - 2i + 1}{2m - i + 1} = \frac{(2m)!}{(2m)! \cdot 2^m}$$
$$= 2^{-m}.$$

If $n = 2m + 1$, then we derive that

$$\prod_{i=1}^{m} \left(1 - \frac{i}{2m - i + 2}\right) = \prod_{i=1}^{m} \frac{2m - 2i + 2}{2m - i + 2} = \frac{2^m m!(m+1)!}{(2m+1)!}$$
$$= 2^m \Big/ \binom{2m+1}{m+1}.$$

For the cops to capture the robber, at least one surviving cop must remain after m moves; this requires at least 2^m total cops when $n = 2m$ and at least $\binom{2m+1}{m+1} 2^{-m}$ when $n = 2m + 1$. $\qquad\square$

A similarly randomized strategy for the cops should produce a good upper bound. However, it is difficult to control the deviations from expected; that is, over all the cops together. Our strategy will group the play of the game into phases that enable us to give essentially the same bound on undesirable deviations in each phase.

If there are enough cops to cover the entire middle level, then the robber cannot sneak through. The size of the middle level is asymptotic to $2^n/\sqrt{\pi n/2}$ (see the discussion after Lemma 1.3.2). This trivial upper bound is roughly the square of the lower bound in Theorem 3.4.1. If n is odd, then a slight improvement follows by observing that one only needs to block each $(n+1)/2$-set by reaching some $(n-1)/2$-set under it. More substantial improvements use the fact that as the robber starts to move, the family of sets needing to be protected shrinks.

The following upper bound on c_n, proved in [122], is $O(\log n)$ times the lower bound in Theorem 3.4.1. In the proof, a randomized strategy was used for the cops; it may or may not succeed in capturing the robber. However, with sufficiently many cops, the strategy succeeds a.a.s. Consequently, some deterministic strategy for the cops (in response to the moves by the robber) wins the game.

Theorem 3.4.2 ([122]).

$$c_n = \begin{cases} O(2^m \log n), & n = 2m; \\ O(2^{-m}\binom{2m+1}{m+1}) \log n), & n = 2m+1. \end{cases}$$

Proof. We consider the case $n = 2m$ first, returning later to the case $n = 2m+1$. We will specify the number of cops later. All the cops begin at $\varnothing$. Let R be the current set occupied by the robber. On their kth turn, for $1 \le k \le m$, each surviving cop at set C chooses the next element for their set uniformly at random from among $R - C$. We claim that, regardless of how the robber moves, this cop strategy succeeds a.a.s.

To facilitate analysis of the cops' strategy, we introduce some notation and terminology. Consider an instance of the game. We say that this instance satisfies *property* $P(t, a)$ if, after t rounds, every m-set below the robber also has at least a cops at or below it.

Intuitively, the m-sets below the robber are the places where the robber can potentially be captured; property $P(t, a)$ means that each of them can be reached by at least a cops.

To show that the cop strategy a.a.s. captures the robber, we will show that, no matter how the robber plays, a.a.s. property $P(t_i, a_i)$ holds for specific choices of t_i and a_i. Let $r = \lceil \log_2 \log_2 n \rceil$, and for $i \in \{0, \dots, r\}$ let $s_i = 2^{r-i}$ and $t_i = m - s_i$. Furthermore, let

$$a_i = 1600 \left(\prod_{j=1}^{i} (1 - \varepsilon_j) \right) 2^{s_i} \log n,$$

where $\varepsilon_j = \sqrt{s_j / 2^{s_j}}$. In particular, $a_0 = 1600 \cdot 2^{2^r} \log n$. Note that always

$$\prod_{j=1}^{i} (1 - \varepsilon_j) \geq \prod_{j=1}^{r} (1 - \varepsilon_j)$$

$$\geq \exp\left(-2 \sum_{j=1}^{r} \varepsilon_j \right)$$

$$\geq \exp\left(-2 \left(\sqrt{2^0 / 2^{2^0}} + \sqrt{2^1 / 2^{2^1}} \right.\right.$$

$$\left.\left. + \sqrt{2^2 / 2^{2^2}} (1 + 1/2 + 1/4 + \dots) \right) \right)$$

$$= \exp(-2\sqrt{2} - 2) > 1/200,$$

and hence, $a_i \geq 8 \cdot 2^{s_i} \log n$. (Above, the second inequality uses the fact that $1 - x \geq \exp(-2x)$ whenever $0 \leq x \leq 1/\sqrt{2}$, while the third inequality uses the observation that $\varepsilon_{j-1} \leq \varepsilon_j / 2$ for $0 \leq j \leq r - 2$.)

We play the game with $\lceil 3200 \cdot 2^m \log n \rceil$ cops. We claim that a.a.s. property $P(t_i, a_i)$ holds for all i in $\{0, \dots, r\}$. We also claim that a.a.s. property $P(m, 1)$ holds. This ensures that in the final round the cops can cover all vertices where the robber can move; hence, they win.

We break the game into $r+2$ *phases*. Phase 0 consists of rounds 1 through t_0. For $i \in \{1, \dots, r\}$, Phase i consists of rounds $t_{i-1}+1$

through t_i. Phase $r + 1$ consists of the single round $t_r + 1$. Our analysis is inductive. For Phase 0, we show that a.a.s. property $P(t_0, a_0)$ holds. When considering Phase i for $1 \le i \le r$, we assume that property $P(t_{i-1}, a_{i-1})$ holds and show that a.a.s. property $P(t_i, a_i)$ also holds. Finally, for Phase $r + 1$, we assume that property $P(t_r, a_r)$ holds and show that a.a.s. the cops capture the robber.

We begin with Phase 0. We claim that property $P(t_0, a_0)$ holds with probability at least $1 - 1/n$, no matter how the robber moves. Fix a sequence of moves for the robber in the first t_0 rounds of the game, and fix a set S with $|S| = m$ that remains below the robber. A particular cop remains below S if and only if their position contains only elements of S. In round i, each cop below S has already added $i - 1$ such elements, and $m - i + 1$ others remain. Since each surviving cop chooses a new element uniformly from $2m - 2i + 2$ possibilities, the probability that a cop below S remains below S is $\frac{m-i+1}{2m-2i+2}$, which equals $1/2$. Thus, a given cop remains below S after the first t_0 rounds with probability 2^{-t_0}.

Consequently, the number of cops remaining below S after t_0 rounds is a random variable X with the binomial distribution $\mathrm{Bin}(\lceil 3200 \cdot 2^m \log n \rceil, 2^{-t_0})$. Recalling that $t_0 = m - s_0$ and that $s_0 = 2^r \ge \log_2 n$, we have that

$$
\begin{aligned}
\mathbb{E}[X] &\ge 3200 \cdot 2^m \log n \cdot 2^{-t_0} \\
&= 3200 \cdot 2^{s_0} \log n \\
&= 3200 \cdot 2^{2^r} \log n \\
&= 2a_0.
\end{aligned}
$$

The Chernoff bound now yields

$$
\begin{aligned}
\mathbb{P}(X \le a_0) &\le \mathbb{P}\left(X \le \frac{\mathbb{E}[X]}{2}\right) \le \exp\left(-\frac{(1/2)^2 \mathbb{E}[X]}{2}\right) \\
&< \exp(-3n \log n).
\end{aligned}
$$

Thus, the probability that fewer than a_0 cops remain below S is less than $\exp(-3n \log n)$. The number of such sets S below the robber is less than 2^n, which is less than $\exp(n \log n)$. By the union bound, the probability that some m-set below the robber has fewer than a_0 cops below it is thus, less than $\exp(-2n \log n)$. That is, for one sequence of moves by the robber, property $P(t_0, a_0)$ fails to hold with probability at most $\exp(-2n \log n)$. The number of possible move sequences by the robber in Phase 0 is less than n^{t_0}, which in turn is less than $\exp(n \log n)$. By using the union bound, the probability that some robber strategy causes property $P(t_0, a_0)$ to fail is less than $\exp(-n \log n)$. Thus, property $P(t_0, a_0)$ holds with probability more than $1 - \exp(-n \log n)$, which is more than $1 - 1/n$.

Next consider Phase i with $1 \leq i \leq r$, consisting of rounds $t_{i-1} + 1$ through t_i. Under the assumption that property $P(t_{i-1}, a_{i-1})$ holds, we claim that property $P(t_i, a_i)$ also holds with probability at least $1 - 1/n$. The argument is similar to that for Phase 0. Fix a sequence of moves for the robber in rounds $t_{i-1} + 1$ through t_i, and fix an m-set S that remains below the robber after round t_i. As before, a cop below S on a given round remains below S after that round with probability $1/2$.

By assumption, at least a_{i-1} cops sat below S at the beginning of Phase i; the number of cops remaining below S at the end of Phase i is thus bounded below by the random variable X with binomial distribution $\mathrm{Bin}(\lceil a_{i-1} \rceil, 2^{-(t_i - t_{t-1})})$. Hence

$$\mathbb{E}[X] \geq a_{i-1} 2^{t_{i-1} - t_i} = a_{i-1} 2^{s_i - s_{i-1}} = a_i \cdot \frac{a_{i-1} \cdot 2^{-s_{i-1}}}{a_i \cdot 2^{-s_i}} = \frac{a_i}{1 - \varepsilon_i}.$$

This time, the Chernoff bound yield

$$\mathbb{P}(X \leq a_i) \leq \mathbb{P}(X \leq (1 - \varepsilon_i)\mathbb{E}[X]) \leq \exp\left(-\frac{\varepsilon_i^2 \cdot \mathbb{E}[X]}{2}\right)$$

$$\leq \exp\left(-\frac{\varepsilon_i^2 \cdot a_i}{2}\right)$$

$$\leq \exp\left(-\frac{\varepsilon_i^2 \cdot 8 \cdot 2^{s_i} \log n}{2}\right) = \exp(-4s_i \log n).$$

At the start of Phase i, the robber occupies level $n - t_{i-1}$. At this time, the number of m-sets that lie below the robber is $\binom{n - t_{i-1}}{m}$. This simplifies to $\binom{m + s_{i-1}}{m}$, which is at most $n^{s_{i-1}}$; since $s_{i-1} = 2s_i$, this is at most $\exp(2s_i \log n)$. Likewise, the number of move sequences available to the robber during Phase i is at most $n^{s_{i-1}-s_i}$, which simplifies to $\exp(s_i \log n)$. Applying the union bound twice, as in Phase 0, we see that property $P(t_i, a_i)$ fails with probability at most $\exp(-s_i \log n)$. Hence $P(t_i, a_i)$ holds with probability at least $1 - \exp(-s_i \log n)$, which is at least $1 - 1/n$.

Finally, we show that if $P(t_r, a_r)$ holds, then $P(m, 1)$ holds with probability at least $1 - 1/n$. Recall that $t_r = m - 1$ and that $a_r \geq 16 \log n$. Each cop chooses from two possible moves, each leading to an m-set. The number of cops that remain below an m-set S is bounded from below by the random variable X with distribution $\text{Bin}(\lceil a_r \rceil, 1/2))$. Now

$$\mathbb{P}(X = 0) = 2^{-\lceil a_r \rceil} \leq 2^{-16 \log n} \leq \frac{1}{n^2},$$

so the probability that no cop reaches S is at most $1/n^2$. There are $m + 1$ choices for S; by the union bound, $P(m, 1)$ fails with probability less than $1/n$. Hence, $P(m, 1)$ holds with probability at least $1 - 1/n$, as claimed.

To complete the proof, we now consider the full game. We want to show that a.a.s. $P(m, 1)$ holds. The probability that $P(m, 1)$ holds is bounded below by the probability that $P(t_0, a_0), \ldots, P(t_r, a_r)$, and $P(m, 1)$ all hold. We have shown that $P(t_0, a_0)$ fails with probability at most $1/n$, that $P(t_i, a_i)$ for $1 \leq i \leq r$ fails with probability at most $1/n$ when $P(t_{i-1}, a_{i-1})$ holds, and that $P(m, 1)$ fails with probability at most $1/n$ when $P(t_r, a_r)$ holds. By the union bound, the probability that some property in this list fails is bounded above by $(r + 2)/n$, which is at most $2 \log_2 \log_2 n/n$ when n is sufficiently large. Thus, the conjunction of these properties (and in particular, property $P(m, 1)$) holds with probability at least $1 - 2 \log_2 \log_2 n/n$. This completes the proof for the case $n = 2m$.

If $n = 2m+1$, then we define the property $P(t, a)$ to mean that after t rounds, at least a cops are positioned below each $(m + 1)$-set that is below the robber. It now suffices to prove that $P(m, 1)$ holds a.a.s., since any cop that remains below the robber at the beginning of round $m + 1$ can capture them. The details of the argument are nearly identical to the previous case, and we omit them. $\qquad\square$

We remark that the cops can play more efficiently by using an appropriate deterministic strategy in round m. This does not improve the asymptotics of our bound, but it does improve the leading constant.

3.5   Containment Game

In this final section of the chapter, we consider a variant of the game of Cops and Robbers called *Containment*, introduced recently by Komarov and Mackey [123]. In this version, cops move from edge to adjacent edge, while the robber moves as in the classic game, from vertex to adjacent vertex (but cannot move along an edge occupied by a cop). The cops win if after some finite number of rounds, all edges incident with the robber are occupied by cops. This is called a *capture*. The robber wins if they can evade capture indefinitely.

If we place a cop at each edge, then the cops are guaranteed to win. Therefore, the smallest number of cops required to win in a graph G is a well-defined positive integer, named the *containability number* of the graph G. Following the notation introduced in [123], we write $\xi(G)$ for the containability number of a graph G and $c(G)$. Note that $\xi(G) \leq |E(G)|$. See Figure 3.4 for an example.

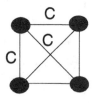

FIGURE 3.4: The graph K_4 has a containability number equaling 3.

In [123], it was proved that for every graph G,

$$c(G) \leq \xi(G) \leq \gamma(G)\Delta(G).$$

It was conjectured that the upper bound can be strengthened and the following holds.

Conjecture 3.5.1 ([123]). *For every graph G,*

$$\xi(G) \le c(G)\Delta(G).$$

This is likely the main question for this variant of the game at the moment. In [162], by investigating expansion properties, asymptotically almost sure bounds on the containability number of binomial random graphs $\mathbb{G}(n, p)$ for a wide range of $p = p(n)$ were obtained. Hence, those results proved Conjecture 3.5.1 holds for some ranges of p (or holds up to a constant or an $O(\log n)$ multiplicative factors for some other ranges of p). We summarize those results from [162] in this section.

Theorem 3.5.2 ([162]). *Let $0 < \alpha < 1$ and $d = d(n) = np = n^{\alpha+o(1)}$.*

(i) *If $\frac{1}{2j+1} < \alpha < \frac{1}{2j}$ for some integer $j \ge 1$, then a.a.s.*

$$\xi(\mathbb{G}(n, p)) = \Theta(d^{j+1}) = \Theta(c(\mathbb{G}(n, p)) \cdot \Delta(\mathbb{G}(n, p))).$$

Hence, a.a.s. Conjecture 3.5.1 holds (up to a multiplicative constant factor).

(ii) *If $\frac{1}{2j} < \alpha < \frac{1}{2j-1}$ for some integer $j \ge 2$, then a.a.s.*

$$\xi(\mathbb{G}(n, p)) = \Omega\left(\frac{n}{d^{j-1}}\right), \quad and$$
$$\xi(\mathbb{G}(n, p)) = O\left(\frac{n \log n}{d^{j-1}}\right)$$
$$= O(c(\mathbb{G}(n, p)) \cdot \Delta(\mathbb{G}(n, p)) \cdot \log n).$$

Hence, a.a.s. Conjecture 3.5.1 holds (up to a multiplicative $O(\log n)$ factor).

(iii) *If $1/2 < \alpha < 1$, then a.a.s.*

$$\xi(\mathbb{G}(n, p)) = \Theta(n) = \Theta(c(\mathbb{G}(n, p)) \cdot \Delta(\mathbb{G}(n, p))/\log n)$$
$$\le c(\mathbb{G}(n, p)) \cdot \Delta(\mathbb{G}(n, p)).$$

Hence, a.a.s. Conjecture 3.5.1 holds.

It follows that a.a.s. $\log_n \xi(\mathbb{G}(n, n^{x-1}))$ is asymptotic to the function $g(x)$ shown in Figure 3.5 (denoted in red). We know that a.a.s. $\log_n c(\mathbb{G}(n, n^{x-1}))$ is asymptotic to the function $f(x)$ shown in Figure 3.5 (denoted in blue).

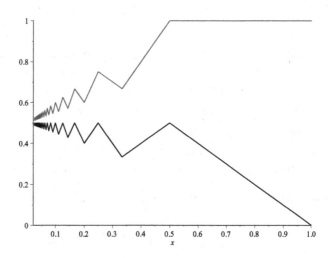

FIGURE 3.5: The zigzag functions representing the ordinary cop number (blue) and the containability number (red).

The fact the conjecture holds is associated with the observation that $g(x) - f(x) = x$, which is equivalent to saying that a.a.s. the ratio $\xi(\mathbb{G}(n, p))/c(\mathbb{G}(n, p)) = dn^{o(1)} = \Delta(\mathbb{G}(n, p)) \cdot n^{o(1)}$. We also mention that Theorem 3.5.2 implies that the conjecture is sharp (again, up to a constant or an $O(\log n)$ multiplicative factors for corresponding ranges of p). Finally, we observe that in the above result we skip the case when $np = n^{1/k+o(1)}$ for some positive integer k or $np = n^{o(1)}$ (see [162] for more details).

To obtain the upper bound in parts (i) and (ii) of Theorem 3.5.2, one needs to combine and adjust ideas from [136] (and [167] to include much sparser graphs). Since we present these results in the next chapter, we omit them here. It remains to concentrate on part (iii) for which we are going to make a simple observation that the containability number is linear if G has a perfect or a near-perfect matching. The result will follow since it is well known that for $p = p(n)$ such that $pn - \log n \to \infty$, $\mathbb{G}(n, p)$ has a perfect (or a near-perfect) matching a.a.s. See [115], for more details.

Lemma 3.5.3 ([162]). *If G is a graph on n vertices with a perfect matching (n is even) or a near-perfect matching (n is odd), then we have that $\xi(G) \leq n$.*

Proof. Suppose first that n is even. The cops start on the edges of a perfect matching; two cops occupy any edge of the matching for a total of n cops. All vertices of G can be associated with unique cops. The robber starts on some vertex v. One edge incident to v (the edge vv' that belongs to the perfect matching used) is already occupied by a cop (by two cops, associated with v and v'). Further, the remaining cops can move so that all edges incident to v are protected and the game ends. Indeed, for each edge vu, the cop associated with u moves to vu.

The case when n is odd is analogous. Two cops start on each edge of a near-perfect matching that matches all vertices but u. If u is isolated, we may simply remove it from G and return to the case when n is even. (Recall that the cops win if all edges incident with the robber are occupied by cops. As this property is vacuously true when the robber starts on an isolated vertex, we may assume that they do not start on u.) Hence, we may assume that u is not isolated. We introduce one more cop on some edge incident to u. The total number of cops is at most $2 \cdot \frac{n-1}{2} + 1 = n$; again, each vertex of G can be associated with a unique cop and the proof goes as before. $\square$

The proof of the lower bound is an adaptation of the proof used for the classic cop number in [136]. We will use the following two technical lemmas.

Lemma 3.5.4 ([162]). *If $d = d(n) = p(n-1) \geq \log^3 n$, then there exists a positive constant c such that a.a.s. the following properties hold in $\mathbb{G}(n, p) = (V, E)$.*

   (i) If $S \subseteq V$ be any set of $s = |S|$ vertices, and let $r \in \mathbb{N}$, then

we have that

$$\left| \bigcup_{v \in S} N_r[v] \right| \geq c \min\{sd^r, n\}.$$

In addition, if s and r are such that $sd^r < n/\log n$, then

$$\left| \bigcup_{v \in S} N_r[v] \right| \sim sd^r.$$

(ii) $\mathbb{G}(n, p)$ *is connected.*

(iii) *If $r = r(n)$ is the largest integer such that $d^r \leq \sqrt{n \log n}$, then for every vertex $v \in V$ and $w \in N_{r+1}(v)$, the number of edges from w to $N_r(v)$ is at most b, where*

$$b = \begin{cases} 250 & \text{if } d \leq n^{0.49} \\ \frac{3 \log n}{\log \log n} & \text{if } n^{0.49} < d \leq \sqrt{n}. \end{cases}$$

Lemma 3.5.5 ([162]). *Let ε and α be constants such that $0 < \varepsilon < 0.1$, $\varepsilon < \alpha < 1 - \varepsilon$, and let $d = d(n) = p(n-1) = n^{\alpha+o(1)}$. Let $\ell \in \mathbb{N}$ be the largest integer such that $\ell < 1/\alpha$. Then a.a.s. for every vertex v of $\mathbb{G}(n, p)$ the following properties hold.*

(i) *If $w \in N_i[v]$ for some i with $2 \leq i \leq \ell$, then $P_i(v, w) \leq \frac{3}{1-i\alpha}$.*

(ii) *If $w \in N_{\ell+1}[v]$ and $d^{\ell+1} \geq 7n \log n$, then $P_{\ell+1}(v, w) \leq \frac{6}{1-\ell\alpha} \frac{d^{\ell+1}}{n}$.*

(iii) *If $w \in N_{\ell+1}[v]$ and $d^{\ell+1} < 7n \log n$, then $P_{\ell+1}(v, w) \leq \frac{42}{1-\ell\alpha} \log n$.*

Further, a.a.s. we have the following.

(iv) *Every edge of $\mathbb{G}(n, p)$ is contained in at most εd cycles of length at most $\ell + 2$.*

The two bounds, corresponding to parts (i) and (ii) in Theorem 3.5.2, are proved independently in the following two lemmas.

Lemma 3.5.6 ([162]). *If $\frac{1}{2j+1} < \alpha < \frac{1}{2j}$ for some integer $j \geq 1$, $c = c(j, \alpha) = \frac{3}{1-2j\alpha}$, and $d = d(n) = np = n^{\alpha+o(1)}$, then a.a.s.*

$$\xi(\mathbb{G}(n, p)) > K = \left(\frac{d}{30c(2j+1)} \right)^{j+1}.$$

Proof. Since our aim is to prove that the desired bound holds a.a.s. for $\mathbb{G}(n, p)$, we may assume, without loss of generality, that a graph G on which the players play satisfies the properties stated in Lemmas 3.5.4 and 3.5.5. Suppose that the robber is chased by K cops. Our goal is to provide a winning strategy for the robber on G. For vertices $x_1, x_2, \ldots, x_s$, let $C_i^{x_1, x_2, \ldots, x_s}(v)$ denote the number of cops in $E_i(v)$ (that is, at distance i from v) in the graph $G \setminus \{x_1, x_2, \ldots, x_s\}$.

Right before the robber makes their move, we say that the vertex v occupied by the robber is *safe*, if for some neighbor x of v we have $C_1^x(v) \leq \frac{d}{30c(2j+1)}$, and

$$C_{2i}^x(v), C_{2i+1}^x(v) \leq \left(\frac{d}{30c(2j+1)} \right)^{i+1}$$

for $i = 1, 2, \ldots, j - 1$ (such a vertex x will be called a *deadly neighbor* of v). The reason for introducing deadly neighbors is to deal with a situation in which many cops apply a greedy strategy and always decrease the distance between them and the robber. As a result, there might be many cops right behind the robber but they are not so dangerous unless they make a step backwards by moving to a vertex they came from in the previous round, a deadly neighbor. Note that a vertex is called safe for a reason: if the robber occupies a safe vertex, then the game is definitely not over since the condition for $C_1^x(v)$ guarantees that at most a small fraction of incident edges are occupied by cops.

Since a.a.s. G is connected (see Lemma 3.5.4(ii)), without loss of generality we may assume that at the beginning of the game all cops begin at the same edge, e. Subsequently, the robber may choose a vertex v so that e is at distance $2j + 2$ from v (see Lemma 3.5.4(i) applied with $r = 2j + 1$ to see that almost all vertices are at distance $2j + 1$ from both endpoints of e). Hence, even if all cops will move from e to $E_{2j+1}(v)$ after this move, v will remain safe as no bound is required for $C_{2j+1}^x(v)$. (Of course, again, without loss of generality we may assume that all cops pass for the next round and stay at e before starting to apply their best strategy against the robber.) Hence, in order to prove the lemma, it is enough to show that if the robber's current vertex v is safe, then they can move along an unoccupied edge to a neighbor y so that no matter how the cops move in the next round, y remains safe.

For $0 \leq r \leq 2j$, we say that a neighbor y of v is *r-dangerous* if

(i) an edge vy is occupied by a cop (for $r = 0$), or

(ii) $C_r^{v,x}(y) \geq \frac{1}{3}\left(\frac{d}{30c(2j+1)}\right)^i$ (for $r = 2i$ or $r = 2i - 1$, where $i = 1, 2, \ldots, j$),

where x is a deadly neighbor of v. We will check that for every $r \in \{0, 1, \ldots, 2j\}$, the number of r-dangerous neighbors of v, which we denote by $\mathrm{dang}(r)$, is smaller than $\frac{d}{2(2j+1)}$. Since v is safe,

$$\mathrm{dang}(0) \leq C_1^x(v) \leq \frac{d}{30c(2j+1)} \leq \frac{d}{2(2j+1)}.$$

Suppose then that $r = 2i$ or $r = 2i - 1$ for some $i \in \{1, 2, \ldots, j\}$. Every r-dangerous neighbor of v has at least $\frac{1}{3}\left(\frac{d}{30c(2j+1)}\right)^i$ cops occupying $E_{\leq(r+1)}(v)$. However, every edge from $E_{\leq(r+1)}(v)$ is incident to at most 2 vertices at distance at most r from v. Note that Lemma 3.5.5(i) implies that there are at most c paths between v and any $w \in N_{\leq r}(v)$. Finally, by the assumption that v is safe, we have $C_{2i}^x(v), C_{2i+1}^x(v) \leq \left(\frac{d}{30c(2j+1)}\right)^{i+1}$, provided that $i \leq j - 1$; the corresponding conditions for $C_{2j}^x(v)$ and $C_{2j+1}^x(v)$ hold, since both can be bounded from above by K, the total number of cops. Combining all of these yields

$$\begin{aligned}
\frac{1}{3}\left(\frac{d}{30c(2j+1)}\right)^i \cdot \mathrm{dang}(r) &\leq 2c \cdot C_{\leq(r+1)}^x(v) \\
&\leq 2c \cdot (2 + o(1))C_{r+1}^x(v) \\
&\leq 5c \cdot \left(\frac{d}{30c(2j+1)}\right)^{i+1},
\end{aligned}$$

and consequently $\text{dang}(r) \leq \frac{d}{2(2j+1)}$, as required. Thus, there are at most $d/2$ of neighbors of v are r-dangerous for some $r \in \{0, 1, \ldots, 2j\}$.

Since we have $(1 + o(1))d$ neighbors to choose from (see Lemma 3.5.4(i)), there are many neighbors of v that are not r-dangerous for any $r = 0, 1, \ldots, 2j$ and the robber might want to move to one of them. However, there is one small issue we have to deal with. In the definition of being dangerous, we consider the graph $G \setminus \{v, x\}$, whereas in the definition of being safe we want to use $G \setminus \{v\}$ instead. Fortunately, Lemma 3.5.5(iv) implies that we can find a neighbor y of v that is not dangerous but also x does not belong to the $2j$-neighborhood of y in $G \setminus \{v\}$. It follows that vy is not occupied by a cop and $C_r^v(y) < \frac{1}{3} \left(\frac{d}{30c(2j+1)} \right)^i$ for $r = 2i$ or $r = 2i - 1$, where $i = 1, 2, \ldots, j$. We move the robber to y.

Now, it is time for the cops to make their move. Because of our choice of the vertex y, we can assure that the desired upper bound for $C_r^v(y)$ required for y to be safe will hold for $r \in \{1, 2, \ldots, 2j-1\}$. Indeed, the best that the cops can do to try to fail the condition for $C_r^v(y)$ is to move all cops at distance $r - 1$ and $r + 1$ from y to the r-neighborhood of y, and to make cops at distance r stay put, but this would not be enough. Thus, regardless of the strategy used by the cops, y is safe and the proof follows. $\qquad\square$

Lemma 3.5.7 ([162]). *If $\frac{1}{2j} < \alpha < \frac{1}{2j-1}$ for some integer $j \geq 1$, $\bar{c} = \bar{c}(\alpha) = \frac{6}{1-(2j-1)\alpha}$ and $d = d(n) = np = n^{\alpha+o(1)}$, then a.a.s.*

$$\xi(\mathbb{G}(n, p)) \geq \bar{K} = \left(\frac{d}{30\bar{c}(2j + 1)} \right)^{j+1} \frac{n}{d^{2j}}.$$

Proof. The proof is very similar to that of Lemma 3.5.6. The only difference is that checking the desired bounds for $\text{dang}(2j-1)$ and $\text{dang}(2j)$ is more complicated. As before, we do not control the number of cops in $E_{2j}(v)$ and $E_{2j+1}(v)$. We note that $C_{2j}^x(v)$ and $C_{2j+1}^x(v)$ are bounded from above by $\bar{K}$, the total number of cops.

We derive that

$$\frac{1}{3}\left(\frac{d}{30\bar{c}(2j+1)}\right)^{j} \cdot \mathrm{dang}(2j-1) \leq 2\bar{c} \cdot \mathrm{C}^{x}_{\leq(2j)}(v)$$

$$\leq 2\bar{c} \cdot (2+o(1))\bar{K}$$

$$\leq 5\bar{c} \cdot \left(\frac{d}{30\bar{c}(2j+1)}\right)^{j+1},$$

and consequently $\mathrm{dang}(2j-1) \leq \frac{d}{2(2j+1)}$, as required. (Note that we have room to spare here but we cannot take advantage of it, so we do not modify the definition of being $(2j-1)$-dangerous.)

Let us now notice that a cop at distance $2j+1$ from v can contribute to the dangerousness of more than $\bar{c}$ neighbors of v. However, the number of paths of length $2j$ joining v and w is bounded from above by $\bar{c}d^{2j}/n$ (see Lemma 3.5.5(ii) and note that $d^{2j} = n^{2j\alpha+o(1)} \geq 7n\log n$, since $2j\alpha > 1$). Hence,

$$\frac{1}{3}\left(\frac{d}{30\bar{c}(2j+1)}\right)^{j} \cdot \mathrm{dang}(2j) \leq \frac{2\bar{c}d^{2j}}{n} \cdot \mathrm{C}^{x}_{\leq(2j+1)}(v)$$

$$\leq \frac{2\bar{c}d^{2j}}{n} \cdot (2+o(1))\bar{K}$$

$$\leq 5\bar{c} \cdot \left(\frac{d}{30\bar{c}(2j+1)}\right)^{j+1},$$

and, as desired, $\mathrm{dang}(2j) \leq \frac{d}{2(2j+1)}$. Besides this modification, the argument remains basically the same. $\qquad\square$

Chapter 4

Zombies and Survivors

So far we have discussed many natural variants of Cops and Robbers in Chapter 3. In the present, shorter chapter, we concentrate on another recent variant that deserves a separate treatment. One novel feature of the game we introduce here is that randomness is embedded within its definition.

For simplicity, we will only consider connected graphs. In the game of *Zombies and Survivors* suppose that k zombies (analogous to the cops) start the game on random vertices of G; each zombie, independently, selects a vertex uniformly at random to begin the game. Then the *survivor* (analogous to the robber) occupies some vertex of G. As zombies have limited intelligence, in each round, a given zombie moves toward the survivor along a shortest path connecting them. In particular, the zombie decreases the distance from their vertex to the survivor's. If there is more than one neighbor of a given zombie that is closer to the survivor than the zombie is, then they move to one of these vertices chosen uniformly at random. Each zombie moves independently of all other zombies. As in Cops and Robbers, the survivor may move to another neighboring vertex, or *pass* and not move. The zombies win if one or more of them *eat* the survivor; that is, land on the vertex that the survivor currently occupies. The survivor, as survivors should do in the event of a zombie attack, attempts to survive by applying an optimal strategy; that is, a strategy that minimizes the probability of being eaten. Note that there is no strategy for the zombies; they merely move on geodesics toward the survivor in each round. In this sense, Zombies and Survivors is a one-person game. Note also that since zombies always move toward the survivor, they can pass at most $\mathrm{diam}(G)$ times before being eaten by some zombie.

The probabilistic version of Zombies and Survivors was first introduced in [50] and subsequently studied in the preprint [164].

The random zombie model was inspired by a deterministic version of this game (with similar rules, but the zombies may choose their initial positions, and also choose which shortest path to the survivor they will move on) first considered in [89].

Let $s_k(G)$ be the probability that the survivor wins the game, provided that they follow the optimal strategy. It is evident that $s_k(G) = 1$ for $k < c(G)$, where $c(G)$ is the cop number of G. Note that $s_k(G) < 1$, provided that there is a strategy for $k \geq c(G)$ cops in which the cops always try to get closer to the robber, since with positive probability, the zombies may follow such a strategy. Usually, $s_k(G) > 0$ for any $k \geq c(G)$; however, there are some examples of graphs for which $s_k(G) = 0$ for every $k \geq c(G)$ (consider, for example, trees). Further, note that $s_k(G)$ is a nondecreasing function of k; that is, for every $k \geq 1$, we have that $s_{k+1}(G) \leq s_k(G)$, and $s_k(G) \to 0$ as $k \to \infty$. The latter limit follows since the probability that each vertex is initially occupied by at least one zombie tends to 1 as $k \to \infty$.

Define the *zombie number* of a graph G by

$$z(G) = \min\{k \geq c(G) : s_k(G) \leq 1/2\}.$$

Note that the zombie number is well-defined as

$$\lim_{k \to \infty} s_k(G) = 0.$$

In particular, $z(G)$ is the smallest number of zombies such that the probability that they eat the survivor is at least $1/2$. The ratio $Z(G) = z(G)/c(G) \geq 1$ is what we refer to as the *cost of being undead*. Note that there are examples of families of graphs for which there is no cost of being undead; that is, $Z(G) = 1$ (as is the case if G is a tree), and, as we show in the next section, there are examples of graphs with $Z(G) = \Theta(n)$.

Results in this chapter focus on the analysis of the zombie number for several graph classes, such as cycles, hypercubes and toroidal grids. Results from Sections 4.1, 4.2, 4.3, and 4.4 were first presented in [50]. Following the work of [164], the final subsection focuses on the interesting open question of finding the zombie number of the toroidal grid.

4.1 The Cost of Being Undead Can Be High

In this section, we consider a family of connected graphs $(G_n)_{n \in \mathbb{N}}$ for which $Z(G_n)$ is of order as large as the number of vertices. Notice how this already contrasts strongly with the behavior of the cop number on connected graphs.

Let G_n be the graph consisting of a 5-cycle with vertices v_i, where $1 \leq i \leq 5$, and $n - 5$ end-vertices attached to v_1, as shown in Figure 4.1. Although two cops suffice to capture the robber in this graph, a linear number of zombies are needed to eat the survivor with probability at least $1/2$, as shown in the following result. Simple modifications of this example provide many graphs each with a linear zombie number (we leave the reader to consider other examples).

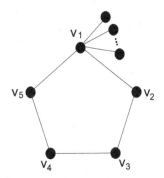

FIGURE 4.1: The graph G_n has a large cost of being undead.

Theorem 4.1.1 ([50]). *For G_n as defined above, we have that*

$$z(G_n) \sim \log(1/\alpha)n \approx 0.2180n,$$

where α is the only solution of $\alpha^4(2 - \alpha) = 1/2$ in $(0, 1)$. In particular, $Z(G_n) \sim \frac{\log(1/\alpha)}{2}n$.

Proof. A quick check demonstrates that the function $f(x) = x^4(2 - x) - 1/2$ is increasing and has a unique root in the interval

$(0,1)$. We refer to this unique root as α. Suppose that the game is played against $k \sim \log(1/\hat{\alpha})n$ zombies, for some fixed $\hat{\alpha} \in (0,1)$. Letting $S = V(G_n) \setminus \{v_1, v_2, v_3, v_4, v_5\}$, the probability that either $S \cup \{v_2\}$ or $S \cup \{v_5\}$ contain the starting vertices of all zombies is

$$2\left(1 - \frac{4}{n}\right)^k - \left(1 - \frac{5}{n}\right)^k$$

$$= \left(1 - \frac{4}{n}\right)^k \left(2 - \left(1 - \frac{1}{n-4}\right)^k\right) \sim \hat{\alpha}^4 \left(2 - \hat{\alpha}\right),$$

which is strictly greater than $1/2$ if $\hat{\alpha} > \alpha$. If this occurs, then we claim that the survivor has a deterministic winning strategy. Indeed, they can start at vertex v_2 (or v_5, whichever is initially free of zombies). Then all zombies immediately move to v_1. The survivor continues to walk around the cycle, and all zombies chase them indefinitely without eating them. The probability that the previous property fails (that is, neither $S \cup \{v_2\}$ nor $S \cup \{v_5\}$ contain all zombies initially) and additionally at least one zombie starts in S is at least

$$\left(1 - 2\left(1 - \frac{4}{n}\right)^k + \left(1 - \frac{5}{n}\right)^k\right) - \left(\frac{5}{n}\right)^k \sim 1 - \hat{\alpha}^4 \left(2 - \hat{\alpha}\right),$$

which is strictly greater than $1/2$ if $\hat{\alpha} < \alpha$. If this occurs, then there is at least one zombie initially in S and either some zombie in $\{v_1, v_3, v_4\}$ or some in both v_2 and v_5. A simple case analysis shows that from such an initial configuration, the zombies will eat the survivor regardless of their strategy. Putting the two cases together and since $\hat{\alpha}$ can be arbitrarily close to α, we obtain that $z(G_n) \sim \log(1/\alpha)n$, and so $Z(G_n) \sim \frac{\log(1/\alpha)}{2}n$. □

4.2   Cycles

We analyze the case of cycles C_n of length n next. Even for elementary graphs such as cycles, unusual situations may arise that never occur in Cops and Robbers, as illustrated by the following statement.

Lemma 4.2.1 ([50]). *The survivor wins on C_n against $k \geq 2$ zombies if and only if all zombies are initially located on an induced subpath containing at most $\lceil n/2 \rceil - 2$ vertices.*

Hence, zombie hordes with cardinality a little less than half the cycle will (depending on their arrangement on the cycle) not be able to eat the survivor! In view of this, we have the following lemma that gives the probability that the survivor wins against $k \geq 2$ zombies. For $k = 1$, trivially $s_1(C_3) = 0$, and $s_1(C_n) = 1$ for $n \geq 4$.

Lemma 4.2.2 ([50]). *For any natural numbers $k \geq 2$ and $n \geq 9$, we have that*

$$k \left(\frac{1}{2} - \frac{4}{n} \right)^{k-1} \leq s_k(C_n) < k \left(\frac{1}{2} \right)^{k-1}.$$

In particular, $s_k(C_n) \sim k(1/2)^{k-1}$, as $n \to \infty$.

Proof. By Lemma 4.2.1, the survivor has a strategy to remain un-eaten indefinitely if and only if all zombies initially lie in a subpath of $r \leq \lceil n/2 \rceil - 2$ vertices. To bound $s_k(C_n)$ from above, we first overcount these configurations, by distinguishing two zombies, one at each end of the subpath. There are $k(k-1)$ ways to select the two distinguished zombies that are placed at the ends of a sub-path consisting of $r \leq \lceil n/2 \rceil - 2$ vertices, and n choices for the position of the path. The two zombies start at the right place with probability $(1/n)^2$, the remaining ones must start on the subpath, which happens with probability $(r/n)^{k-2}$. Since we are overcounting configurations, it follows that

$$
\begin{aligned}
s_k(C_n) &\leq \sum_{r=1}^{\lceil n/2 \rceil - 2} k(k-1)n(1/n)^2(r/n)^{k-2} \\
&< k(k-1)n^{1-k} \int_1^{n/2} x^{k-2} dx \\
&< k \left(\frac{1}{2} \right)^{k-1}.
\end{aligned}
$$

To derive a lower bound on $s_k(C_n)$, we consider only those configurations in which the two end-vertices of the subpath are occupied by exactly one zombie. These configurations form a subset of all those configurations that allow the survivor to win, but the advantage is that the corresponding probabilities can be computed exactly without overcounting. Therefore, we have that

$$s_k(C_n) \geq \sum_{r=3}^{\lceil n/2 \rceil - 2} k(k-1)n(1/n)^2((r-2)/n)^{k-2}$$

$$\geq k(k-1)n^{1-k} \left(\int_0^{n/2-4} x^{k-2} dx \right)$$

$$= k \left(\frac{1}{2} - \frac{4}{n} \right)^{k-1}.$$

The proof of the lemma follows. In the special cases of $k = 2$ and $k = 3$, it is straightforward to find the winning probability precisely:

$$s_2(C_n) = \sum_{r=2}^{\lceil n/2 \rceil - 2} 2 \cdot n \cdot \frac{1}{n^2} + n \cdot \frac{1}{n^2}, \text{ and}$$

$$s_3(C_n) = \sum_{r=3}^{\lceil n/2 \rceil - 2} 3 \cdot 2 \cdot n \cdot \frac{1}{n^2} \cdot \frac{r-2}{n} + 2 \sum_{r=2}^{\lceil n/2 \rceil - 2} 3 \cdot n \cdot \frac{1}{n^3}$$

$$+ n \cdot \frac{1}{n^3}. \quad \square$$

As an immediate consequence of Lemma 4.2.2, we deduce that $z(C_n) = 4$ for all $n \geq 44$. Combining this with a direct examination of the smaller values of n, we derive the zombie number and the cost of being undead for the cycle C_n of length $n \geq 3$.

Theorem 4.2.3 ([50]).

$$z(C_n) = \begin{cases} 4 & \text{if } n \geq 27 \text{ or } n = 23, 25, \\ 3 & \text{if } 11 \leq n \leq 22 \text{ or } n = 9, 24, 26, \\ 2 & \text{if } 4 \leq n \leq 8 \text{ or } n = 10, \\ 1 & \text{if } n = 3; \end{cases}$$

and therefore,

$$Z(C_n) = \begin{cases} 2 & \text{if } n \geq 27 \text{ or } n = 23, 25, \\ 3/2 & \text{if } 11 \leq n \leq 22 \text{ or } n = 9, 24, 26, \\ 1 & \text{if } 3 \leq n \leq 8 \text{ or } n = 10. \end{cases}$$

4.3 Hypercubes

We know from [137] that $c(Q_n) = \lceil \frac{n+1}{2} \rceil$. Our aim in this section is to show that approximately $4/3$ times more zombies are needed to eat the survivor. We remark that some of the ideas in this section have a similar flavor to (but were derived independently from) some observations by Offner and Ojakian [153], who studied some variations of the Cops and Robbers game on the hypercube and proved that the cop number is $\lceil 2n/3 \rceil$.

Theorem 4.3.1 ([50]). *The zombie number of the hypercube satisfies* $z(Q_n) = \frac{2n}{3} + \Theta(\sqrt{n})$, *as* $n \to \infty$. *Hence,* $Z(Q_n) \sim \frac{4}{3}$, *as* $n \to \infty$.

We concentrate on an upper bound and, for simplicity, we show a weaker bound.

Lemma 4.3.2 ([50]). *If* $k = \frac{2}{3}n + \omega\sqrt{n}$, *where* $\omega = \omega(n)$ *is any function tending to infinity as* $n \to \infty$, *then* $s_k(Q_n) \to 0$, *as* $n \to \infty$.

Proof. Our goal is to show that a.a.s. k zombies can win. It follows from the Chernoff bounds that a.a.s. at least $n/3$ zombies start in both positions having an even number of ones and an odd number of ones, and this property remains true throughout the game. Denote by $d(j,t)$ the graph distance between the jth zombie and the survivor after the t-th round, and let $\vec{d}(t)$ be the corresponding k-dimensional vector of distances at time t. Since $\vec{d}(t)$ is coordinate-wise nonincreasing, it suffices to show that given any starting position (for both the survivor and the zombies) there is a positive probability that after a finite number of moves the distance vector decreases in at least one coordinate. Indeed, suppose that, regardless of the starting position, with probability $\delta > 0$ (observe that δ might be a function of n that tends to zero as $n \to \infty$) after $T(n)$ steps the distance vector decreases, where $T(n)$ is some function of n. By concatenating disjoint intervals of length $T(n)$, the probability of having a strictly decreasing distance vector can be boosted as high as desired.

To show this, observe the following: if immediately after the zombies' move there is no coordinate in which all zombies have the same binary value as the survivor, then, regardless of what the survivor does in the next round, at least one zombie will become closer to the survivor "for free" (that is, there exists $1 \leq j \leq k$ such that after the t-th round we have $d(j, t) < d(j, t-1)$). Otherwise, suppose that there exist $1 \leq C \leq n-1$ coordinates such that all zombies have the same binary value as the survivor in this coordinate, and in the latter we can maintain the distances to all zombies by flipping the bit corresponding to any such coordinate. In the next round, consider the following strategy: all but one zombie flip the bit recently flipped by the survivor, and the remaining zombie flips a coordinate in which they are not the only zombie differing from the survivor in that bit. Note that this is indeed possible, since by our assumption, after the survivor's move, at least $n/3$ zombies differ in least one bit (other than the last one flipped by the survivor), and at least $n/3$ zombies differ in least two bits (again, other than the last one flipped by the survivor).

Therefore, the total number of bits in which zombies differ is at least n, and so, by the pigeonhole principle, there exists a bit (one more time, other than the last one flipped by the survivor) in which at least two zombies differ. With probability at least $(1/n)^k > 0$, the zombies choose this strategy, and if they do so, in the next round there are less coordinates in which all zombies have the same binary value as the survivor. It follows that with probability at least $(1/n)^{kC} \geq (1/n)^{k(n-1)} > 0$, the zombies follow this sequence of strategies, and then the survivor is forced to choose a coordinate in which the distance to at least one zombie decreases. Since this holds independently of the distance vector, the distances eventually decrease, and the survivor is eaten with probability 1.

□

4.4   Toroidal Grids

We now consider the zombie number of grids formed by products of cycles. This family of graphs seems to be the most interesting one to analyze and is poorly understood at present. Let T_n be the *toroidal grid* $n \times n$, which is isomorphic to $C_n \square C_n$. For simplicity, we take the vertex set of T_n to consist of $\mathbb{Z}_n \times \mathbb{Z}_n$, where $\mathbb{Z}_n$ denotes the ring of integers modulo n.

Lower Bound

The following lower bound for the zombie number of T_n was proved in [50].

Theorem 4.4.1 ([50]). *If $\omega = \omega(n)$ is a function tending to infinity as $n \to \infty$, then a.a.s. $z(T_n) \geq \sqrt{n}/(\omega \log n)$.*

To prove this lower bound on $z(T_n)$, we assume henceforth that there are $k = \lfloor \sqrt{n}/(\omega \log n) \rfloor$ zombies, for any given $\omega = \omega(n)$ that tends to infinity as $n \to \infty$. We will find a strategy for the survivor that a.a.s. allows the survivor to avoid being eaten indefinitely.

It is convenient for the analysis to assume that the game runs over infinitely many rounds, even if some zombie catches the survivor (in which case they will remain together indefinitely). A *trajectory* is a sequence $\boldsymbol{u} = (u_t)_{t \in I}$ of vertices of T_n, where I is an interval (finite or infinite) of nonnegative integers corresponding to time-steps. We say that the survivor (or one zombie) follows a trajectory $\boldsymbol{u} = (u_t)_{t \in I}$ if, for each $t \in I$, u_t denotes the position of that survivor or zombie at time-step t. Recall that zombies move first, so a zombie with zombie trajectory $\boldsymbol{v}$ catches the survivor with trajectory $\boldsymbol{u}$ at time-step t if $v_t = u_{t-1}$. (If $v_t = u_t$ because the survivor moves to the zombie's location, then $v_{t+1} = u_t$, so we may interpret this as if the zombie catches the survivor at time-step $t+1$.) Sometimes it is useful to imagine that the survivor and the zombies move simultaneously, but the zombies observe the position of the survivor at time t to decide their new position at time $t + 1$, whereas the survivor looks at the positions of the zombies at time $t + 1$ to decide their new position at time $t + 2$. Since the zombies' trajectories may depend on the survivor's trajectory and vice versa, it is convenient to formulate the strategy of each player a priori in a way that does not depend on the other player's choices.

A *zombie strategy* is given by $(v_0, \boldsymbol{\sigma})$, where $v_0 \in \mathbb{Z}_n \times \mathbb{Z}_n$, $\boldsymbol{\sigma} = (\sigma_t)_{t \in \mathbb{N}}$, and each σ_t is a permutation of the symbols U, D, L, R (up, down, left, right). Each zombie will choose a zombie strategy $(v_0, \boldsymbol{\sigma})$ uniformly at random and independently from everything else, and this will determine the zombie's decisions throughout the process in the following manner. Initially, the zombie starts at position v_0. At each step $t \in \mathbb{N}$, the zombie moves from v_{t-1} to v_t (before the survivor moves). To do so, the zombie picks the first direction in the permutation σ_t that decreases their distance to the survivor, and takes a step in that direction. This determines the new position v_t. Sometimes we will use modular notation to describe these steps. For instance, if the t-th step is taken to the right (that is, in the direction R), we may write $v_t = v_{t-1} + (1, 0)$, where the sum is to be interpreted coordinatewise modulo n.

A *survivor strategy* is given by $(u_0, \boldsymbol{m})$, where $u_0 : (\mathbb{Z}_n \times \mathbb{Z}_n)^k \to \mathbb{Z}_n \times \mathbb{Z}_n$, $\boldsymbol{m} = (m_t)_{t \in \mathbb{N}}$ and $m_t : (\mathbb{Z}_n \times \mathbb{Z}_n)^{k+1} \to \{\text{U}, \text{D}, \text{L}, \text{R}\}$. This strategy is chosen deterministically by the survivor before the zombie strategies have been exposed, and will determine the decisions of the survivor during the game as follows. Initially, the survivor starts at vertex u_0, which is a function of the zombies' initial configuration. At each time-step $t \in \mathbb{N}$, after the zombies move, the survivor moves from u_{t-1} to u_t. The direction of this move is determined by m_t, which is a function of the positions of the zombies and the survivor right before their move. Note that the strategy of the survivor depends not only on the positions of all players at a given time-step t, but also on t. Possibly, the dependency on t does not provide an essential advantage for the survivor, but makes the description of the argument easier.

The formulation above is useful, since it allows us to decouple all decisions by the survivor and the zombies prior to the start of the game. Note that the final trajectory of each player (either the survivor or zombie) will depend not only on their own strategy, but will be a deterministic function of all strategies together.

Throughout the section, let B be a fixed $\lfloor K \log n \rfloor \times \lfloor K \log n \rfloor$ box contained in the toroidal grid, where $K = 5 \cdot 10^4$. A survivor strategy is *B-boxed* during the time period $[0, 4n]$ if the following hold: the initial position u_0 belongs to the box B and is chosen independently of the positions of the zombies; the sequence of moves $\boldsymbol{m} = (m_t)_{t \in [1, 4n]}$ is such that the survivor always stays inside of B, regardless of the positions of the zombies in that period; each move m_t ($t \in [1, 4n]$) does not depend on the positions of the zombies that lie outside of B at that given step t (that is, any two configurations of the zombies at time t that only differ in the positions of some zombies not in B must yield the same value of m_t).

Later in this section, we will specify a particular B-boxed strategy for the survivor that will allow them to survive indefinitely a.a.s. The next two lemmas describe the locations of the zombies before they reach the box B, by only assuming that the survivor's strategy is B-boxed during the time period $[0, 4n]$.

Lemma 4.4.2 ([50]). *Assume that the survivor's strategy is B-boxed during the time period $[0, 4n]$, and pick any zombie strategy for all but one distinguished zombie. For any $t \in [1, 4n]$, the probability that this zombie is initially outside of the box B and arrives at B at the t-th step of the game is at most $20Kt \log n / n^2$.*

Lemma 4.4.3 ([50]). *Consider $k = \lfloor \sqrt{n}/(\omega \log n) \rfloor$ zombies on T_n, for any given $\omega = \omega(n)$ that tends to infinity as $n \to \infty$. Assume that the survivor follows a B-boxed strategy during the time period $[0, 4n]$. Then a.a.s. the following hold.*

(i) There is no zombie in B initially.

(ii) All zombies arrive at B within the first $3n$ steps.

(iii) No two zombies arrive at B less than $M \log n$ steps apart, where $M = 12K$.

A zombie strategy $(v_0, \boldsymbol{\sigma})$ is called *regular* if, for any direction $\alpha \in \{L, R, U, D\}$ and any interval of consecutive steps $I \subseteq [1, 4n]$ of length $\lfloor 20 \log n \rfloor$, there is a subset of steps $J \subseteq I$ (not necessarily consecutive) with $|J| = \lceil \log n \rceil$ such that, for every $i \in J$, σ_i has α as the first symbol in the permutation. Informally, for every $\lfloor 20 \log n \rfloor$ consecutive steps in $[1, 4n]$, there are at least $\log n$ steps in which the zombie tries to move in the direction of α if that decreases the distance to the survivor.

Lemma 4.4.4 ([50]). *Consider $k = \sqrt{n}/(\omega \log n)$ zombies on T_n, for any given $\omega = \omega(n)$ that tends to infinity as $n \to \infty$. Then a.a.s. every zombie has a regular zombie strategy.*

Given a trajectory $\boldsymbol{u} = (u_t)_{t \in [a,b]}$, any integer $a < j < b$ such that $u_j - u_{j-1} \neq u_{j+1} - u_j$ is called a *turning point* (that is, the direction of the trajectory changes at time-step $j + 1$). A turning point j is *proper* if additionally $u_j - u_{j-1} \neq -(u_{j+1} - u_j)$ (that is, the trajectory does not turn $180°$). For convenience, we also consider the first and last indices a and b of the trajectory to be proper turning points.

A trajectory is *stable* if all its turning points are proper and, for any two different turning points j and j', we have $j - j' \geq \lfloor 20 \log n \rfloor$.

Lemma 4.4.5 ([50]). *Suppose that a zombie has a fixed regular zombie strategy $(v_0, \boldsymbol{\sigma})$ and that the survivor follows a stable trajectory $\boldsymbol{u}$ during the time interval $[a, b]$. Let $\boldsymbol{v}$ be the trajectory of the zombie (determined by their strategy and the survivor's trajectory). Suppose moreover that v_a and u_a are at distance $d \in \{2, 3\}$ and that v_{a+1} and u_{a+1} are also at distance d (that is, the first move of the survivor is not toward the zombie). Then deterministically, v_t and u_t are at distance d for all $t \in [a, b]$.*

Now, we are ready to prove the lower bound.

Proof of Theorem 4.4.1. We will describe a B-boxed strategy for the survivor during the time period $[0, 4n]$ that a.a.s. succeeds at attracting all the zombies to two vertices at distances 2 and 3 from the survivor's position and on the same side of that position. Once that is achieved, the survivor can simply keep moving in a straight line around the toroidal grid, staying away from all the zombies forever; see Figure 4.2.

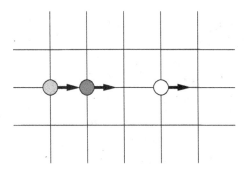

FIGURE 4.2: If at some point the survivor is on the white vertex and all the zombies are at distance 2 or 3 on the grey vertices, and the survivor is moving away from the zombies, then they can keep the same direction indefinitely on T_n and survive.

All geometric notions in the argument that may be ambiguous in the torus (for example, top, bottom, left, right, above, below, and so on) should be interpreted with respect to the box B where the survivor plays their strategy. Let C be a smaller $\lfloor 20 \log n \rfloor \times \lfloor 20 \log n \rfloor$ box centered at the center of box B. The survivor starts at the top left corner u_0 of C, and will always follow a stable trajectory $\boldsymbol{u}$ during the time period $[0, 4n]$ inside B. The survivor's decisions regarding what trajectory to follow, will depend on the positions of the zombies inside B, but not on those outside of B. Therefore, the survivor strategy is B-boxed during the time period $[0, 4n]$.

In our description of the survivor's strategy, we will only consider situations that are achievable assuming that the conclusions of Lemmas 4.4.3 and 4.4.4 hold. That is, we assume that initially there is no zombie in B; they all arrive at B within the first $3n$ steps; no two zombies arrive at B less than $M \log n$ steps apart; and all zombies have regular strategies. One key consequence of this is that the survivor will be able to handle one zombie at a time. If at some step the survivor has to face a situation not covered by our description (because, for instance, two zombies arrived at B at the same time), then they give up and simply defaults to any arbitrary fixed B-boxed strategy, ignoring the zombies from then on. As a result the survivor will probably be eaten, but a.a.s. this situation does not occur.

The survivor starts at the top left corner u_0 and starts going in circles clockwise around C until the time a first zombie arrives at B. Let a be the time this situation occurs. The survivor keeps going in circles around C until the zombie is at distance less than or equal to $542 \log n$. From our assumption on the regularity of the zombies' strategies, this takes at most $(K/2)\lfloor 20 \log n \rfloor \leq 10K \log n$ steps from time a (since there will be at least $(K/2) \log n$ steps among those in which the zombie tries to move in the direction of α, for each $\alpha \in \{L, R, U, D\}$). Then the survivor keeps going until the next corner u of C. At that point the zombie is at a distance between $500 \log n$ and $542 \log n$ from the survivor. The survivor makes that corner u their next proper turning point in their trajectory, and changes the direction so that they are not moving toward the zombie. The survivor can always do so by choosing between a $90°$ or a $-90°$ turn. (Notice that they might leave C at u.)

Without loss of generality, we may suppose that this direction is to the right (the description of their strategy in any other case is analogous). The survivor keeps moving right for $1000 \lfloor 20 \log n \rfloor \leq 2 \cdot 10^4 \log n$ time-steps. During those steps, the zombie gets horizontally aligned with the survivor (and it is still at the same distance), since their zombie strategy is regular and at least $1000 \log n$ of those steps decrease the vertical distance between the two individuals. Then the survivor moves down for $\lfloor 20 \log n \rfloor$ steps, left for $\lfloor 20 \log n \rfloor$ steps and up again for $\lfloor 20 \log n \rfloor$ steps. The zombie is still to the left of the survivor (at horizontal distance of between $440 \log n$ and $542 \log n$) and either horizontally aligned or below (at vertical distance between 0 and $\lfloor 20 \log n \rfloor$). Next, the survivor moves to the left $20 \lfloor 20 \log n \rfloor \leq 400 \log n$ steps. Both individuals must now be horizontally aligned and at a distance between $40 \log n$ and $143 \log n$. The survivor keeps on moving left until they are at distance 2 or 3 from the zombie. Let b be the time-step when this happens. Finally, the survivor moves down for $\lfloor 20 \log n \rfloor$ steps, and then moves back to the top left corner u_0 of C using a stable trajectory. He chooses the shortest stable trajec-

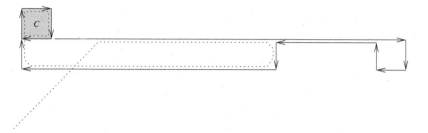

FIGURE 4.3: Approximate depiction of the survivor's strategy when a new zombie approaches. The black arrows describe the trajectory of the survivor and the dotted lines the trajectory of the zombie.

tory that allows the survivor to reach u_0 by a step up. Let c be the time they get back at u_0. Then they resume the strategy of going around C clockwise until the next zombie arrives at B. See Figure 4.3 for a visual representation of the above description.

Note that the survivor's trajectory described so far is stable, since all the turning points are proper and are at least $\lfloor 20 \log n \rfloor$ steps apart. Further, the survivor's move down at step $b + 1$ is not toward the zombie (which is at time b horizontally aligned with the survivor and to their left). Therefore, by Lemma 4.4.5, as long as the survivor maintains a stable trajectory over the whole time period $[0, 4n]$, then the trajectory of this zombie will keep a constant distance (either 2 or 3) to the survivor's trajectory during all steps in $[b, 4n]$. Also, observe that the whole process between time a and time c takes at most $11K \log n < M \log n$ steps, so by assumption there was no other zombie in B during that period. The survivor's strategy does not depend on zombies outside of B and, in spite of their long excursion of around $2 \cdot 10^4 \log n$ steps to the right from one corner of C, the survivor never abandons the box B, as required by the definition of B-boxed strategy.

The survivor can proceed analogously each time a new zombie arrives to B, ignoring all zombies that are already at distance 2 or 3 from them. By construction, this defines a B-boxed strategy during the time period $[0, 4n]$ (recall that for any configuration not covered by our previous description, the survivor just adopts any default B-strategy). If all our assumptions on the zombies hold (which occurs a.a.s. by Lemmas 4.4.3 and 4.4.4), the survivor will follow a stable trajectory during the time period $[0, 4n]$ with the following properties: at step $4n$ all zombies are at distance 2 or 3 from the survivor, and the survivor is moving away from all of them. Then from that moment on, the survivor can keep going in the same direction and survive forever. The proof of the theorem follows.   □

Upper Bound

Observe that $z(T_n) = O(n^2 \log n)$. To see this, suppose that the game is played against $k = 3n^2 \log n$ zombies. It is straightforward to see that a.a.s. every vertex is initially occupied by at least one zombie and if so the survivor is eaten immediately. Indeed, the probability that at least one vertex is not occupied by a zombie is at most $n^2(1 - 1/n^2)^k \leq n^2 \exp(-k/n^2) = 1/n = o(1)$.

In this subsection, we provide a general approach that gives immediately a small improvement, an upper bound of $O(n^2)$. The results here derive from the preprint [164]. We investigate this technique by analyzing a few natural strategies and show that this approach has no hope of giving an upper bound better than $O(n^{3/2})$. Determining the actual upper bound on the zombie number of toroidal grids remains open.

Consider the family $\mathcal{F}$ of strategies of the survivor for the first $M = M(n) = \lfloor n/4 \rfloor$ moves of the game. Here, a *strategy* is simply a sequence of moves of the survivor, which is not affected neither by the initial distribution of zombies or by location during the game. We may assume that zombies do not eat the survivor immediately when they land on them; instead, they walk with them to the end of this sequence of T moves and then do their job. Hence, since there are n^2 vertices to choose from for the starting position of the survivor and give options in each round (U, D, L, R, and "stay put"), the number of strategies in $\mathcal{F}$ is $n^2 5^M = n^2 5^{\lfloor n/4 \rfloor}$. Finally, let $\mathcal{F}_0 \subseteq \mathcal{F}$ be a subfamily of $5^{\lfloor n/4 \rfloor}$ strategies of the survivor that finish their walk at vertex $(0, 0)$ of T_n.

We concentrate on a given strategy $\mathcal{S} \in \mathcal{F}$ finishing at vertex

(a, b) of T_n. For all x and $y \in \mathbb{Z}$ such that $|x| + |y| \leq M$, let $p_{\mathcal{S}}(x, y)$ be the probability that a zombie starting at vertex $(a+x, b+y)$ eats the survivor using strategy $\mathcal{S}$ for the first M moves. It is evident that if a zombie starts at a vertex at distance larger than M from (a, b), then it is impossible for them to eat the survivor (even a player with access to all strategies would not be able to reach (a, b) in M moves!); hence, in this situation $p_{\mathcal{S}}(x, y) = 0$. Note also that restricting starting positions for zombies to the subgraph around (a, b) guarantees that the survivor and zombies starting at this subgraph do not leave it during the first M rounds. As a result, the game during these M rounds is played as if it was played on the square grid $(P_n \square P_n)$ centered at (a, b), not the toroidal one. Since T_n is a vertex transitive graph, without loss of generality, we may assume that the survivor finishes their walk at vertex $(0, 0)$. These little observations will simplify the analysis below. Finally, let

$$t(\mathcal{S}) = \sum_{x=-M}^{M} \sum_{y=-M}^{M} p_{\mathcal{S}}(x, y).$$

As mentioned earlier, due to the fact that T_n is vertex-transitive, we have that

$$t_n = \min_{\mathcal{S} \in \mathcal{F}} t(\mathcal{S}) = \min_{\mathcal{S} \in \mathcal{F}_0} t(\mathcal{S}).$$

Now, we are ready to state our first observation.

Theorem 4.4.6 ([164]). *A.a.s.* $k = n^3/t_n$ *zombies eat the survivor on* T_n. *Hence,* $z(T_n) = O(n^3/t_n)$.

Proof. Let $X_1, X_2, \ldots$ be a sequence of independent random variables, each of them being the Bernoulli random variable with parameter $p = 1/2$ (that is, $\Pr(X_i = 1) = \Pr(X_i = 0) = 1/2$ for each $i \in \mathbb{N}$). This (random) sequence will completely determine the location of all the zombies. Formally, we first fix a permutation π of k zombies. Then in each round, we consider all zombies, one by one, using permutation π. If there is precisely one shortest path between a given zombie and the survivor, the next move is determined; otherwise, the next random variable X_i from the sequence guides it. For example, if $X_i = 0$, the zombie moves horizontally; otherwise, they move vertically.

Our goal is to show that a.a.s. the survivor is eaten during the first $M = \lfloor n/4 \rfloor$ rounds, regardless of the strategy they use. To show this property, let us pretend that the game is played on a *real board* but, at the same time, there are $n^2 5^M$ *auxiliary boards* where the game is played against all strategies from $\mathcal{F}$. An important assumption is that the same sequence $X_1, X_2, \ldots$ is used for all the boards.

Since zombies select initial vertices uniformly at random, the probability that a given zombie wins against a given strategy $\mathcal{S} \in \mathcal{F}$ is at least

$$\sum_{x=-M}^{M} \sum_{y=-M}^{M} \frac{p_{\mathcal{S}}(x, y)}{n^2} \geq \frac{t_n}{n^2}.$$

Since zombies play independently, the probability the survivor using strategy $\mathcal{S}$ is not eaten during the first M rounds is at most

$$\left(1 - \frac{t_n}{n^2} \right)^k \leq \exp\left(-\frac{t_n k}{n^2} \right) = \exp(-n) = o\left(\frac{1}{n^2 5^{n/4}} \right).$$

(Note that, in particular, if two zombies start at the same vertex (x, y), each of them catches the survivor with probability $p_{\mathcal{S}}(x, y)$ and the corresponding events are independent.) It follows that the expected number of auxiliary games where the survivor is not eaten is $o(1)$ and so a.a.s. they lose on all auxiliary boards. As a result, a.a.s. zombies win the real game, regardless of the strategy used by the survivor. The survivor should make decisions based on the locations of the zombies (who make decisions at random and so very often behave in a sub-optimal way); however, if the survivor wins the real game, at least one auxiliary game is also won by some specific strategy which we showed cannot happen a.a.s. That is why we needed the trick with the sequence $X_1, X_2, \ldots$ guiding all the games considered.   □

It is possible to derive a recursive formula for calculating $p_{\mathcal{S}}(x, y)$ for a given strategy $\mathcal{S} \in \mathcal{F}$. (See [164].) It follows that $t_n \geq 4M = 4\lfloor n/4 \rfloor \sim n$. As a result, from Theorem 4.4.6 we obtain the following upper bound for $z(T_n)$.

Corollary 4.4.7 ([164]). $z(T_n) = O(n^2)$.

This improves the trivial upper bound of $O(n^2 \log n)$ but more work remains to determine the zombie number of toroidal grids. Is there an upper bound of $O(n^{2-\varepsilon})$ for some $\varepsilon > 0$?

Now, let us present a few natural strategies from family $\mathcal{F}_0$ for various kinds of movement of the survivor.

(a) The survivor does not move.

(b) The survivor performs a random walk.

(c) The survivor moves straight down.

(d) The survivor moves along the diagonal (that is, moves down and then immediately left in each pair of the two consecutive rounds).

(e–h) The survivor moves along edges of a square (k times): $k = 1, 2, 3$ or 8.

For each of them, we calculated $p_S(x, y)$ for various values of n to estimate $t(\mathcal{S})$. Functions $p_S(x, y)$ are presented visually in Figures 4.4 and 4.5. In the figure, darker colors correspond to values of probabilities that are close to 1, while lighter ones to values close to 0.

We might conjecture that $t_n = \Theta(n^2)$ and this in turn would imply $z(T_n) = O(n)$. Unfortunately, this is not true. Based on simulations, we were able to identify that strategy (d) might create a problem and serve as a counterexample, which turned out to be the case. We will show that the conjecture was too optimistic and $t_n = O(n^{3/2})$. But there remains the possibility that $t_n = \Theta(n^{3/2})$ and so it may hold that $z(T_n) = O(n^{3/2})$.

Theorem 4.4.8 ([164]). $t(\mathcal{S}) = O(n^{3/2})$ *for strategy (d).*

Proof. To estimate $t(\mathcal{S})$, we need to estimate $p_S(x, y)$ for each $x, y \in \mathbb{Z}$ such that $|x| + |y| \leq M$. Due to the symmetry, we may assume that $x \geq y$. Suppose that a given zombie starts the game at vertex (x, y). We will partition the part of the grid we investigate into 4 regions and deal with each of them separately. See Figure 4.6. Let $c \in \mathbb{R}_+$ be some fixed, sufficiently large constant.

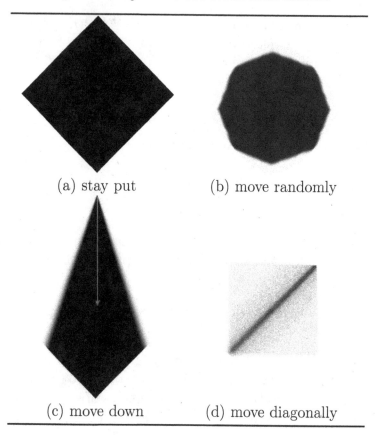

(a) stay put   (b) move randomly

(c) move down   (d) move diagonally

FIGURE 4.4: Examples of four strategies for the first M moves.

Region **R1**: Suppose that $x \geq M/2 + c\sqrt{n}$. The zombies move randomly in the directions U or L until their position and a position of the survivor match horizontally or vertically. If they match horizontally, then they will never be able to eat the survivor—the distance will be preserved till the end of the game. However, if coordinates are matched vertically, then there is a chance. The probability of this event can be estimated (see discussion for Region **R3** below) but we do not need it here (we may use a trivial upper bound of 1 for this conditional probability). It is enough to notice that if the vertical match occurs, then at some point of the game a zombie moved $x - \lfloor M/2 \rfloor$ more often L than they moved

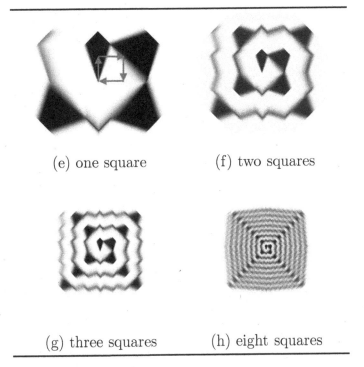

(e) one square   (f) two squares

(g) three squares   (h) eight squares

FIGURE 4.5: Examples of four more strategies for the first M moves.

U. Hence,

$$p_S(x, y) \leq \mathbb{P}\Big(S_t \geq t/2 + (x - \lfloor M/2 \rfloor) \text{ for some } t \leq M\Big),$$

where $X_1, X_2, \ldots$ is a sequence of independent random variables, each of them being the Bernoulli random variable with parameter $p = 1/2$, and $S_t = \sum_{i=1}^{t} X_t$. It follows that

$$p_S(x, y) \leq \exp\left(-\Omega\left(\frac{(x - \lfloor M/2 \rfloor)^2}{M}\right)\right).$$

(For example, by converting S_t into a martingale and then using Azuma-Hoeffding inequality. See Section 6.5 for more background, and Theorem 6.5.1 for the Azuma-Hoeffding inequality.) The con-

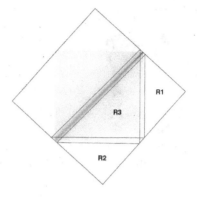

FIGURE 4.6: A closer look at Strategy (d).

tribution to $t(\mathcal{S})$ from Region **R1** is then at most

$$\sum_{x \geq M/2 + c\sqrt{n}} O(n) \exp\left(-\Omega\left(\frac{(x - \lfloor M/2 \rfloor)^2}{M}\right)\right)$$

$$= O(n) \sum_{x \geq c\sqrt{n}} \exp\left(-\Omega\left(\frac{x^2}{n}\right)\right) = O(n^{3/2}),$$

provided that c is large enough.

Region **R2**: Suppose now that $y \leq -M/2 - c\sqrt{n}$. The zombie moves U or R, and it is expected that players match horizontally after r rounds at which point the zombie occupies vertex $(\hat{x}, \hat{y})$, where

$$
\begin{aligned}
\hat{x} &= \frac{M/2 + x}{2} + O(1) = \frac{M}{4} + \frac{x}{2} + O(1) \\
r &= 2(\hat{x} - x) + O(1) = 2\left(\frac{M}{4} - \frac{x}{2}\right) + O(1) \\
\hat{y} &= y + \frac{r}{2} = y + \left(\frac{M}{4} - \frac{x}{2}\right) + O(1) = \frac{M}{4} - \frac{x}{2} + y + O(1).
\end{aligned}
$$

The distance from $(\hat{x}, \hat{y})$ to $(0,0)$ is $d = \hat{x} - \hat{y} = x - y + O(1)$ and so $r + d = M/2 - y + O(1) \geq M + c\sqrt{n} + O(1)$. It follows that in order for the zombie to have a chance to win, a horizontal match has to occur much later than expected. Arguing as before, we can estimate the probability of this event and show that the contribution to $t(\mathcal{S})$ from Region **R2** is $O(n^{3/2})$, provided that c is large enough.

Region **R3**: Suppose now that $x \geq y + c\sqrt{n}$, $y \geq -M/2 + c\sqrt{n}$, and $x \leq M/2 - c\sqrt{n}$. This case seems to be the most interesting. As for the previous region, it is expected that players match horizontally when the zombie occupies vertex $(\hat{x}, \hat{y})$ and the distance between players is

$$k = \left(\frac{M}{2} - y\right) - \left(\frac{M}{2} - x\right) = x - y \geq c\sqrt{n}.$$

Arguing as before, we can show that with probability $1 - \exp(-\Omega(-k^2/M))$ not only does this happens, but at that time the distance between players, Y, is at least $k/2$.

Conditioning on $Y = y \geq k/2$, we aim now to estimate the probability that the survivor is eaten. Assume that y is even; the odd case can be handled similarly. Consider a sequence of two consecutive rounds. At the beginning of each pair of rounds, before the survivor moves in the D direction, we measure the absolute difference between the corresponding x-coordinates of the players, to obtain a sequence $Z_0, Z_1, \ldots, Z_{y/2}$ of random variables. It is straightforward to see that $Z_0 = 0$ and the survivor is eaten if and only if $Z_{y/2} = 0$. Indeed, if $Z_{y/2} > 0$, then the zombie ends up lined up horizontally before getting close to the survivor and from that point on they will continue keeping the distance. We use the notation w.p. to denote "with probability." If $Z_t > 0$, then we have that:

$$Z_{t+1} = \begin{cases} Z_t + 1 & \text{w.p. } 1/4 \text{ (zombie moves U twice)} \\ Z_t - 1 & \text{w.p. } 1/4 \text{ (zombie moves L twice)} \\ Z_t & \text{w.p. } 1/2 \text{ (otherwise).} \end{cases}$$

However, if $Z_t = 0$, then the first move of the zombie is forced (they move U) and so

$$Z_{t+1} = \begin{cases} 1 & \text{w.p. } 1/2 \text{ (the second move is U)} \\ 0 & \text{w.p. } 1/2 \text{ (the second move is L).} \end{cases}$$

We can couple this process with a lazy random walk and show that $\mathbb{P}(Z_{y/2} = 0) = \Theta(1/\sqrt{y}) = \Theta(1/\sqrt{k})$. The contribution to $t(\mathcal{S})$ from Region **R3** is

$$O(n) \sum_{k=c\sqrt{n}}^{O(n)} \left(\exp\left(-\Omega\left(\frac{k^2}{n}\right)\right) + \frac{1}{\sqrt{k}} \right)$$

$$= O(n^{3/2}) + O(n) \int_{\sqrt{n}}^{O(n)} \frac{dx}{x} = O(n^{3/2}),$$

provided that c is large enough.

The number of vertices not included in the three regions we considered is $O(n^{3/2})$ and so the total contribution is $O(n^{3/2})$ and the proof follows.   □

Chapter 5

Large cop number and Meyniel's conjecture

Meyniel's conjecture is now a central topic in the fields of graph searching and Cops and Robbers. The conjecture claims that if G is a connected graph of order n, then

$$c(G) = O(\sqrt{n}). \tag{5.1}$$

Rephrasing this, there is a universal constant $d > 0$ such that for every connected graph G of order n we have that

$$c(G) \leq d\sqrt{n}.$$

The inequality (5.1) is called the *Meyniel bound*.

The conjecture has an interesting history. It was mentioned in Frankl's paper [91] as a personal communication to him by Henri Meyniel in 1985 (see page 301 of [91] and reference [8] in that paper; see Figure 5.1 for a rare photograph of Meyniel). Since 1985 the conjecture lay mostly dormant, until a new upper bound on the cop number was provided in 2008 [65].

Meyniel's conjecture definitely stands out as one of the deepest problems on the cop number. Many new works supplying solutions of partial cases have emerged; see [13, 36, 65, 95, 133, 167, 168, 174]. See also the survey [14] and Chapter 3 of [52].

For a positive integer n, let $c(n)$ be the maximum value of $c(G)$, where G is a connected graph of order n. For example, $c(1) = c(2) = c(3) = 1$, while $c(4) = c(5) = 2$. Note that $c(n)$ is a nondecreasing function (to see this, note that adding a vertex of degree one does not change the cop number). Using this function, we can rephrase Meyniel's conjecture as

$$c(n) = O(\sqrt{n}).$$

FIGURE 5.1: Henri Meyniel. Photo courtesy of Geňa Hahn.

As a first step toward proving Meyniel's conjecture, Frankl [91] proved that $c(n) = O(n \log \log n / \log n) = o(n)$. Recent work has improved this upper bound, but only somewhat; see Section 5.1 for more details. To further highlight how far we are from proving the conjecture, even the so-called *soft Meyniel's conjecture* is open, which states that for some fixed constant $\varepsilon > 0$,

$$c(n) = O(n^{1-\varepsilon}).$$

We note that Meyniel's conjecture was proved, using the probabilistic method, in the classes of diameter 2 and bipartite diameter 3 graphs by Lu and Peng [133].

Theorem 5.0.1 ([133]). *If G is a connected graph of order n and either diameter 2 or bipartite diameter 3, then*

$$c(G) \leq 2\sqrt{n} - 1.$$

The upper bound in Theorem 5.0.1 was recently improved to $\sqrt{2n}$ in [178] without using randomized arguments.

Our goal in this chapter is to summarize some results on partial solutions of Meyniel's conjecture, especially those emphasizing random techniques. In Section 5.1 we provide the best currently known upper bound on $c(n)$, and supply a proof due to Frieze, Krivelevich, and Loh [95] using expansion and the probabilistic method. We discuss the proof of Meyniel's conjecture for binomial random graphs in Section 5.2; random d regular graphs are discussed in Section 5.3. We finish by discussing families of graphs realizing the tightness of the Meyniel bound (5.1) in Section 5.4.

5.1 Upper Bounds for $c(n)$

For many years, the best known upper bound on the cop number of general graphs was the one proved by Frankl [91] in 1987.

Theorem 5.1.1 ([91]). *For a positive integer n,*

$$c(n) = O\left(n\frac{\log\log n}{\log n}\right).$$

The proof of Theorem 5.1.1 used a greedy approach and Moore's theorem; see also Chapter 3 of [52]. It was many years later, in 2008, that Chinifooroshan gave an improved bound removing the $\log\log n$ term.

Theorem 5.1.2 ([65]). *For a positive integer n,*

$$c(n) = O\left(\frac{n}{\log n}\right). \tag{5.2}$$

An even more recent improvement exists to the bound (5.2) in Theorem 5.1.2. Interestingly, the following theorem was proved independently by three sets of authors (each using different methods).

Theorem 5.1.3 ([95, 133, 174]). *For a positive integer n, we have that*

$$c(n) \le \frac{n}{2^{(1-o(1))\sqrt{\log_2 n}}}. \tag{5.3}$$

The bound in (5.3) is currently the best upper bound for general graphs that is known, but it is still far from proving Meyniel's conjecture or even the soft version of the conjecture. Indeed, the bound in (5.3) only gives $c(n) \leq n^{1-o(1)}$. We note that the proofs of Theorem 5.1.3 in [133, 174] use the greedy approach as in the proofs of Theorems 5.1.1 and 5.1.2, while expansion properties are used in [95].

We give the proof of Theorem 5.1.3 as presented in [95] using expansion properties. We will use the following well-known theorem of Hall that provides a sufficient and necessary condition for a bipartite graph to have a perfect matching. Hall's theorem will be also be useful in the next section when we prove Meyniel's conjecture for random graphs.

If a bipartite graph $G = (X \cup Y, E)$ contains a perfect matching, then $|X| = |Y|$. Also, every subset of X must have enough neighbors in Y; that is, $|N(S)| \geq |S|$ for all $S \subseteq X$. The following theorem says that this obvious necessary condition (which we call *Hall's condition*) is sufficient.

Theorem 5.1.4 ([107]). (Hall's theorem) *A bipartite graph $G = (X \cup Y, E)$ with $|X| = |Y|$ contains a perfect matching if and only if $|N(S)| \geq |S|$ for all $S \subseteq X$.*

We also need the notion of the boundary of a set of vertices. For a set of vertices S in G, the *boundary* of S, written ∂S, is the set of vertices not in S but which are adjacent to some vertex in S. See Figure 5.2.

Proof of Theorem 5.1.3. Throughout, we assume G is connected with order n. We prove the theorem by a series of claims.

Claim 1: If G has a set of vertices S with $|S| < n/2$ and $|\partial S| \leq p|S|$ for some $0 < p < 1$, then $c(G) \leq p|S| + c(G')$, where G' is a connected graph with at most $n - |S|$ vertices.

To see this, place a cop on each vertex of ∂S. Then the robber is forced to remain within a single component X of $G[V(G) \setminus \partial S]$. Note that each such component is either a subgraph of $G[S]$ or a subgraph of $G[V(G) \setminus (S \cup \partial S)]$. By hypothesis, $|S| < n/2 < n - |S|$.

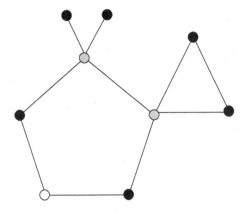

FIGURE 5.2: The boundary of the gray vertices are the black vertices.

Observe that all components in $G[V(G)\setminus(S\cup\partial S)]$ have cardinality at most $n - |S| - |\partial S| \leq n - |S|$, and the claim follows.

Claim 2: If G is bipartite with parts A and B and k is a constant, then we may partition A as $S\cup T$ such that $|N(S)| \leq k|S|$ and for every subset $U \subseteq T$, we have that $|N(U)| \geq k|U|$,

Begin with $S = \emptyset$ and $T = S$. If there exists $U \subseteq T$ with $|N(U)| \leq k|U|$, then add elements of U to S and delete them from T. Repeat this process so long as such a U exists. At the end of the process, in the graph G, we have that $|N(S)| \leq k|S|$ and for all $U \subseteq T$, we have that $|N(U)| > k|U|$. The claim follows.

Claim 3: Fix $p \in (0, 1)$, and let n and r be positive integers. For np sufficiently large, it is possible to place $2pn$ cops in G such that for every S with $|N_r(S)| \geq \frac{16}{p}|S|\log n$, there are at least $|S|$ cops in $N_r(S)$.

For the proof of Claim 3, place a cop on each vertex independently with probability p. With S as in the statement of the claim, the expected number of cops in $N_r(S)$ is at least $16|S|\log n$. The Chernoff bound shows that the probability that this quantity is below half of its expectation is at most

$$\exp(-1/8 \cdot 16|S|\log n) \leq n^{-2|S|}.$$

An application of the union bound over all subsets S of cardinality s implies that with probability at least $1 - n^{-s}$, every $N_r(S)$ contains at least $8|S|\log n \geq |S|$ cops. By a second application of the union bound over all $s \in [n]$, we have that with probability at least $1 - 2/n$, this holds for every $1 \leq |S| \leq n$. As $\text{Bin}(n, p)$ is at most $2pn$ a.a.s. by the Chernoff bounds, we derive that with positive probability we may distribute $2pn$ cops with the desired property, and the claim follows.

Claim 4: For a fixed $p \in (0, 1)$, we have that $c(G) \leq pn + c(G')$, where G' is a connected graph with $m \leq n$ vertices and maximum degree at most $1/p$. Further, G' is a *p-expander*: every subset $V(G)$ of at most $m/2$ vertices has $|\partial S| > p|S|$.

If there is a vertex v of degree at least $1/p$, then by placing one cop on v, we can reduce the number of vertices by at least $1/p + 1$. By Claim 1, if there is a set S of with $|S| < n/2$, but $|\partial S| \leq p|S|$, then we can reduce the number of vertices by at least $|S|$ using $p|S|$ cops. Hence, by repeating this process as long as possible, we will end up with an induced subgraph G' with the stated properties in the claim, such that

$$c(G) \leq p(n - m) + c(G') \leq pn + c(G').$$

The claim follows.

Claim 4 now plays a major role in the proof, as using it we can force the robber into a subgraph with expansion properties. The following claim, in conjunction with Claim 4, will finish the proof of the theorem. A subgraph H is *guarded* by cops if whenever the robber enters H, it is captured in the next round. We say that the cops *guard* H.

Claim 5: There is a function $p = p(n) = 2^{-(1-o(1))\sqrt{\log_2 n}}$ such that for all G with order $m \leq n$, maximum degree less than $1/p$ and expansion at least p, we can guard G with at most $(1 + o(1))\sqrt{\log_2 n} \cdot 2pn$ cops.

We may assume, without loss of generality, that $m > pn$, or else the result follows. It can be shown (we omit details here; see

[95]) that there is a function $p = 2^{-(1-o(1))\sqrt{\log_2 n}}$ and a positive integer $t \sim \sqrt{\log_2 n}$ such that if $k = \frac{16}{p} \log n$, then we have that:

$$k^{t+1} \leq (1+p)^{-2^t} np/2, \quad \text{and} \quad (1+p)^{2^t} \geq k. \quad (5.4)$$

We place the cops into $(t+1)$-many groups $C_0, C_1, \ldots, C_t$ each of cardinality $2pn$. We choose the initial positions of the cops in C_i by applying Claim 3 with $r = 2^i$.

Suppose now that the robber's initial position is v. Let $N_0 = N(v)$. Hence, $|N_0| = |N(v)| \leq 1/p < k$. Consider the auxiliary bipartite graph formed with parts $A = N_0$, $B = V(G)$, and $a \in A$ adjacent to $b \in B$ if and only they are at distance at most one in G. Then Claim 2 implies that we may partition N_0 as $S_0 \cup T_0$ such that in G, $|N_1(S_0)| \leq k|S_0|$, and for every subset $U \subseteq T_0$ we have that $|N(U)| \geq k|U|$. Hence, by Hall's theorem, we may send a distinct cop from C_0 to each vertex in T_0 in the first move of the cops, thereby preventing the robber from moving to T_0.

We therefore have that the robber, after their second move, must remain in $M_1 = N_1(S_0)$. Further, $|N_1(S_0)| \leq k|S_0| \leq k^2$. We may then partition $M_1 = S_1 \cup T_1$, and have the cops in C_1 guard T_1. Also, we have that $|N_2(S_1)| \leq k|S_1| \leq k^3$. The cops have now restricted the robber to $M_2 = N_2(S_1)$.

We may now iterate. In each iteration of this procedure of restricting the possible positions of the robber, the balls double in radii. Thus, in their 2^t-th move, the robber is restricted within a set $M_t = N_{2^{t-1}}(S_{t-1})$ of cardinality at most k^{t+1}. In the final iteration, we must have that $M_t = S_t \cup T_t$ with $S_t = \emptyset$. Indeed, as G is a p-expander, every nonempty set S of vertices of cardinality at most

$$(1+p)^{-(r-1)}m/2 > (1+p)^{-(r-1)}np/2$$

has $|N_r(S)| \geq (1+p)^r|S|$. As the inequalities in (5.4) imply that $|M_t| \leq k^{t+1} \leq (1+p)^{-2^t}np/2$ and that $(1+p)^{2^t} \geq k$, we derive that S_t is empty.

Hence, the cops in C_t can cover all vertices of N_t in 2^t moves. As N_t consists of all possible positions of the robber after their 2^t-th move, the robber must eventually be captured. $\qquad\square$

5.2   Binomial Random Graphs

Our goal in this section is to give a sketch of the proof of Meyniel's conjecture for binomial random graphs. We first summarize some earlier results on the cop number of $\mathbb{G}(n,p)$. Bonato, Wang, and Prałat started investigating Cops and Robbers in $\mathbb{G}(n,p)$ random graphs and generalizations used to model complex networks with a power-law degree distribution (see [53, 55]). From their results it follows that if $2\log n/\sqrt{n} \le p < 1 - \varepsilon$ for some $\varepsilon > 0$, then a.a.s. for $G \in \mathbb{G}(n,p)$

$$c(G) = \Theta(\log n/p).$$

Hence, Meyniel's conjecture holds a.a.s. for such p. A simple argument using dominating sets shows that Meyniel's conjecture also holds a.a.s. if p tends to 1 as n goes to infinity (see [166] for this and stronger results). Bollobás, Kun, and Leader [36] showed that if $p(n) \ge 2.1\log n/n$, then a.a.s. for $G \in \mathbb{G}(n,p)$

$$\frac{1}{(pn)^2}n^{1/2-9/(2\log\log(pn))} \le c(G) \le 160000\sqrt{n}\log n.$$

From these results, if $np \ge 2.1\log n$ and either $np = n^{o(1)}$ or $np = n^{1/2+o(1)}$, then a.a.s. $c(G) = n^{1/2+o(1)}$, where $G \in \mathbb{G}(n,p)$. Surprisingly, between these values it was shown by Łuczak and Prałat [136] that the cop number has more complicated behavior. It follows from [136] that a.a.s. $\log_n c(G)$, where $G \in \mathbb{G}(n,n^{x-1})$ is asymptotic to the function $f(x)$ shown in Figure 5.3. From the above results, we know that Meyniel's conjecture holds a.a.s. for random graphs except perhaps when $np = n^{1/(2k)+o(1)}$ for some $k \in \mathbb{N}$, or when $np = n^{o(1)}$. Wormald and Prałat showed recently [167] that the conjecture holds a.a.s. in $\mathbb{G}(n,p)$ for all $np > (1/2+\varepsilon)\log n$ for some $\varepsilon > 0$. Note that Meyniel's conjecture is restricted to connected graphs, but $G \in \mathbb{G}(n,p)$ is a.a.s. disconnected when $np \le (1 - \varepsilon)\log n$ for some $\varepsilon > 0$. Thus, the result says that Meyniel's conjecture holds a.a.s. for $G \in \mathbb{G}(n,p)$ for *all* p, provided the conjecture is stated precisely as follows: if G is connected, then $c(G) = O(\sqrt{n})$.

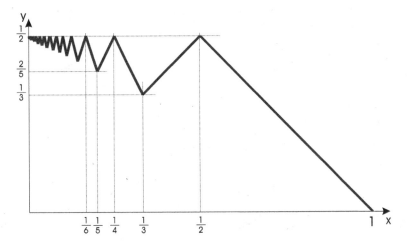

FIGURE 5.3: A graph of the zigzag function f.

In this book, we show that Meyniel's conjecture holds a.a.s. for dense random graphs $(d = p(n-1) \geq \log^3 n)$. This result is from [167], where Meyniel's conjecture is shown to hold for binomial random graphs (including sparser graphs).

We will prove a general purpose result that holds for a family of graphs with some specific expansion properties. After that we will show that dense random graphs a.a.s. fall into this class of graphs and so the conjecture holds a.a.s. for dense random graphs. This more general theorem was used to show that the conjecture holds for random d-regular graphs as well [168].

Let $S_r(v)$ denote the set of vertices whose distance from v is precisely r. We begin with the following expansion-type properties of random graphs.

Theorem 5.2.1 ([167]). *Let $\mathbb{G}_n$ be a set of graphs and $d = d(n) \geq \log^3 n$. Suppose that for some positive constant c, for all $G_n \in \mathbb{G}_n$ the following properties hold.*

    (i) Let $S \subseteq V(G_n)$ be any set of $s = |S|$ vertices, and let $r \in \mathbb{N}$. Then

$$\left| \bigcup_{v \in S} N_r(v) \right| \geq c \min\{sd^r, n\}.$$

Further, if s and r are such that $sd^r < n/\log n$, then

$$\left| \bigcup_{v \in S} N_r(v) \right| \sim sd^r.$$

(ii) Let $v \in V(G_n)$, and let $r \in \mathbb{N}$ be such that $\sqrt{n} < d^{r+1} \le \sqrt{n}\log n$. Then there exists a family

$$\left\{ W(u) \subseteq S_{r+1}(u) : u \in S_r(v) \right\}$$

of pairwise disjoint subsets such that, for each $u \in S_r(v)$,

$$|W(u)| \sim d^{r+1}.$$

Then $c(G_n) = O(\sqrt{n})$.

Before we move to the proof of Theorem 5.2.1, we first provide a useful lemma.

Lemma 5.2.2 ([167]). *Suppose that $d = p(n-1) \ge \log^3 n$. Let G be a graph possessing the properties stated in Theorem 5.2.1. Let $X \subseteq V$ be any set of at most $2\sqrt{n}$ vertices and $r = r(n) \in \mathbb{N}$ is such that $d^r \ge \sqrt{n}\log n$. Let $Y \subseteq V$ be a random set determined by independently choosing each vertex of $v \in V$ to be in Y with probability $C/\sqrt{n}$, where $C \in \mathbb{R}$. Then for sufficiently large constant C, the following statement holds with probability $1 - o(n^{-2})$: it is possible to assign all the vertices in X to distinct vertices in Y such that for each $u \in X$, the vertex in Y to which it is assigned is within distance r of u.*

Proof. To show that the required assignment exists with probability $1 - o(n^{-2})$, we show that with this probability the random choice of vertices in Y satisfies Hall's condition for matchings in bipartite graphs. Set

$$k_0 = \max\{k : kd^r < n\}.$$

Let $K \subseteq X$ with $|K| = k \le k_0$. We may apply the condition in Theorem 5.2.1(i) to bound the cardinality of $\bigcup_{u \in K} N_r(u)$.

From the definition of k we have $k\sqrt{n}\log n \leq kd^r < n$, and hence, the number of vertices of Y in $\bigcup_{u \in K} N_r(u)$ can be stochastically bounded from below by the binomial random variable $\mathrm{Bin}(\lfloor ck\sqrt{n}\log n \rfloor, C/\sqrt{n})$, whose expected value is asymptotic to $Cck\log n$. Using the Chernoff bounds we find that the probability that there are fewer than k vertices of Y in this set of vertices is less than $\exp(-4k\log n)$ when C is a sufficiently large constant. Hence, the probability that Hall's condition fails for at least one set K with $|K| \leq k_0$ is at most

$$\sum_{k=1}^{k_0} \binom{|X|}{k} \exp(-4k\log n) \leq \sum_{k=1}^{k_0} n^k \exp(-4k\log n) = o(n^{-2}).$$

Now consider any set $K \subseteq X$ with $k_0 < |K| = k \leq |X| \leq 2\sqrt{n}$ (if such a set exists). Note that the condition in Theorem 5.2.1(i) implies that the cardinality of $\bigcup_{u \in K} N_r(u)$ is at least cn, so we expect at least $Cc\sqrt{n}$ vertices of Y in this set. Again using the Chernoff bounds, we deduce that the number of vertices of Y in this set is at least $2\sqrt{n} \geq |X| \geq |K|$ with probability at least $1 - \exp(-4\sqrt{n})$, by taking the constant C to be large enough. Since

$$\sum_{k=k_0+1}^{|X|} \binom{|X|}{k} \exp(-4\sqrt{n}) \leq 2\sqrt{n}2^{2\sqrt{n}} \exp(-4\sqrt{n}) = o(n^{-2}),$$

Hall's condition holds with probability $1 - o(n^{-2})$. $\qquad\square$

We now return to the proof of Theorem 5.2.1.

Proof of Theorem 5.2.1. We need to introduce two independent teams of cops that are distributed at random. In Case 1 described below, one team is enough. Each team of cops is determined by independently choosing each vertex of $v \in V(G_n)$ to be occupied by such a cop with probability $C/\sqrt{n}$, where C is a (large) constant to be determined soon. The total number of cops is $\Theta(\sqrt{n})$ a.a.s.

The robber appears at some vertex $v \in V(G_n)$. Let $r = r(d)$ be the smallest integer such that $d^{r+1} \geq \sqrt{n}$. Note that it follows from $d^r < \sqrt{n}$, and assumption (i) that $|N_r(v)| < 2\sqrt{n}$. We consider two cases, depending on the value of d^{r+1}, and in each case we give a strategy which permits the cops to win a.a.s. The first case is based on the idea used in [136].

Case 1. $d^{r+1} \geq \sqrt{n} \log n$.

Since $|N_r(v)| < 2\sqrt{n}$ and $d^{r+1} \geq \sqrt{n} \log n$, it follows from Lemma 5.2.2 that with probability $1 - o(n^{-1})$ it is possible to assign distinct cops from the first team to all vertices u in $N_r(v)$ such that a cop assigned to u is within distance $(r+1)$ of u. Observe that here, the probability refers to the randomness in distributing the cops; the graph G_n is fixed. If this can be done, then after the robber appears, these cops can begin moving straight to their assigned destinations in $N_r(v)$. Since the first move belongs to the cops, they have $(r+1)$ steps, after which the robber must still be inside $N_r(v)$, which is fully occupied by cops.

Hence, the cops will win with probability $1 - o(n^{-1})$, for each possible starting vertex $v \in V(G_n)$. It follows that the strategy gives a win for the cops a.a.s.

Case 2. $d^{r+1} = \sqrt{n} \cdot \omega$, where $1 \leq \omega = \omega(n) < \log n$.

Suppose that $r \geq 1$ (the case $r = 0$ has to be treated differently and will be discussed at the end of the proof). This time, the first team cannot necessarily capture the robber a.a.s., since we cannot hope to find cops inside all the required neighborhoods. Instead, the first team of cops will be given the task of densely covering the sphere $S_r(v)$. Not every $u \in S_r(v)$ can have a distinct cop assigned. For convenience, we restrict ourselves to trying to find one in each set $W(u) \subseteq S_{r+1}(u)$, as defined in condition (ii). Using the estimate on $|W(u)|$ given there, the probability that $W(u)$ contains no cop from the first team is bounded from above by

$$\left(1 - \frac{C}{\sqrt{n}}\right)^{|W(u)|} < \exp\left(-\frac{C}{\sqrt{n}} \cdot \frac{1}{2}\sqrt{n} \cdot \omega\right)$$

$$= \exp\left(-\frac{C}{2} \cdot \omega\right) < \frac{1}{10\omega},$$

for C sufficiently large.

There is a complication, that the robber might try to hide inside $N_r(v)$. We can use an auxiliary team of $C\sqrt{n}$ cops, randomly placed, and for every $u \in N_r(v)$, we search for one of them within distance $(r + 2)$ of u. Since $|N_r(v)| < 2\sqrt{n}$ and $d^{r+2} = d\sqrt{n}\omega \geq \sqrt{n} \log n$, it follows from Lemma 5.2.2 that these cops will capture the robber if they take at least $(r + 1)$ steps to reach $S_r(v)$. Thus, we may assume that the robber reaches $S_r(v)$ in precisely r steps.

The second main team of cops is released when the robber is at $z \in S_{\lfloor r/2 \rfloor}(v)$. (Note that for $r = 1$, we have $z = v$ and both teams start at the same time.) We may assume that after r steps from the start, the robber must be located in the set $S \subseteq S_{\lceil r/2 \rceil}(z) \cap S_r(v)$ that is not covered by cops from the first team. Note that the robber can see both teams of cops right from the start, so the robber can try to use a strategy by selecting appropriately the pair of vertices v and z. However, since there are $O(n^2)$ possible pairs to consider, it is enough to show that, for a given starting vertex v, and assuming the robber appears at z after $\lfloor r/2 \rfloor$ steps, the second team of cops can then capture the robber in the next $(r + 1)$ steps with probability $1 - o(n^{-2})$. Note that the cops get an extra initial step when the robber appears at this place, so they can have $(r + 2)$ steps available for this purpose.

From condition (i) we have $|S_{\lceil r/2 \rceil}(z)| \leq 2d^{\lceil r/2 \rceil}$. Hence, from above, the expected cardinality of S is at most

$$2d^{\lceil r/2 \rceil} \cdot \frac{1}{10\omega},$$

and $|S| \leq \frac{d^{\lceil r/2 \rceil}}{4\omega}$ with probability at least $1 - o(n^{-2})$. We can therefore, assume that this bound on $|S|$ holds.

Since we can assume (as noted above) that the robber will be at a vertex in $S \subseteq S_r(v)$ after taking $\lceil r/2 \rceil$ steps from z, the second team of cops has the assignment of covering the set of vertices

$$U = \bigcup_{s \in S} S_{\lfloor r/2 \rfloor + 1}(s),$$

of cardinality at most

$$2d^{\lfloor r/2 \rfloor + 1}|S| \leq \frac{d^{r+1}}{2\omega} < \sqrt{n},$$

by condition (i). Thus, for each $u \in U$, we would like to find a cop within distance $(r + 2)$. Since $d^{r+2} \geq \sqrt{n} \log n$, it follows from

Lemma 5.2.2 that the second team of cops with probability at least $1 - o(n^{-2})$ can occupy all of U in $(r + 2)$ steps after the robber reaches z.

The argument above showed that the robber reached vertex $s \in S \subseteq S_r(v)$ after r steps, and thus, the robber must be on or inside one of the spheres in the definition of U when the second team of cops are in place. If the robber is not already caught, we can use a "clean-up" team (disjoint from the two main teams and the auxiliary one), of $C\sqrt{n}$ cops, and send them to $N(s, \lfloor r/2 \rfloor + 1)$ to capture the robber while the other cops remain at their final destinations. (Note that it follows from condition (i) that $|N_{\lfloor r/2 \rfloor + 1}(s)| = o(\sqrt{n})$ so an even smaller team would do the job.)

Now suppose that $r = 0$; that is, $d \in [\sqrt{n}, \sqrt{n} \log n)$. The first team of cops is generated randomly, exactly the same way as before. However, in this case, if the first team of cops were to try to cover $N_1(v)$ in one step, since the expected number of them is $\Theta(\sqrt{n})$, they could normally occupy only a negligible part of the neighborhood of v if d is substantially larger than $\sqrt{n}$. So instead, we hold the first team at their original positions for their first move. For a sufficiently large constant C, the first team a.a.s. dominates all of $S(v, 1)$ except for a set $S \subseteq S_1(v)$ with $|S|/|S_1(v)| < e^{-C\omega/2} < 1/(2\omega)$. Hence, if the robber moves in the first step, they have to go to S, and $|S| < \sqrt{n}$. The second team is generated the same way as the first team, but independently, and is instructed to try to cover all of $S \cup \{v\}$ in two steps. Since $d^2 \geq n > \sqrt{n} \log n$, Lemma 5.2.2 implies that with probability $1 - o(n^{-2})$ we can find cops, one in $S_2(s)$ for each $s \in S \cup \{v\}$, that are distinct. In this case the second team can capture the robber after the robber's first step, on their second move. This strategy gives a win for the cops a.a.s., regardless of the initial choice for $v \in V$. □

Now we will verify that dense random graphs a.a.s. satisfy the conditions in the hypotheses of Theorem 5.2.1.

Lemma 5.2.3 ([167]). *If $d = p(n - 1) \geq \log^3 n$ and $G = (V, E) \in \mathbb{G}(n, p)$, then conditions (i) and (ii) in the hypotheses of Theorem 5.2.1 hold a.a.s.*

Theorem 5.2.1, combined with the lemma, immediately shows that Meyniel's conjecture holds for dense random graphs.

Theorem 5.2.4 ([167]). *If $d = p(n-1) \geq \log^3 n$ and $G = (V, E) \in \mathbb{G}(n, p)$, then a.a.s.*
$$c(G) = O(\sqrt{n}).$$

Before we prove Lemma 5.2.3, we need two definitions: $N[S]$ denotes $\bigcup_{v \in S} N(v)$, the closed neighborhood of S, and $N(S) = N[S] \setminus S$ denotes the (open) neighborhood of S.

Proof of Lemma 5.2.3. Let $S \subseteq V$, $s = |S|$, and consider the random variable $X = X(S) = |N(S)|$. For (i), we will bound X from below in a stochastic sense. There are two things that need to be estimated: the expected value of X, and the concentration of X around its expectation.

It is clear that

$$
\begin{aligned}
\mathbb{E}[X] &= \left(1 - \left(1 - \frac{d}{n-1}\right)^s\right)(n-s) \\
&= \left(1 - \exp\left(-\frac{ds}{n}(1 + O(d/n))\right)\right)(n-s) \\
&= \frac{ds}{n}(1 + O(ds/n))(n-s) \\
&= ds(1 + O(\log^{-1} n)),
\end{aligned}
$$

provided $ds \leq n/\log n$. We next use the Chernoff bounds which implies that the expected number of sets S that have $\big||N(S)| - d|S|\big| > \varepsilon d|S|$ and $|S| \leq n/(d\log n)$ is, for $\varepsilon = 2/\log n$, at most

$$
\begin{aligned}
\sum_{s \geq 1} 2n^s \exp\left(-\frac{\varepsilon^2 s \log^3 n}{3 + o(1)}\right) \\
= \sum_{s \geq 1} 2 \exp\left(\left(1 - \frac{4}{3 + o(1)}\right) s \log n\right) = o(1).
\end{aligned}
\tag{5.5}
$$

So a.a.s. if $|S| \leq n/d\log n$, then

$$|N(S)| = d|S|(1 + O(1/\log n)),$$

where the constant implicit in $O(\cdot)$ does not depend on the choice of S. Since $d \geq \log^3 n$, for such sets we have

$$|N[S]| = |N(S)|(1 + O(1/d)) = d|S|(1 + O(1/\log n)).$$

We may assume this statement holds.

Given this assumption, we have good bounds on the ratios of the cardinalities of $N[S]$, $N[N[S]] = \bigcup_{v \in S} N_2(v)$, and so on. We consider this up to the r'th iterated neighborhood provided $sd^r \leq n/\log n$ and therefore, $r = O(\log n/\log \log n)$. Then it follows that the cumulative multiplicative error term is $(1 + O(\log^{-1} n))^r \sim 1$; that is,

$$\left| \bigcup_{v \in S} N_r(v) \right| \sim sd^r \tag{5.6}$$

for all s and r such that $sd^r \leq n/\log n$. This establishes (i) in this case.

Suppose now that $sd^r = c'n$ with $1/\log n < c' = c'(n) \leq 1$. Using (5.6), $U = \bigcup_{v \in S} N_{r-1}(v)$ has cardinality $(1 + o(1))sd^{r-1}$. Now $N[U]$ has expected cardinality

$$n - e^{-c'}n(1 + o(1)) \geq \frac{1}{2}c'n(1 + o(1)) \sim \frac{1}{2}sd^r,$$

since $c' \leq 1$. The Chernoff bounds can be used again in the same way as before to show that with high probability $|N[U]|$ is concentrated near its expected value, and hence, that a.a.s. $|N[U]| > \frac{4}{9}sd^r$ for all sets S and all r for which $n/\log n < sd^r \leq n$, where $s = |S|$. Thus, (i) holds also in this case.

Finally, if $sd^r > n$, consider the maximum r_0 such that $sd^{r_0} \leq n$. From the penultimate conclusions of the previous two cases, it follows that for $U = \bigcup_{v \in S} N_{r_0}(v)$ we have $|U| \geq \frac{4}{9}sd^{r_0} \geq \frac{4}{9}n/d$. Next, we show that taking one more step, that is, at most r steps in total, will reach at least cn vertices for some universal constant $c > 0$, as required. Applying another version of the Chernoff bounds (see Theorem 1.6.3) and taking the union bound over all S with $|S| = s$, we deduce that for any $\varepsilon > 0$, with probability conservatively at least $1 - o(n^{-3})$ we have

$$\left| \bigcup_{v \in S} N_{r_0+1}(v) \right| = |N[U]| > n(1 - e^{-4/9} - \varepsilon),$$

which is at least $n/3$ for small enough ε. Of course, for any $v \in S$, we note that $N_{r_0+1}(v) \subseteq N_r(v)$. We conclude that (i) holds a.a.s. for all sets S and all r, where $s = |S|$, in this case. Hence, (i) holds in full generality a.a.s.

To prove (ii) holds a.a.s., note first that if such an r exists, then it follows from $d^{r+1} \leq \sqrt{n} \log n$ and $d \geq \log^3 n$ that $d^r \leq \sqrt{n}/\log^2 n$, and in particular r is uniquely determined from d and n.

Let us say that $v \in V$ is *erratic* if $|S_r(v)| > 2d^r$. From (5.6), we see that the expected number of erratic vertices is $o(1)$. Let us take any fixed v, expose the first r neighborhoods of v out to $S_r(v)$, condition on v not being erratic (which can be determined at this point), and then fix $u \in S_r(v)$. Set $U = V \setminus S_r(v)$. We now expose the iterated neighborhoods of u, but restricting ourselves to the graph induced by U. Since v is not erratic, $|U| = n - o(\sqrt{n})$. Vertices found at distance $(r + 1)$ from u form a set $W(u) \subseteq S_{r+1}(u)$. We now argue as in the derivation of (5.5), but with $\varepsilon = 4/\log n$, and note that we are searching within a set of $n - o(\sqrt{n})$ vertices. In this way it is straightforward to see that, with probability at least $1 - o(n^{-2})$, we have that $|W(u)| \sim d^{r+1}$.

Next we iterate the above argument, redefining U to be the vertices not explored so far. In each case where the bounds on $|W(u)|$ hold for all steps, we have $|U| = n - o(n)$, since we stop when we have treated all of the at most $2d^r$ vertices in $S_r(v)$, and $2d^{2r+1} \leq 2n/\log n$. Hence, (ii) holds a.a.s. $\qquad\square$

5.3 Random d-Regular Graphs

In the work [168], it was shown that Meyniel's conjecture holds a.a.s. for the random d-regular graph $\mathbb{G}_{n,d}$. Results for the two main random graph models (binomial and random regular graphs) support the conjecture regardless of the fact that there remains a large gap in the deterministic bounds; as already mentioned earlier in this chapter, even the soft Meyniel's conjecture remains open.

Our focus here is on dense random d-regular graphs; that is,

$n^{1/3}/\log^3 n \leq d = d(n) < \sqrt{n}\log n$. Proving the conjecture for sparse graphs is the main challenge faced in [168] but dealing with dense case here gives us an opportunity to present an interesting technique—the switching method. See [184] for more on this technique.

We will verify that for this range of d, random d-regular graphs a.a.s. satisfy the conditions (i) and (ii) in the hypotheses of Theorem 5.2.1. Note that only the condition (i) needs to be verified as the condition (ii) does not apply for this range of d, since $r \geq 1$. Moreover, as we will only need part (i) to be verified for $r = O(1)$, the following lemma will imply the result.

Lemma 5.3.1 ([168]). *Suppose that $n^{1/3}/\log^3 n \leq d = d(n) < \sqrt{n}\log n$. Let $G = (V, E) \in \mathbb{G}_{n,d}$. Then there exists some positive constant c such that the following properties hold a.a.s. For any set $S \subseteq V$ of $s = |S| \leq cn/d$ vertices*

$$|N[S]| \geq csd. \tag{5.7}$$

Moreover, if s is such that $sd < n/\log n$, then we have that

$$|N[S]| \sim sd. \tag{5.8}$$

Proof. In [22] there are bounds on the number of edges within a given set of vertices in $\mathbb{G}_{n,d}$. Some of these bounds are obtained by using switchings. However, the results obtained there do not suffice for our present needs. The main additional information we need is a bound on the number of edges between two sets of certain sizes.

We will show the required expansion of a set S to its neighbors. We do this by showing first that there cannot be too many edges within S, and second, that there cannot be too many edges from S to an "unusually" small set T, where often sd is approximately the "usual" size. It follows that if the neighborhood of S is too small, by setting it equal to T, we see that there is not enough room for the edges incident with S (each vertex of which must have degree d). Hence, the neighborhood must be large.

We start with the more difficult issue: edges from S to T. The approach is similar to some of the results in [22] and similar papers on random regular graphs of high degree. Suppose $|S| = s$ and $|T| = t$ where S and T are disjoint subsets of V. Let $U = V \backslash (S \cup T)$ and put $u = |U| = n - s - t$. Moreover, assume that $s + t \leq n/3$ and $s < u - n/2$.

In $\mathbb{G}_{n,d}$, consider the set of graphs $\mathcal{C}_i$ with exactly i edges from S to T. Since this is a uniform probability space, we may bound $\Pr(\mathcal{C}_i)$ via the inequality

$$\Pr(\mathcal{C}_i) \leq \frac{|\mathcal{C}_i|}{|\mathcal{C}_{i_1}|}, \tag{5.9}$$

which holds for any i_1 such that $\mathcal{C}_{i_1} \neq \emptyset$. We will do this for all $i > (1 - \alpha)sd$, where $\alpha > 0$ is sufficiently small. Let G be a member of $\mathcal{C}_i$ where $i > 0$. Consider a "switching" applied to G, which is an operation consisting of the following steps. First, select one of the i edges ab in G with $a \in S$ and $b \in T$, and some other edge $a'b'$ such that aa' and bb' are not edges of G and $a' \in U$, $b' \notin S$. Then replace the edges ab and $a'b'$ by new edges aa' and bb'. Call the resulting graph G'. Then G' is clearly a d-regular graph and must lie in $\mathcal{C}_{i-1}$ since the edge ab is removed and none of $a'b'$, aa' or bb' can join S to T. There are iud ways to choose one of the i edges for ab and one of the ud pairs of vertices $a'b'$ where $a' \in U$ and b' is adjacent to it. Hence, the number N of ways to choose a graph in $\mathcal{C}_i$ and perform this switching is

$$N = |\mathcal{C}_i|(iud - A),$$

where A denotes the average number of choices excluded by the other constraints. The number of exclusions due to aa' or bb' being an edge of G is $O(id^2)$. The number excluded because $b' \in S$ is at most isd since there are at most sd edges from S to U. Therefore, $A \leq isd + O(id^2)$, and we obtain $N \geq |\mathcal{C}_i|(i(u - s)d + O(id^2))$.

The number of ways to arrive at a given graph in $\mathcal{C}_{i-1}$ after applying such a switching to a graph in $\mathcal{C}_i$ is at most $(sd - i + 1)td$ since this is a clear upper bound on the number of choices for aa' and bb'. Thus, $N \leq |\mathcal{C}_{i-1}|(sd - i + 1)td$ and we derive that

$$\frac{|\mathcal{C}_i|}{|\mathcal{C}_{i-1}|} \leq \frac{(sd - i + 1)t}{i(u - s) + O(id)}.$$

For the present lemma, $n^{1/3+o(1)} \leq d \leq n^{1/2+o(1)}$. Since $u-s > n/2$, we have that

$$\frac{|\mathcal{C}_i|}{|\mathcal{C}_{i-1}|} \leq \frac{(sd - i + 1)t}{i(u - s)}(1 + O(n^{-1/4})).$$

By (5.9) and omitted calculations (taking $\beta = \frac{t}{u-s}$), we may derive that

$$\begin{aligned}
\Pr(\mathcal{C}_{i_0}) &= O(sd)\left(\frac{\beta^{1-\alpha}(1 + O(n^{-1/4}))}{(1 + \beta)\alpha^\alpha(1 - \alpha)^{(1-\alpha)}}\right)^{sd} \\
&\leq \left(\frac{\beta^{1-\alpha}(1 + O(n^{-1/4}))}{\alpha^\alpha(1 - \alpha)^{(1-\alpha)}}\right)^{sd}.
\end{aligned}$$

In turn, this gives

$$\Pr(\mathcal{C}_{i_0}) \leq \left(\beta^{1-\alpha}(1 + O(n^{-1/4}))\right)^{sd}.$$

By the union bound, the probability there exists such a pair of sets S and T of sizes s and $t = csd$, respectively, with precisely i_0 edges joining them is at most

$$\begin{aligned}
p &= \binom{n}{t}\binom{n - t}{s}\Pr(\mathcal{C}_{i_0}) \leq (en/t)^t(en/s)^s \Pr(\mathcal{C}_{i_0}) \\
&= \left(\left(\frac{en}{csd}\right)^c \left(\frac{en}{s}\right)^{1/d}\right)^{sd} \Pr(\mathcal{C}_{i_0}) \\
&= \left(\left(\frac{en}{csd}\right)^c (1 + O(n^{-1/4}))\right)^{sd} \Pr(\mathcal{C}_{i_0}).
\end{aligned}$$

Assuming that $sd \leq cn$, which implies that $t \leq c^2 n$, we note that

$$\begin{aligned}
\beta^{1-\alpha} &= \left(\frac{t}{u - s}\right)^{1-\alpha} \\
&\leq \left(\frac{csd}{n - c^2 n - O(n/d)}\right)^{1-O(d^{-1})} \\
&= \frac{csd}{(1 - c^2)n}(1 + O(n^{-1/4})),
\end{aligned}$$

and thus, we derive that

$$
\begin{aligned}
p^{1/sd} &\le \left(\frac{en}{csd}\right)^c \frac{csd}{(1-c^2)n}(1+O(n^{-1/4})) \\
&\le \frac{ce^c}{(1-c^2)c^c}\left(\frac{sd}{n}\right)^{1-c}(1+O(n^{-1/4})) \\
&\le \frac{c^2e^c}{(1-c^2)c^{2c}}(1+O(n^{-1/4})) \le 1/2
\end{aligned}
$$

for c small enough. Finally, we take the union bound over all s such that $n/\log n \le sd \le cn$ and all $i_0 = ds - O(s\log n)$, to conclude that (5.7) holds a.a.s.

Let $\varepsilon = 1/\log\log\log n$. By a similar argument, the probability there exists such a pair of sets S and T of sizes s and $t = sd(1-\varepsilon)$, respectively, with precisely i_0 edges joining them is at most

$$
p = \left(\left(\frac{en}{sd(1-\varepsilon)}\right)^{1-\varepsilon}(1+O(n^{-1/4}))\right)^{sd}\Pr(\mathcal{C}_{i_0}).
$$

In this case, we assume that $sd \le n/\log n$, which implies that $t = O(n/\log n)$. Therefore, we find that

$$
\beta^{1-\alpha} = \frac{sd}{n}(1-\varepsilon)(1+O(1/\log n)),
$$

which can be shown (by omitted calculations) that $p^{1/sd} = o(1)$. As in the previous case, we take the union bound over all s such that $sd \le n/\log n$ and all $i_0 = ds - O(s\log n)$, to conclude that (5.8) holds a.a.s.

It remains to show that a.a.s. each set of $s \le cn/d$ vertices induces at most $s\log n$ edges. This can be shown using an approach similar to the above argument for edges out of S, but is significantly simpler and so omitted here (full details may be found in [168]). □

5.4   Meyniel Extremal Families

How close can we get from below to the Meyniel bound? We finish this chapter by supplying deterministic examples of graphs whose cop number is as large as possible (assuming the veracity of Meyniel's conjecture). It is interesting that the known explicit families of such graphs derive in some way from designs and finite geometries.

An *incidence structure* consists of a set P of points, and a set L of lines along with an incidence relation consisting of ordered pairs of points and lines. Given an incidence structure S, we define its *incidence graph* $G(S)$ to be the bipartite graph whose vertices consist of the points (one partite set), and lines (the second partite set), with a point adjacent to a line if two are incident in S. Projective planes are some of the most well-studied examples of incidence structures. A *projective plane* consists of a set of points and lines satisfying the following axioms.

(i) There is exactly one line incident with every pair of distinct points.

(ii) There is exactly one point incident with every pair of distinct lines.

(iii) There are four points such that no line is incident with more than two of them.

Finite projective planes possess $q^2 + q + 1$ points for some integer $q > 0$ (called the *order* of the plane). Projective planes of order q exist for all prime powers q, and an unsettled conjecture claims that q must be a prime power for such planes to exist. For more on projective planes, see for example, [75]. See Figure 5.4 for $G(P)$, where P is the Fano plane (that is, the projective plane of order 2). We note the incidence graph of the Fano plane is isomorphic to the well-known *Heawood graph*.

Note that the girth of every incidence graph of a projective

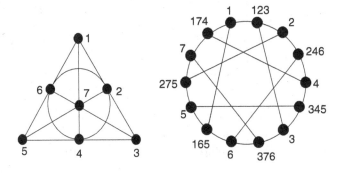

FIGURE 5.4: The Fano plane and its incidence graph.

plane is at least 6. The lower bound for the cop number in Theorem 2.0.3 from Chapter 2 proves that $c(G(P)) \geq q+1$. As proven first in [166], we actually have that $c(G(P)) = q + 1$. However, the orders of $G(P)$ depend on the orders of projective planes. The only orders where projective planes are known to exist are prime powers. What about integers which are not prime powers? An infinite family of graphs $(G_n)_{n \in \mathbb{N}}$ is *Meyniel extremal* if there is a positive constant d such that for all n, $c(G_n) \geq d\sqrt{|V(G_n)|}$.

Recall the *Bertrand-Chebyshev theorem* about prime numbers.

Theorem 5.4.1 ([63, 81])**.** *For all integers $x > 1$, there is a prime in the interval $(x, 2x)$.*

In [166], a Meyniel extremal family was given using incidence graphs of projective planes and Theorem 5.4.1. Using Theorem 5.4.1, it was shown that

$$c(n) \geq \sqrt{\frac{n}{8}}.$$

Using this theorem and a result from number theory, it was shown in [166] that for sufficiently large n,

$$c(n) \geq \sqrt{\frac{n}{2}} - n^{0.2625}. \tag{5.10}$$

A graph is (a, b)-*regular* if each vertex has degree either a or b.

We next give infinitely many Meyniel extremal families consisting of graphs which are (a, b)-regular for certain a and b. An *affine plane of order q* has q^2-many points, each line has q points, and each pair of distinct points is on a unique line. In an affine plane, there are $q^2 + q$ lines, and each point is on $q + 1$ lines. Affine planes exist for all prime power orders q, and it is conjectured they exist only for these orders (as with projective planes); see [75]. The relation of parallelism on the set of lines is an equivalence relation, and the equivalence classes are called *parallel classes*. Note that each parallel class contains q lines, and there are $q + 1$ parallel classes. Note that asymptotics are now as a function of q, not n.

Theorem 5.4.2 ([14]). *For prime powers q and all $k = o(q)$, there exist graphs of order $2q^2 + (1 - k)q$ which are $(q + 1 - k, q)$-regular and have cop number in the interval $[q + 1 - k, q]$.*

Proof. Consider an affine plane $\mathcal{A}$ with order q. The incidence graph $G(\mathcal{A})$ has order $2q^2 + q$ and is $(q + 1, q)$-regular. As $G(\mathcal{A})$ has girth at least 6, we have that $\{G(\mathcal{A}) : \mathcal{A}$ is an affine plane of order $q\}$ forms a Meyniel extremal family.

Affine planes of order q may be partitioned into $(q + 1)$-many parallel classes, each containing q lines. Form the partial planes $\mathcal{A}^{-k}$ by deleting the lines in some fixed set of $k > 0$ parallel classes. For a given $\mathcal{A}^{-k}$, the bipartite graph $G(\mathcal{A}^{-k})$ is then $(q + 1 - k, q)$-regular, and has order $2q^2 + (1 - k)q$. As the girth is at least 6, we have by Theorem 2.0.3 that

$$c(G) \geq q + 1 - k.$$

We claim that

$$c(G) \leq q. \tag{5.11}$$

To prove (5.11), we play with q cops $c_1, c_2, \ldots, c_q$. Fix a parallel class which was not deleted, say ℓ, and place one cop on each line of the parallel class. As each point is on some line in ℓ, the robber must move to some line $L \notin \ell$ to avoid being captured in the first round. See Figure 5.5.

Fix a point P of L, and let L' be the line of ℓ which intersects L at P. Move the cop c_i on L' to P. Now the robber cannot remain

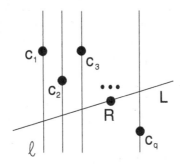

FIGURE 5.5: The robber must move to L or be captured.

on L without being captured by c_i, and so must move to some point. However, each point not on L' is adjacent to some cop, so the robber must move to a point of L'. But the unique point on L' adjacent to L is P, which is occupied by a cop. $\qquad\square$

Note that the graphs described in Theorem 5.4.2 have order $(1 - o(1))q^2$ with cop number $(1 - o(1))q$ and so are Meyniel extremal.

Recent work in [41] provides constructions of new Meyniel extremal families from designs and geometries, and we finish by presenting that work. The following lemma is crucial to proving the existence of new Meyniel extremal families. A graph G is *H-free* if it does not contain H as a subgraph.

Lemma 5.4.3 ([41]). *Let $t \geq 1$ be an integer. If G is $K_{2,t}$-free, then $c(G) \geq \delta(G)/t$.*

Proof. Suppose there are k cops playing, where $k < \delta(G)/t$. We show first that the robber may choose a vertex that is not adjacent to any cop in the first round. Let S be the set of vertices occupied by the cops, and let u be a vertex not in S. If no vertex of S is adjacent to u, then the robber can pick vertex u. Now assume that $N(u) \cap S \neq \emptyset$. Since G is $K_{2,t}$-free, a cop C which is adjacent to u can be adjacent to at most $t - 1$ other neighbors of u. Hence, C is equal or adjacent to at most t neighbors of u. Similarly, a cop not adjacent to u can be adjacent to at most $t - 1$ neighbors of u. It follows that the number of neighbors of u which are equal or adjacent to a cop is at most kt. But since $kt < \delta(G)$, this means that there must be some neighbor v of u which is not adjacent to any cop, and the robber can begin the game on v.

Now assume that in round $r \geq 0$ of the game, the cops and robber have moved and the robber is on some vertex u such that no cop is adjacent to u. Suppose in round $r + 1$ the cop "attacks" the robber by moving to some vertex of their neighbor set. A similar argument to that in the initial round shows that there is a neighbor v of u which is not adjacent to any cop, so the robber can move to v in round $r + 1$ and avoid capture. $\square$

From Lemma 5.4.3 the following holds.

Corollary 5.4.4 ([41]). *If G is C_4-free, then $c(G) \geq \delta(G)/2$.*

We now describe a family of Meyniel extremal graphs which are diameter 2. Fix q as a prime power. The *Erdős-Rényi graphs,* written $\mathrm{ER}(q)$, have vertices that are the points of a projective plane of order q, with u adjacent to v if $u^T v = 0$ (where we identify vertices with 1-dimensional subspaces of $\mathrm{GF}(q)^3$). These are well-known examples of graphs which are C_4-*free extremal*, in the sense that they have the largest possible number of edges in a C_4-free graph (which is asymptotic to $(1/2)n^{3/2}$); see [58, 83].

The Erdős-Rényi graphs are part of the more general family of polarity graphs. For a given projective plane of order q with points P and lines L, a *polarity* $\pi : P \to L$ is a bijection such that for all points p_1 and p_2, $p_1 \in \pi(p_2)$ if and only if $p_2 \in \pi(p_1)$. The *polarity graphs* are formed on vertex set P by joining distinct vertices u and v if $u \in \pi(v)$. For example, the orthogonal polarity (which exists for all $\mathrm{PG}(2, q)$) gives rise to the Erdős-Rényi graphs. For more on polarity graphs, see [149]. Polarity graphs have order $q^2 + q + 1$, with $q(q + 1)^2$ edges, and each vertex has degree $q + 1$ or q. These graphs are C_4-free and have diameter 2.

We have the following theorem that is crucial in proving that polarity graphs are Meyniel extremal.

Theorem 5.4.5 ([41]). *Suppose that G satisfies the following properties.*

(i) The graph G has order $q^2 + q + 1$, with each vertex of degree $q + 1$ or q.

(ii) The graph G is C_4-free and has diameter 2.

Then
$$q/2 \leq c(G) \leq q + 1.$$

Proof. The proof of the lower bound follows by Corollary 5.4.4. For the upper bound, we play with $q + 1$ cops. Fix a vertex u, and start the cops on $N(u)$. The robber is captured in the next round. □

We now provide a general technique to provide Meyniel extremal families via infinite families that lack some orders. We summarize the method in the following lemma, which may serve useful in finding other Meyniel extremal families.

Lemma 5.4.6 ([41]). *Let X be a set of positive integers with the property that for all integers $y \geq 1$, there is an $x \in X$ such that $y \leq x \leq 2y$. Suppose that for all $x \in X$ there exists a graph G_x with order $ax^2 + bx + d$ for rationals $a > 0$, b, d with either $a \geq 1$ or $b = 0$ (note that b and d can be.negative), such that $c(G_x) \geq mx$, for some fixed rational $m \in (0, 1]$. Then for $n \geq 1$, an integer, there exists a graph G_n of order n with the property that for n sufficiently large,*

$$c(G_n) \geq \frac{m}{2}\sqrt{\frac{n}{a}}(1 - o(1)).$$

Proof. Assume first that $b \geq 0$. Let $y = \left\lfloor \frac{1}{2}\sqrt{\left(\frac{n}{a} - \frac{b}{a}\sqrt{n} - \frac{d}{a}\right)} \right\rfloor$ and choose $x \in X$ such that $y \leq x \leq 2y$ (we choose n so that y is positive). Let $z = \frac{n}{a} - \frac{b}{a}\sqrt{n} - \frac{d}{a}$. Then we have that for large enough n,

$$
\begin{aligned}
ax^2 + bx + d &\leq az + b\sqrt{z} + d \\
&\leq n - b\sqrt{n} - d + b\sqrt{n} + d \\
&= n,
\end{aligned}
$$

where the second inequality follows since either

$$b\sqrt{\frac{n}{a} - \frac{b}{a}\sqrt{n} - \frac{d}{a}} \le b\sqrt{\frac{n}{a}} \le b\sqrt{n}$$

in the case $a \ge 1$, or trivially if $b = 0$. Form G_n from the graph G_x by attaching k-many vertices of degree 1 to a fixed vertex, where $k = n - (ax^2 + bx + d)$. We therefore have that

$$
\begin{aligned}
c(G_n) \;&=\; c(G_x) \\
&\ge\; mx \\
&\ge\; m\left\lfloor \frac{1}{2}\sqrt{\frac{n}{a} - \frac{b}{a}\sqrt{n} - \frac{d}{a}} \right\rfloor \\
&=\; \frac{m}{2}\sqrt{\frac{n}{a}}(1 - o(1)).
\end{aligned}
$$

If $b < 0$, then we write $b = -b'$, with $b' > 0$. Since $ax^2 + bx + d \le ax^2 + b'x + d$, the proof now follows from the case $b > 0$, replacing b with b'.   □

Corollary 5.4.7 ([41]). *The following holds.*

(i) *There exists a Meyniel extremal family whose graphs have diameter 2.*

(ii) *There exists a Meyniel extremal family whose graphs are C_4-free.*

Proof. For a positive integer n, let q be the least prime power such that $q^2 + q + 1 \le n$. If $n = q^2 + q + 1$, then we are done by using the polarity graphs and Theorem 5.4.5. If $q^2 + q + 1 < n$, then we form graphs G_1 and G_2 from a polarity graph G with order $q^2 + q + 1$ as follows. The graphs G_i will constitute the Meyniel extremal families in item (1) in the statement of the corollary, for $i = 1, 2$.

For item (2), let $m = n - (q^2 + q + 1)$. For G_1, fix a vertex x and add m new vertices x_i (where $1 \le i \le m$) adjacent to all the vertices in $N[x]$ and to no other vertices. The vertices x_i are corners, and the addition of corners does not change the cop number; see [1] or Chapter 2 of [52]. Hence, $c(G_1) = c(G)$, and it is straightforward to see that the diameter of G_1 is 2.

For G_2, attach m-many vertices y_i (where $1 \leq i \leq m$) of degree 1 to a fixed vertex x. As the vertices y_i are corners, we have that $c(G_2) = c(G)$. It is straightforward to see that G_2 does not contain 4-cycles.

We now apply Lemma 5.4.6 to polarity graphs with $a = b = d = 1$, $m = 1/2$, and X the set of all primes. $\qquad\square$

Examples first given in [96] provide another new family of Meyniel extremal graphs, this time containing no $K_{2,t+1}$. In the finite field of order q a prime power, written $GF(q)$, fix h an element of order t, and let $H = \{1, h, \ldots, h^{t-1}\}$. The vertices of G are the t-element orbits of $(GF(q) \times GF(q))\backslash\{(0,0)\}$ under the action of multiplication by powers of H. Two classes $\langle a, b\rangle$ and $\langle c, d\rangle$ are adjacent if $ac + bd \in H$. We call these the t-*orbit graphs*. Every vertex of G is degree q or $q+1$. By Theorem 5.4.5, $c(G) \geq q/(t+1)$. Hence, by Lemma 5.4.6 (with $a = 1/t$, $b = 0$, $d = -1/t$, $m = 1$, and X the set of all primes) we have the following result.

Corollary 5.4.8 ([41]). *Let $t \geq 1$ be an integer. There exists a Meyniel extremal family whose graphs are $K_{2,t+1}$-free.*

Chapter 6

Graph cleaning

The graph cleaning model, introduced in [141, 144], is a graph searching game that considers the decontamination of a graph by sending brushes from vertex to vertex. Graph cleaning may be viewed as a combination of the chip-firing game and edge searching on graphs. See also [98] where the parallel version of the process is studied, [56, 59, 145, 160] for the closely related processes of cleaning with brooms and with additional cleaning restrictions, [146] for connections to ordered sets, and [103] for a combinatorial game. The brush number of a graph G defined below corresponds to the minimum total imbalance of G, which has applications in graph drawing theory [97].

We now define graph cleaning, and give an example. Initially, every edge and vertex of a graph is *dirty*, and a fixed number of brushes start on a set of vertices. At each time-step, a vertex v and all its incident edges that are dirty may be *cleaned* if there are at least as many brushes on v as there are incident dirty edges. When a vertex is cleaned, every incident dirty edge is traversed (that is, cleaned) by one and only one brush, and brushes cannot traverse a clean edge. See Figure 6.1 for an example of this cleaning process. The initial configuration has only two brushes, both at one vertex. The solid edges are dirty and the dashed edges are clean. The circle indicates which vertex is cleaned next. The assumption in [144], and made here also, is that a graph is cleaned when every vertex has been cleaned. If every vertex has been cleaned, then it follows that every edge has been cleaned. It may be that a vertex v has no incident dirty edges at the time it is cleaned, in which case no brushes move from v. Although this viewpoint might seem unnatural, it simplifies much of the analysis in [144].

FIGURE 6.1: A cleaning process with brushes represented by the stars.

Formally, at each time-step t, $\omega_t(v)$ denotes the number of brushes at vertex v (here $\omega_t : V(G) \to \mathbb{N} \cup \{0\}$) and D_t denotes the set of dirty vertices. An edge $uv \in E(G)$ is dirty if and only if both u and v are dirty: $\{u, v\} \subseteq D_t$. Finally, let $D_t(v)$ denote the number of dirty edges incident to v at step t:

$$
D_t(v) = \begin{cases} |N(v) \cap D_t| & \text{if } v \in D_t, \\ 0 & \text{otherwise.} \end{cases}
$$

The *cleaning process* $\mathfrak{P}(G, \omega_0) = (\omega_t, D_t)_{t=0}^{T}$ of an undirected graph $G = (V, E)$ with an *initial configuration of brushes* ω_0 is as follows.

(**0**) Initially, all vertices are dirty: $D_0 = V$; set $t = 0$.

(**1**) Let α_{t+1} be any vertex in D_t such that $\omega_t(\alpha_{t+1}) \geq D_t(\alpha_{t+1})$. If no such vertex exists, then stop the process, set $T = t$, and return the *cleaning sequence* $\alpha = (\alpha_1, \alpha_2, \ldots, \alpha_T)$, the *final set of dirty vertices* D_T, and the *final configuration of brushes* ω_T.

(**2**) Clean α_{t+1} and all incident dirty edges by moving a brush from α_{t+1} to each dirty neighbor. More precisely, $D_{t+1} = D_t \setminus \{\alpha_{t+1}\}$,

$$
\omega_{t+1}(\alpha_{t+1}) = \omega_t(\alpha_{t+1}) - D_t(\alpha_{t+1}),
$$

and for every $v \in N(\alpha_{t+1}) \cap D_t$, $\omega_{t+1}(v) = \omega_t(v) + 1$ (the other values of ω_{t+1} remain the same as in ω_t).

(**3**) Set $t = t + 1$ and go back to *(1)*.

Note that for a graph G and initial configuration ω_0, the cleaning process can return different cleaning sequences and final configurations of brushes; consider, for example, an isolated edge uv and $\omega_0(u) = \omega_0(v) = 1$. It has been shown (see Theorem 2.1 in [144]), however, that the final set of dirty vertices is determined by G and ω_0. Thus, the following definition is natural. A graph $G = (V, E)$ *can be cleaned* by the initial configuration of brushes ω_0 if the cleaning process $\mathfrak{P}(G, \omega_0)$ returns the empty final set of dirty vertices (that is, $D_T = \emptyset$). Let the *brush number of G*, written $b(G)$, be the minimum number of brushes needed to clean G; that is,

$$b(G) = \min_{\omega_0:V\to\mathbb{N}\cup\{0\}} \left\{ \sum_{v\in V} \omega_0(v) : G \text{ can be cleaned by } \omega_0 \right\}.$$

Similarly, $b_\alpha(G)$ is defined as the minimum number of brushes needed to clean G using the cleaning sequence α.

It is evident that for every cleaning sequence α, $b_\alpha(G) \geq b(G)$ and

$$b(G) = \min_\alpha b_\alpha(G).$$

The last relation can be used as an alternative definition of $b(G)$. In general, it is difficult to find $b(G)$, but $b_\alpha(G)$ can be easily computed. For this, it seems better not to choose the function ω_0 in advance, but to run the cleaning process in the order α, and compute the initial number of brushes needed to clean a vertex. We can adjust ω_0 along the way by

$$\begin{aligned} \omega_0(\alpha_{t+1}) &= \max\{D_t(\alpha_{t+1}) - \big(d(\alpha_{t+1}) - D_t(\alpha_{t+1})\big), 0\} \\ &= \max\{2D_t(\alpha_{t+1}) - d(\alpha_{t+1}), 0\}, \end{aligned} \tag{6.1}$$

for $t = 0, 1, \ldots, |V| - 1$, since that is the number of brushes we have to add over and above what we obtain for free.

We now present a few examples. It is straightforward to see that $b(P_n) = 1$ for any $n \geq 2$, and $b(C_n) = 2$ for any $n \geq 3$. Since each brush used during the cleaning process follows a path (we call it the *brush path*), we derive that for every graph G,

$$b(G) \geq \frac{d_o(G)}{2},$$

where $d_o(G)$ denotes the number of vertices of odd degree in G. For any tree T, this trivial lower bound is attained; that is, $b(T) = d_o(T)/2$ (see Theorem 5.1 in [144]). However, the brush number of the complete graph K_n is much larger (see Theorem 5.2 in [144]):

$$b(K_n) = \left\lfloor \frac{n^2}{4} \right\rfloor = \begin{cases} \frac{n^2}{4} & \text{if } n \text{ is even,} \\ \frac{n^2-1}{4} & \text{otherwise.} \end{cases}$$

We will show which values can be obtained for the brush number of a graph on n vertices. We will use the following elementary property (left as an exercise, or see [144]).

Lemma 6.0.1 ([144]). *For any graph G with $e \notin E(G)$, we have that*

$$b_\alpha(G) - 1 \le b_\alpha(G + e) \le b_\alpha(G) + 1$$

for any cleaning sequence α.

Note that Lemma 6.0.1 implies that adding an edge can only change the brush number by 1. Using this property and the one that will be observed in Lemma 6.0.4, we can show the next theorem.

Theorem 6.0.2 ([145]). *If $n \in \mathbb{N}$, then for each $k = 0, 1, \ldots, \lfloor n^2/4 \rfloor$, there exists a graph G on n vertices with $b(G) = k$. No other value can be obtained for an n-vertex graph.*

Proof. Let $N = \binom{n}{2}$. Consider a sequence $(G_t = (V, E_t))_{t=0}^{N}$ of simple graphs on n vertices. Let G_0 be the empty graph on n vertices. For $1 \le t \le N$ we form G_t from G_{t-1} by adding any edge; hence, G_N is a clique. Now consider a sequence of numbers $(b(G_t))_{t=0}^{N}$; $b(G_0) = 0$ and

$$b(G_N) = b(K_n) = \lfloor n^2/4 \rfloor$$

by Theorem 5.2 in [144]. Additionally, it follows from Lemma 6.0.1 that any two consecutive numbers in the sequence can differ by at most one. Therefore, for each k (where $0 \le k \le \lfloor n^2/4 \rfloor$) there exists t, where $0 \le t \le N$, such that $b(G_t) = k$. Note that we do not construct this graph, we only prove it exists!

Finally, it follows from Theorem 3.7 in [7] (see also Lemma 6.0.4) that

$$b(G) \leq \frac{|E|}{2} + \frac{|V|}{4} - \frac{1}{4} \sum_{v \in V(G),\, \deg(v)\, \text{even}} \frac{1}{\deg(v)+1}$$

$$\leq \frac{n(n-1)}{4} + \frac{n}{4}$$

$$= \frac{n^2}{4},$$

so no other value can be obtained, which finishes the proof. □

We present a universal upper bound for the brush number from [7], whose proof uses the probabilistic method.

Theorem 6.0.3 ([7]). *Let G be a d-regular graph on n vertices. If d is even, then*

$$b(G) \leq \frac{n}{4}\left(d+1-\frac{1}{d+1}\right),$$

and if d is odd, then

$$b(G) \leq \frac{n}{4}(d+1).$$

Proof. Let π be a random permutation of the vertices of G taken with uniform distribution. We clean G according to this permutation to obtain the value of $b_\pi(G)$ (note that $b_\pi(G)$ is a random variable now). For a given vertex $v \in V$, we have to assign to v exactly

$$X(v) = \max\{d^+(v) - d^-(v), 0\} = \max\{0, 2d^+(v) - d\}$$

brushes in the initial configuration, where $d^+(v)$ is the number of neighbors of v that follow it in the permutation (that is, the number of dirty neighbors of v at the time when v is cleaned); see (6.1). The random variable $d^+(v)$ attains each of the values $0, 1, \ldots, d$ with probability $1/(d+1)$. Indeed, this follows from the fact that the random permutation π induces a uniform random permutation on the set of $d + 1$ vertices consisting of v and its neighbors (to see this, we can fix the positions of nonneighbors of

v first, and then fixing one of $d + 1$ free positions for the vertex v will yield desired values of $d^+(v)$ with uniform distribution). Therefore, the expected value of $X(v)$, for even d, is

$$\frac{d + (d - 2) + \cdots + 2}{d + 1} = \frac{d + 1}{4} - \frac{1}{4(d + 1)},$$

and for odd d it is

$$\frac{d + (d - 2) + \cdots + 1}{d + 1} = \frac{d + 1}{4}.$$

Thus, by the linearity of expectation,

$$
\begin{aligned}
\mathbb{E}\big[b_\pi(G)\big] &= \mathbb{E}\left[\sum_{v \in V} X(v)\right] \\
&= \sum_{v \in V} \mathbb{E}\big[X(v)\big] \\
&= \begin{cases} \frac{n}{4}\left(d + 1 - \frac{1}{d+1}\right), & \text{if } d \text{ is even;} \\ \frac{n}{4}(d + 1), & \text{otherwise.} \end{cases}
\end{aligned}
$$

This discussion implies that there is a permutation π_0 such that $b(G) \le b_{\pi_0}(G) \le \mathbb{E}\big[b_\pi(G)\big]$ and the assertion holds. $\square$

This result can be easily generalized to any (not necessarily regular) graph G.

Theorem 6.0.4 ([7]). *For a graph* $G = (V, E)$,

$$b(G) \le \frac{|E|}{2} + \frac{|V|}{4} - \frac{1}{4} \sum_{v \in V(G), d(v) \text{ is even}} \frac{1}{d(v) + 1}.$$

6.1 Tools: Convergence of Moments Method

We now take a brief interlude to discuss the convergence of moments method, known also as Brun's sieve. This method will be used in the next section and later in the book. For more details about this method, see, for example, Section 8.3 in [8]. The first and second moment methods can be used when the expectation tends to zero and to infinity, respectively. We next introduce another powerful method that is used when the expectation tends to a constant other than zero.

Let X be a random variable whose range is the set $\{0, 1, \ldots, n\}$ such that $\{\omega \in \Omega : X(\omega) = j\} \in \mathcal{F}$ for every $j \in \{0, 1, \ldots, n\}$. The k-th *binomial moment* B_k is defined by

$$B_k = \mathbb{E}\left[\binom{X}{k}\right].$$

Note that $B_1 = \mathbb{E}[X]$. Moments are particularly important when the random variable X counts certain objects. Consider some events $A_1, A_2, \ldots, A_n$, and let X count how many events actually occur. There is an obvious connection between the binomial moment and the inclusion-exclusion principle.

Theorem 6.1.1.

$$B_k = \mathbb{E}\left[\binom{X}{k}\right] = \sum_{1 \leq i_1 < i_2 < \cdots < i_k \leq n} \mathbb{P}\left(\bigcap_{j=1}^{k} A_{i_j}\right).$$

Proof. Note that $X = \sum_{i=1}^{n} I_{A_i}$, where I_{A_i} is the indicator random variable of the event A_i. Interpreting $\binom{X}{k}$ as a combinatorial object, we derive that

$$\binom{X}{k} = \sum_{1 \leq i_1 < i_2 < \cdots < i_k \leq n} \prod_{j=1}^{k} I_{A_{i_j}}.$$

After taking expectations we obtain the result. $\qquad \square$

Now, we are ready to state the method.

Lemma 6.1.2 (The convergence of moments method). *Consider a sequence of probability spaces* $(\Omega_n, \mathcal{F}_n, \mathbb{P}_n)$. *Let* $A_1^n, A_2^n, \ldots, A_{r_n}^n \in \mathcal{F}_n$,

$$B_\ell^n = \sum_{1 \leq j_\ell < \cdots < j_\ell \leq r_n} \mathbb{P}\left(\bigcap_{i=1}^{\ell} A_{j_i}^n\right),$$

and let S_n be the number of events among $A_1^n, A_2^n, \ldots, A_{r_n}^n$ that actually occur; that is,

$$S_n = \sum_{i=1}^{r_n} I_{A_i^n}.$$

Suppose that for some λ, $\lim_{n\to\infty} B_k^n = \frac{\lambda^k}{k!}$ for all fixed positive integers k. Then S_n tends to the Poisson distribution with parameter λ; that is,

$$\lim_{n\to\infty} \mathbb{P}(S_n = \ell) = e^{-\lambda}\frac{\lambda^\ell}{\ell!}.$$

In particular,

$$\lim_{n\to\infty} \mathbb{P}(S_n = 0) = e^{-\lambda}.$$

Proof. We will show the case $\ell = 0$ only. Fix $\varepsilon > 0$. Choose s so that

$$\left| \sum_{r=0}^{2s}(-1)^r\frac{\lambda^r}{r!} - e^{-\lambda} \right| \le \frac{\varepsilon}{2}.$$

The Bonferroni inequalities (see Theorem 1.2.8) state that the inclusion-exclusion formula alternately over- and under-estimates $\mathbb{P}(S_n = 0)$. In particular,

$$\mathbb{P}(S_n = 0) = 1 - \mathbb{P}(S_n \ge 1) = 1 - \mathbb{P}\left(\bigcup_{i=1}^{r_n} A_i^n\right)$$

$$\le \sum_{r=0}^{2s}(-1)^{r-1}B_r^n.$$

Select n_0 so that for $n \ge n_0$,

$$\left| B_r^n - \frac{\lambda^r}{r!} \right| \le \frac{\varepsilon}{2(2s+1)}$$

for $0 \le r \le 2s$. For such n,

$$\mathbb{P}(S_n = 0) \le e^{-\lambda} + \varepsilon.$$

Similarly, taking the sum to $2s + 1$ we find n_0 so that for $n \ge n_0$,

$$\mathbb{P}(S_n = 0) \ge e^{-\lambda} - \varepsilon.$$

As ε was arbitrary, $\mathbb{P}(S_n = 0)$ tends to $e^{-\lambda}$. $\qquad\square$

6.2 Binomial Random Graphs

One of the most intensively studied phenomena in the field of random graphs is the behavior of a random graph $\mathbb{G}(n, p)$ when $p = d/n$ for d near 1. A graph G is called *unicyclic* if it contains exactly one cycle.

(i) In the *very subcritical phase* (that is, when $d < 1$), a.a.s. $\mathbb{G}(n, p)$ consists of small trees and unicyclic components. The cardinality of the largest component is of order $\log n$.

(ii) In the *barely subcritical phase* (that is, when $d = 1 - \varepsilon$ and $\varepsilon = \varepsilon(n) = \lambda n^{-1/3}$ such that $\varepsilon = o(1)$ and $\lambda = \lambda(n) \to \infty$), a.a.s. all components of $\mathbb{G}(n, p)$ are *simple*: acyclic or unicyclic. The cardinality of the largest component is of order $\varepsilon^{-2} \log \lambda = n^{2/3} \lambda^{-2} \log \lambda$.

(iii) We know that the giant component (that is, a component containing $O(n)$ vertices) is formed from smaller ones during the so-called *critical phase* (that is, when $d = 1 + \varepsilon$, $\varepsilon = \varepsilon(n) = \lambda n^{-1/3}$, and λ is a real constant). The largest component has cardinality of order $n^{2/3}$.

(iv) In the *barely supercritical phase* (that is, when $d = 1 + \varepsilon$ and $\varepsilon = \varepsilon(n) = \lambda n^{-1/3}$ such that $\varepsilon = o(1)$ and $\lambda = \lambda(n) \to \infty$), a.a.s. $\mathbb{G}(n, p)$ consists of one *complex* (that is, not simple) component of cardinality asymptotic to $2\varepsilon n = 2\lambda n^{2/3}$, and some number of small trees and unicyclic components. The cardinality of the second-largest component is of order $\varepsilon^{-2} \log \lambda = n^{2/3} \lambda^{-2} \log \lambda$. (In particular, since $\varepsilon = o(1)$, then the cardinality of the largest component and the number of edges in this component are still $o(n)$.)

(v) For $d > 1$, a.a.s. the cardinality of the largest component is $(1 + o(1))\beta n$, where $\beta = \beta(d)$ is the positive real number satisfying

$$\beta + e^{-\beta d} = 1.$$

The second-largest component is simple and has cardinality of order $\log n$. This phase is called the *very supercritical phase*.

The behavior for $d < 1$, $d = 1$, and $d > 1$ was first studied in the famous paper of Erdős and Rényi [82]. The behavior when $d = 1 + \varepsilon$ and $\varepsilon = \varepsilon(n) = o(1)$ was studied by Bollobás [32] and Łuczak [135]. It was recently shown by Bollobás and Riordan [40] that the case $d \to 1$ can be studied via branching processes but much more is known about this fascinating phenomena.

In this section, we investigate the brush number of a random graph in the subcritical phase (that is, when $p = (1 - s)/n$ and $s = s(n) \gg n^{-1/3}$). Let us recall that $d_o(G)$ denotes the number of vertices of G of odd degree. This section will be devoted to proving the following result from [161].

Theorem 6.2.1 ([161]). *For $G \in \mathbb{G}(n, p)$, $p = p(n) = d/n$, and $1 - d \gg n^{-1/3}$ (thus, $d \leq 1 - o(1)$),*

$$b(G) = \frac{d_o(G)}{2} + 2X + Y,$$

where X tends to the Poisson distribution $\mathrm{Po}(\lambda_X)$ with parameter λ_X, and Y tends to the Poisson distribution $\mathrm{Po}(\lambda_Y)$ with parameter λ_Y, with

$$\lambda_X = -\frac{1}{2}\log(1 - d\exp(-d)) - \frac{(d\exp(-d))^2}{4} - \frac{d\exp(-d)}{2},$$

and

$$\lambda_Y = \frac{1 - \exp(-d)}{2(1 - d\exp(-d))}d^3\exp(-2d).$$

In particular, a.a.s. we have that

$$b(G) = \frac{d_o(G)}{2} + O(\log\log n) \sim n\frac{1 - e^{-2d}}{4},$$

and $b(G) = d_o(G)/2$ with probability tending to $\exp(-\lambda_X - \lambda_Y)$ as $n \to \infty$.

Before we prove Theorem 6.2.1, we need one more lemma whose proof is omitted but can be found in [161]. We already know that $b(G) \geq d_o(T)/2$ for any graph G and that $b(T) = d_o(T)/2$ for any tree T (see Theorem 5.1 in [144]). There are some unicyclic graphs that need a little bit more than $d_o(G)/2$ brushes, but we can show that two additional brushes always suffice.

Theorem 6.2.2 ([161]). *For any unicyclic graph G with $d_o(G)$ vertices of odd degree and $d_t(G)$ vertices on the cycle with degree at least three, we have that*

$$
b(G) = \begin{cases}
d_o(G)/2, & \text{if } d_t(G) \geq 2, \\
d_o(G)/2 + 1, & \text{if } d_t(G) = 1, \\
d_o(G)/2 + 2, & \text{if } d_t(G) = 0.
\end{cases}
$$

We now move to the proof of Theorem 6.2.1.

Proof of Theorem 6.2.1. Recall that in the subcritical phase all components are either trees or unicyclic graphs. Thus, by Theorem 6.2.2 we derive that

$$
b(G) = \frac{d_o(G)}{2} + 2X(G) + Y(G),
$$

where $X(G)$ denotes the number of components that are cycles in G, and $Y(G)$ denotes the number of components that are unicyclic graphs with exactly one tree attached to the cycle. Note that $d_o(G)$, $X(G)$, and $Y(G)$ are random variables, and our goal is to investigate their asymptotic distributions.

We know that a.a.s. there is no component of cardinality more than, say, $\log^2 n$. By computing the probability that a given set of k vertices induces a cycle, it is not difficult to see that

$$
\mathbb{E}[X] \sim \sum_{k=3}^{\lfloor \log^2 n \rfloor} \binom{n}{k} \frac{(k-1)!}{2} p^k (1-p)^{k(n-3)}
$$

$$
\sim \sum_{k=3}^{\lfloor \log^2 n \rfloor} \frac{(d \exp(-d))^k}{2k}
$$

$$= \frac{1}{2} \sum_{k=1}^{\lfloor \log^2 n \rfloor} \frac{(d \exp(-d))^k}{k} - \frac{(d \exp(-d))^2}{4} - \frac{d \exp(-d)}{2}$$

$$\sim -\frac{1}{2} \log(1 - d \exp(-d)) - \frac{(d \exp(-d))^2}{4} - \frac{d \exp(-d)}{2}$$

$$= \lambda_X,$$

and

$$\mathbb{E}[Y] \sim \sum_{k=3}^{\lfloor \log^2 n \rfloor} \binom{n}{k} \frac{(k-1)!}{2} p^k (1-p)^{(k-1)(n-3)} k(1 - \exp(-d))$$

$$\sim \frac{1 - \exp(-d)}{2 \exp(-d)} \sum_{k=3}^{\lfloor \log^2 n \rfloor} (d \exp(-d))^k$$

$$\sim \frac{1 - \exp(-d)}{2 \exp(-d)} \frac{(d \exp(-d))^3}{1 - d \exp(-d)}$$

$$= \lambda_Y.$$

We can also check that, for a given $r \geq 2$, the rth binomial moments of $X(G)$ and $Y(G)$ tend to λ_X^r and λ_Y^r, respectively. Thus, both variables $X(G), Y(G)$ tend to a Poisson distribution with parameters λ_X, λ_Y, respectively, by the convergence of moments method. Further, these two random variables are asymptotically independent.

It follows from Markov's inequality that a.a.s. $X(G) + Y(G)$ is at most, say, $\log \log n$. Hence, in order to finish the proof, we have to estimate the number of vertices of odd degree. For any $k = k(n) < \log n$, let p_k denote the probability that a given vertex has degree k. (It is known that a.a.s. there are no vertices of degree more than $\log n$.) It is clear that

$$p_k = \binom{n-1}{k} p^k (1-p)^{n-1-k} \sim \frac{d^k}{k!} e^{-d}. \tag{6.2}$$

Thus, the expected number of vertices of odd degree is given by

$$\mathbb{E}[d_o(G)] = n \sum_{1 \le k < \log n, k \text{ is odd}} p_k$$

$$\sim ne^{-d} \sum_{k, k \text{ is odd}} \frac{d^k}{k!}$$

$$= ne^{-d} \frac{e^d - e^{-d}}{2}$$

$$= n \frac{1 - e^{-2d}}{2}.$$

It can be shown that $d_o(G)$ is well concentrated around its expectation (we omit the proof here), which finishes the proof of the theorem. $\qquad\square$

We next focus on finding an upper bound for the brush number of a random graph $\mathbb{G}(n, p)$ for $p = d/n$ and $d > 1$. For this range of the parameter p, we already stated the asymptotic cardinality of the giant component. Note that it is possible to investigate the structure of the graph formed by deleting the giant component. We mentioned that the cardinality of the giant component is a.a.s. $\beta n + o(n)$, where $\beta = \beta(d) \in (0, 1)$ is the unique solution to

$$\beta + e^{-\beta d} = 1.$$

It is known that the graph formed by deleting the giant component is essentially equivalent to $\mathbb{G}(n', p')$, where $n' \sim (1 - \beta)n$ and

$$p' = \frac{d}{n} \sim \frac{d(1 - \beta)}{n'}.$$

(Note that $d(1 - \beta) < 1$ so, indeed, a.a.s. all other components are of order $O(\log n)$, since the structure of the graph after removing the giant component corresponds to the very subcritical phase.)

Knowledge about the order of the giant component helps us to obtain a stronger upper bound for the brush number (but a numerical one). We also present an exact formula that works well for relatively large values of d.

Theorem 6.2.3 ([161]). *For $G \in \mathbb{G}(n,p)$, with $p = p(n) = d/n$ and $d > 1$, a.a.s.*

$$b(G) \leq \frac{n}{4}\left(d + 1 - \frac{1 - e^{-2d}}{2d}\right)(1 + o(1)),$$

and

$$b(G) \leq \frac{n}{4}\left(1 + d\beta(2 - \beta) - e^{-d(2-\beta)} + \frac{e^{-2d} - e^{-2d(1-\beta)}}{2d}\right)$$
$$\cdot(1 + o(1))$$

We give a plot of both upper bounds of $b(G)/dn$ in Figure 6.2.

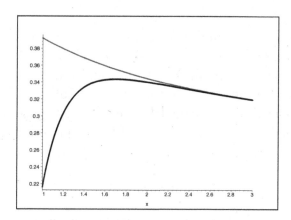

FIGURE 6.2: Upper bounds of $b(G)/dn$ in the very supercritical phase.

Proof of Theorem 6.2.3. Using the Chernoff bounds, we can show that the total number of edges is well-concentrated around $dn/2$. As already mentioned in (6.2), the probability that a given vertex has degree k can be estimated as follows:

$$p_k = \binom{n-1}{k}p^k(1-p)^{n-1-k} \sim \frac{d^k}{k!}e^{-d},$$

for any $k = k(n) < \log n$. (It is known that a.a.s. there are no vertices of degree more than $\log n$.) Finally, it follows from Theorem 3.7 in [7] that

$$b(G) \leq \frac{|E|}{2} + \frac{|V|}{4} - \frac{1}{4} \sum_{v \in V(G), d(v) \text{ is even}} \frac{1}{d(v) + 1}. \tag{6.3}$$

Thus, we have that

$$
\begin{aligned}
b(G) &\leq \frac{n}{4} \left(d + 1 - \sum_{k \geq 0, k \text{ is even}} \frac{1}{k+1} \cdot \frac{d^k}{k!} e^{-d} \right) (1 + o(1)) \\
&\sim \frac{n}{4} \left(d + 1 - \frac{e^{-d}}{d} \sum_{k \geq 1, k \text{ is odd}} \frac{d^k}{k!} \right) \\
&= \frac{n}{4} \left(d + 1 - \frac{1 - e^{-2d}}{2d} \right).
\end{aligned}
$$

To find a better (but numerical) upper bound, we can use (6.3) one more time to derive a bound on the number of brushes needed to clean the giant component and give an asymptotically almost sure value for small components.

Similar calculations to ones we had before (see (6.2)) can be used to show that the probability that a given vertex outside the giant component has degree k is equal to

$$p_k' \sim \frac{(d(1 - \beta))^k}{k!} e^{-d(1-\beta)}. \tag{6.4}$$

Since a.a.s. all small components are trees or unicyclic, and the total number of unicyclic components is negligible, the total number of brushes needed to clean all small components is equal to

$$
\begin{aligned}
(1 + o(1)) &\frac{1}{2} (1 - \beta) n \sum_{1 \leq k < \log n, k \text{ is odd}} p_k' \\
&\sim \frac{1}{2} e^{-d\beta} n e^{-d(1-\beta)} \sum_{k \geq 1, k \text{ is odd}} \frac{(d(1 - \beta))^k}{k!} \\
&= \frac{1}{2} n e^{-d} \frac{e^{d(1-\beta)} - e^{-d(1-\beta)}}{2} \\
&= \frac{n}{4} \left(1 - \beta - e^{-d(2-\beta)} \right). \tag{6.5}
\end{aligned}
$$

(Recall that $b(T) = d_o(T)/2$ for any tree T and so the task here is to estimate the number of vertices of odd degree; see Theorem 5.1 in [144].)

Since the number of edges in the giant component is equal to

$$(1 + o(1)) \left(\binom{n}{2} p - \binom{n'}{2} p' \right) \sim \frac{dn}{2} \beta(2 - \beta),$$

we obtain an upper bound on the number of brushes needed to clean the giant component G' of G; in particular,

$$
\begin{aligned}
b(G') &\leq \frac{|E(G')|}{2} + \frac{|V(G')|}{4} - \frac{1}{4} \sum_{v \in V(G'), d(v) \text{ is even}} \frac{1}{d(v) + 1} \\
&\sim \frac{n}{4} \left(d\beta(2 - \beta) + \beta - \sum_{k \geq 0, k \text{ is even}} \frac{1}{k + 1} \right. \\
&\qquad \left. \cdot \left(\frac{d^k}{k!} e^{-d} - (1 - \beta) \frac{(d(1 - \beta))^k}{k!} e^{-d(1-\beta)} \right) \right) \\
&\sim \frac{n}{4} \left(d\beta(2 - \beta) + \beta + \frac{e^{-2d} - e^{-2d(1-\beta)}}{2d} \right). \qquad (6.6)
\end{aligned}
$$

To derive a new upper bound and finish the proof, it is enough to add together (6.5) and (6.6). $\qquad\qquad\square$

6.3   Tools: Pairing Model

We take another brief interlude to introduce the pairing model and to discuss a few important properties. We will use the model in the next section and later in the book. Instead of working directly in the uniform probability space of random regular graphs on n vertices $\mathbb{G}_{n,d}$ (as introduced in Chapter 1), we use the *pairing model* of random regular graphs, first given in its simple explicit form by Bollobás [29] and called the *configuration model*. (A little earlier, Bender and Canfield [21], and Wormald [185] used the model implicitly.) For more about the model, see the survey [184].

Suppose that dn is even, as in the case of random regular graphs, and consider dn points partitioned into n labeled buckets $v_1, v_2, \ldots, v_n$ of d points each. A *pairing* of these points is a perfect matching into $dn/2$ pairs. Given a pairing P, we may construct a multigraph $G(P)$, with loops allowed, as follows: the vertices are the buckets $v_1, v_2, \ldots, v_n$, and a pair $\{x, y\}$ in P corresponds to an edge $v_i v_j$ in $G(P)$ if x and y are contained in the buckets v_i and v_j, respectively.

For a positive integer k, define the *double factorial* by

$$(2k - 1)!! = (2k - 1)(2k - 3) \cdots 3 \cdot 1 = \frac{(2k)!}{2^k k!}.$$

Now, observe that there are $M(dn) = (dn - 1)!!$ different configurations. Let us note that a pairing must be selected uniformly at random. This can be done in many different ways, some of which turn out to be very convenient. In particular, the points in the pairs can be chosen sequentially. At any stage of the process, the first point in the next random pair chosen can be selected using any rule whatsoever, as long as the second point in that pair is chosen uniformly at random from the remaining points.

It is an elementary fact that the probability that a random pairing corresponds to a given simple graph G is independent of the graph; hence, the restriction of the probability space of random pairings to simple graphs is precisely $\mathbb{G}_{n,d}$. We note that it is known that a random pairing generates a simple graph with probability asymptotic to $e^{(1-d^2)/4}$ depending on d but not on n. To demonstrate this fact, we need to analyze a random variable Z_k, the number of cycles of length k in a (multi)graph $G \in G(P)$. (Note that for graphs we let $k = 3, 4, \ldots$ whereas for multigraphs we let $k = 1, 2, \ldots$.) A multigraph is simple if and only if $Z_1 = Z_2 = 0$.

Theorem 6.3.1 ([184]). *Let $\lambda_k = \frac{1}{2k}(d - 1)^k$. The numbers of cycles of length k in $G \in G(P)$, the Z_k, converge in distribution to independent Poisson-distributed random variables $\mathrm{Po}(\lambda_k)$ as $n \to \infty$, jointly for all $k \geq 1$.*

This theorem immediately implies the following two corollaries. The second one was obtained first by Bender and Canfield [21].

They were studying an asymptotic formula for $|\mathbb{G}_{n,d}|$, and Corollary 6.3.3 follows from their result. Using the pairing model, we obtain the corollary.

Corollary 6.3.2. *Let $\lambda_k = \frac{1}{2k}(d-1)^k$. The numbers of cycles of length k in $G \in \mathbb{G}_{n,d}$, the Z_k, converge in distribution to independent Poisson-distributed random variables $\mathrm{Po}(\lambda_k)$ as $n \to \infty$, jointly for all $k \geq 3$.*

Proof. The proof follows directly from Theorem 6.3.1, conditioning on the event that $Z_1 = Z_2 = 0$. $\qquad\square$

Corollary 6.3.3 ([21]). *A graph $G \in \mathbb{G}_{n,d}$ is simple with probability tending to $e^{(1-d^2)/4}$.*

Proof. By Theorem 6.3.1,

$$\mathbb{P}(Z_0 = Z_1 = 0) \to e^{-\lambda_1-\lambda_2},$$

where $\lambda_1 + \lambda_2 = \frac{1}{2}(d-1) + \frac{1}{4}(d-1)^2 = \frac{1}{4}(d^2-1)$. $\qquad\square$

Let us stress the fact that Corollary 6.3.3 holds for constant d. However, Bollobás [29] showed that the formula can be applied for $d = d(n) \leq \sqrt{2\log n} - 1$. McKay [138] used another approach (switchings) to extend the range of d to $d = o(n^{1/3})$. McKay and Wormald [140] then used a new sort of switching to find the formula for $d = o(\sqrt{n})$:

$$\mathbb{P}(G \in G(P) \text{ is simple}) = \exp\left(\frac{1-d^2}{4} - \frac{d^3}{12n} + O\left(\frac{d^2}{n}\right)\right).$$

They also obtained a formula for $d \approx cn$ [139] that may have consequences for quite dense random regular graphs. In this section, we consider graphs with constant degree only.

It follows from Corollary 6.3.3 that any event holding a.a.s. over the probability space of random pairings also holds a.a.s. over the corresponding space $\mathbb{G}_{n,d}$. For this reason, asymptotic results over random pairings suffice for our purposes.

Theorem 6.3.4 ([184]). *Any property that holds a.a.s. for $G(P)$ also holds a.a.s. for $\mathbb{G}_{n,d}$.*

Proof. Suppose that the property P holds a.a.s. for $G(P)$. Let $\Gamma = \mathbb{P}(G \in \mathbb{G}_{n,d}$ does not have $P)$. Then by Corollary 6.3.3, we derive that

$$
\begin{aligned}
\Gamma &= \mathbb{P}(G \in G(P) \text{ does not have } P \mid G \in G(P) \text{ is simple}) \\
&= \frac{\mathbb{P}(G \in G(P) \text{ does not have } P \text{ and is simple})}{\mathbb{P}(G \in G(P) \text{ is simple})} \\
&\leq \frac{\mathbb{P}(G \in G(P) \text{ does not have } P)}{\mathbb{P}(G \in G(P) \text{ is simple})} \to 0,
\end{aligned}
$$

as $n \to \infty$. $\qquad\square$

The converse does not hold, as the trivial example of not containing a loop shows. As we already mentioned, one of the advantages of using the pairing model is that the pairs may be chosen sequentially so that the next pair is chosen uniformly at random over the remaining (unchosen) points. Let us consider the following example.

Lemma 6.3.5. *Let $K \geq 3$, $d \geq 3$ be fixed integers and $G \in \mathbb{G}_{n,d}$. Let $\omega(n)$ be any function tending to infinity. Then a.a.s. the number of vertices that belong to a cycle of length at most*

$$K = \log_{d-1} n - \log_{d-1} \omega(n)$$

is $o(n)$.

Note that this result implies that a.a.s. almost all vertices do not belong to a cycle of length at most K.

Proof. Let $u \in V(G)$ and let $N_i(u)$ denote the set of vertices at distance at most i from u. Let f_i denote the number of vertices in a balanced d-regular tree of depth i; that is,

$$f_i = 1 + d \sum_{j=0}^{i-1} (d-1)^j = 1 + \frac{d((d-1)^i - 1)}{d-2} = O((d-1)^i).$$

For every $u \in V(G)$ we have $|N_i(u)| \leq f_i$.

We will show that in the early stages of this process, the graphs grown from u tend to be trees a.a.s. Hence, if we expose, step by step, the vertices at distance $1, 2, \ldots, i$ from u, then we have to avoid, at step j, edges that induce cycles. Thus, we wish not to find edges between any two vertices at distance j from u or edges that join any two vertices at distance j to the same vertex at distance $j + 1$ from u. We shall refer to edges of this form as "bad edges." Note that the expected number of "bad edges" at step $i+1$ is equal to $O(f_i^2/n) = O((d-1)^{2i}/n)$. Therefore, the expected number of "bad edges" found up to step $i_1 = \lceil K/2 \rceil$ is equal to

$$\sum_{j=0}^{i_1-1} O\big((d-1)^{2j}/n\big) \;=\; O\big((d-1)^{2i_1}/n\big)$$

$$=\; O\big((d-1)^{K}/n\big)$$

$$=\; O\big(1/\omega(n)\big).$$

Thus, the expected number of vertices that belong to a cycle of length at most K is $O(n/\omega(n))$ and the assertion follows from Markov's inequality. □

It is known that a random d-regular graph is a.a.s. connected for any $d \geq 3$. See Wormald [184] for d fixed; see Cooper, Frieze, and Reed [70], and Krivelevich, Sudakov, Vu, and Wormald [126] for large d. It is not difficult to see that a random 2-regular graph is a.a.s. disconnected, and we expect the graph to be a large family of cycles. A lot more is known about random d-regular graphs. The interested reader is directed to the mentioned survey to learn more about the model [184].

6.4   Random Regular Graphs

We showed that for any d-regular graph G, $b(G) \leq \frac{n(d+1)}{4}$ (see Theorem 6.0.3). In this section, we will show that for large d this is a.a.s. very close to the correct value for a random d-regular graph; that is, a.a.s. $b(G) = \frac{n}{4}(d+O(\sqrt{d}))$. We also mention that the upper bound can be improved by analyzing the degree-greedy algorithm using the DE method. The results presented in this section may be found in [7].

Recall that

$$
\begin{aligned}
b(G) \;\; &\geq \;\; \max_{j} \;\; \min_{S \subseteq V, |S|=j} \{jd - 2|E(G[S])|\} \\
&= \;\; \max_{j} \;\; \min_{S \subseteq V, |S|=j} |E(S, V \setminus S)|,
\end{aligned}
\tag{6.7}
$$

where $E(S, V \setminus S)$ is the set of all edges joining S and its complement, and $E(G[S])$ is the set of all edges in the induced subgraph of G on S. The proof is a straightforward corollary of the fact that the minimum above is a lower bound on the number of edges going from the first j vertices cleaned to elsewhere in the graph. Hence, suppose that x and y are functions of n such that the expected number $S(x,y)$ of sets S of xn vertices in $G \in \mathbb{G}_{n,d}$ with at most yn edges to the complement $V(G) \setminus S$ is $o(1)$. Then this theorem, together with the first moment method, gives that the brush number is a.a.s. at least yn.

To find optimal values of x and y, we use the pairing model. This is essentially the same argument used by Bollobás [33] to obtain a lower bound on the isoperimetric number of random regular graphs. It is straightforward to check that

$$
\begin{aligned}
S(x,y) \;\; = \;\; &\binom{n}{xn}\binom{xdn}{yn} M(xdn - yn)\binom{(1-x)dn}{yn} \\
&\times \; (yn)! M((1-x)dn - yn)/M(dn),
\end{aligned}
$$

where $M(i)$ is the number of perfect matchings on i vertices. Recall that

$$
M(i) = i!! = \frac{i!}{(i/2)! 2^{i/2}}.
$$

To see this, we fix xn vertices (xdn points) to form set S (factor $\binom{n}{xn}$) and yn points in S that correspond to yn edges to the complement of S (factor $\binom{xdn}{yn}$). These edges are incident to yn points in $V(G) \setminus S$ (factor $\binom{(1-x)dn}{yn}$). After fixing points in both S and $V(G) \setminus S$, we need to connect them in all possible ways (factor

$(yn)!$). Finally, we need to take a perfect matching of remaining points in S (factor $M(xdn - yn)$) and a perfect matching of remaining points in $V(G) \setminus S$ (factor $M((1-x)dn - yn)$) to consider all possible configurations satisfying our assumption.

After simplification we find that

$$S(x,y) = \frac{n!(xdn)!((1-x)dn)!(dn/2)!2^{yn}}{(xn)!((1-x)n)!(yn)!((xd-y)n/2)!(((1-x)d-y)n/2)!(dn)!}.$$

Using Stirling's formula (see Lemma 1.3.2) and taking the exponential part we obtain that

$$\begin{aligned}
S(x,y) &\leq e^{o(n)} \frac{x^{x(d-1)n}(1-x)^{(1-x)(d-1)n}d^{dn/2}}{y^{yn}(xd-y)^{(xd-y)n/2}((1-x)d-y)^{((1-x)d-y)n/2}} \\
&= e^{f(x,y,d)n+o(n)},
\end{aligned} \tag{6.8}$$

where

$$\begin{aligned}
f(x,y,d) &= x(d-1)\log x + (1-x)(d-1)\log(1-x) \\
&\quad + 0.5d\log d - y\log y - 0.5(xd-y)\log(xd-y) \\
&\quad - 0.5((1-x)d - y)\log((1-x)d - y).
\end{aligned}$$

Thus, if $f(x,y,d) < 0$, then $S(x,y)$ is exponentially small (for large n) and the brush number is at least yn. Not surprisingly, the strongest bound is obtained for $x = 1/2$, in which case $f(x,y,d)$ becomes

$$\begin{aligned}
(d-1)\log(1/2) &+ (d/2)\log d - y\log y - (d/2 - y)\log(d/2 - y) \\
&= -\frac{d}{4}((1+z)\log(1+z) + (1-z)\log(1-z)) + \log 2
\end{aligned}$$

where $y = (d/4)(1-z)$.

It is straightforward to see that this function is decreasing in z for $z \geq 0$. Let ℓ_d/n denote the value of y for which it first reaches 0. Using Stirling's formula, it is also not difficult to see that we can replace $e^{o(n)}$ by $O(n^{-1})$ in (6.8). This gives us the following asymptotically almost sure lower bounds ℓ_d for the brush number of a random d-regular graph: $\ell_4 = 0.22n$, $\ell_5 = 0.36n$, and $\ell_6 = 0.52n$.

To obtain a result useful for all d, it is straightforward to show (since the Taylor expansion of $(1+z)\log(1+z)+(1-z)\log(1-z)$ is $z^2 + z^4/6 + \cdots$) that $l_d/n > (d/4)(1 - 2\sqrt{\log 2}/\sqrt{d})$. This result has the following implication, giving a nontrivial lower bound for $d \geq 3$.

Theorem 6.4.1 ([7]). *For $G \in \mathbb{G}_{n,d}$, a.a.s. we have that*

$$b(G) \geq \frac{dn}{4}\left(1 - \frac{2\sqrt{\log 2}}{\sqrt{d}}\right).$$

6.5 Tools: Differential Equation Method

We next introduce the *Differential Equation method* (or *DE method*), that is powerful and has many applications in analyzing random processes, including random graphs. In the DE method, we derive the expected changes in a sequence of random variables in a given process over time. The variables are taken as continuous, and suggest systems of differential equations based on the expected changes. Concentration results then show that the solutions to the differential equations are close to the values of the (discrete) variables. While we do not provide a comprehensive discussion of the DE method, we note that there is a detailed survey article devoted to this topic [183]; we just scratch the surface by discussing a few simple examples.

We first provide some machinery on martingales. For a detailed discussion of martingales (including the requisite background on conditional expectation, which we do not cover here), see, for example, [104]. A *martingale* is an infinite sequence $X_0, X_1, \ldots$ of random variables defined on a random process such that

$$\mathbb{E}[X_{n+1}|X_0, X_1, \ldots, X_n] = X_n.$$

In many applications we consider, the sequence of random variables is a Markov process. In that case, the martingale satisfies the property that

$$\mathbb{E}[X_{n+1}|X_0, X_1, \ldots, X_n] = \mathbb{E}[X_{n+1}|X_n] = X_n.$$

For example, toss a coin n times. Let S_n be the difference between the number of heads and the number of tails after n tosses. Hence, $S_0 = 0$ and for every $n \in \mathbb{N}$ we have that $S_n = S_{n-1} + 1$ with probability $1/2$; $S_n = S_{n-1} - 1$ otherwise. Therefore, $(S_n)_{n \in \mathbb{N} \cup \{0\}}$ is a random walk on $\mathbb{Z}$. It is not difficult to see that $(S_n)_{n \in \mathbb{N} \cup \{0\}}$ is a martingale. Indeed,

$$\mathbb{E}[S_{n+1}|S_0, S_1, \ldots, S_n] = S_n + \frac{1}{2} \cdot 1 + \frac{1}{2} \cdot (-1) = S_n.$$

For a second example, this time from graph theory, consider a graph process $G_0, G_1, \ldots$, where G_0 is any given graph with $n \geq 3$ vertices. For $t \in \mathbb{N} \cup \{0\}$, we form G_{t+1} from G_t in the following way. Choose three vertices at random from G_t. If they induce a complete graph K_3 or its complement (that is, an independent set), then do nothing. If there is only one edge joining them, then with probability $2/3$ delete it and otherwise add the other two edges to make a K_3. If there are two edges joining them, then with probability $1/3$ delete them and otherwise add the third edge. Set $X_t = |E(G_t)|$. If we condition on the outcome of the random selection of the three vertices, then it is straightforward to see that the expected change of X_t is zero. Thus, $(X_t)_{t \in \mathbb{N} \cup \{0\}}$ is a martingale. Observe that with probability 1 the process eventually becomes constant (we obtain the complete graph or the empty graph). What is the probability p that the final graph G is the complete graph? Since (X_t) is a martingale, it follows that

$$\begin{aligned} |E(G_0)| &= X_0 = \mathbb{E}[X_1] = \cdots = \mathbb{E}[X_t] \\ &\to p \cdot \binom{n}{2} + (1-p) \cdot 0 = p\binom{n}{2}, \end{aligned}$$

as $t \to \infty$. From this it follows that

$$p = |E(G_0)| / \binom{n}{2}.$$

A natural question is the following: how long does it usually take to reach the final graph? To answer this question, we need the following concentration inequality.

Theorem 6.5.1 ([12, 111]). (Azuma-Hoeffding inequality) *Let* $X_0, X_1, \ldots$ *be a martingale. Suppose that there exist constants* $c_k > 0$ *such that*

$$|X_k - X_{k-1}| \le c_k$$

for each $k \le n$. *Then for every* $t > 0$,

$$\mathbb{P}(X_n \ge \mathbb{E}[X_n] + t) \le \exp\left(-\frac{t^2}{2\sum_{k=1}^n c_k^2}\right), \quad and$$

$$\mathbb{P}(X_n \le \mathbb{E}[X_n] - t) \le \exp\left(-\frac{t^2}{2\sum_{k=1}^n c_k^2}\right).$$

Consider the random graph process described above. Since $c_i = 2$ for all $i \in \mathbb{N}$,

$$\mathbb{P}(|X_t - X_0| \ge \alpha) \le 2\exp\left(-\frac{\alpha^2}{8t}\right).$$

From this inequality, the expected time taken for the process to reach an absorbing state is at least $C\alpha^2$, where

$$\alpha = \min\left\{|E(G_0)|, \binom{n}{2} - |E(G_0)|\right\},$$

and C is a constant.

The Azuma-Hoeffding inequality can be generalized to include *supermartingales* ($\mathbb{E}[X_{t+1}|X_t] \le X_t$) and *submartingales* ($\mathbb{E}[X_{t+1}|X_t] \ge X_t$). Further, it is also possible to include random variables close to being martingales. Our proofs use the supermartingale method of Pittel et al. [158]; this method is also described in [183, Corollary 4.1]. We will use the following useful lemma.

Lemma 6.5.2 ([158]). *Let* $G_0, G_1, \ldots, G_L$ *be a random process and* X_t *a random variable determined by* $G_0, G_1, \ldots, G_t$, $0 \le t \le L$. *Suppose that for some real* β *and* γ,

$$\mathbb{E}[X_t - X_{t-1} \mid G_0, G_1, \ldots, G_{t-1}] < \beta$$

and

$$|X_t - X_{t-1} - \beta| \le \gamma$$

for $1 \le t \le L$. Then for all $\varepsilon > 0$, we have that

$$\mathbb{P}(A) \le \exp\left(-\frac{\varepsilon^2}{2L\gamma^2}\right),$$

where A is the event that for some t with $0 \le t \le L$, $X_t - X_0 \ge t\beta + \varepsilon$.

(We note that $L\gamma^2$ can be replaced by $\sum_{t=1}^{L} \gamma_t^2$ in the case when the bound for $|X_t - X_{t-1} - \beta|$ is not uniform.)

A *stopping time* with respect to a random process is a random variable T with values in $\mathbb{N} \cup \{\infty\}$ for which it can be determined whether $T = \hat{t}$, for any time $\hat{t}$, from knowledge of the process up to and including time $\hat{t}$. The name may be misleading, since a process does not necessarily *stop* when it reaches a stopping time. The key result says that if a supermartingale (X_i) stops at a stopping time (that is, (X_i) becomes static for all time after the stopping time), then the result is a supermartingale.

Theorem 6.5.3 ([104]). *If, with respect to some random graph process, (X_i) is a supermartingale and T is a stopping time, then $(X_{\min\{i,T\}})$ is also a supermartingale with respect to the same process.*

Now, as an illustration of the DE method, consider the following problem known as the *coupon collector process* (discussed also in Chapters 2 and 10). There is an urn with n distinct labeled balls in it. We draw a ball at random, observe its label, and put the ball back into the urn. We then keep repeating this experiment until we see all n balls. The question is: How many draws are necessary to finish your task (typically)?

It can be shown that a.a.s. all balls are drawn after

$$N = n(\log n + \omega(n))$$

rounds, where $\omega(n)$ is any function that tends to infinity together with n. For a given time t, let X_t be the random variable counting how many balls have not been drawn up to this point of the process. We would like to investigate the asymptotic behavior of this random variable (as n tends to infinity). Observe that $X_0 = n$. For $t > 0$, we have that X_t decreases by one precisely when, in time step $t + 1$, a new ball is chosen. Therefore, we obtain that

$$\mathbb{E}[X_{t+1} - X_t | X_t] = -\frac{X_t}{n}.$$

To analyze this random variable, we use the Differential Equation method. Defining a real function $z(x)$ to model the behavior of X_{xn}/n, the above relation implies the following differential equation

$$z'(x) = -z(x),$$

with the initial condition $z(0) = 1$. The general solution is $z(x) = \exp(-x + C)$, with $C \in \mathbb{R}$, and the particular solution is $z(x) = \exp(-x)$. This *suggests* that the random variable X_t should behave similarly to the deterministic function $n \exp(-t/n)$. The following theorem precisely states the conditions under which this holds.

Theorem 6.5.4. *A.a.s., for every t in the range*

$$0 \le t \le t_f = \frac{1}{2} n \log n - 2n \log \log n,$$

we have

$$X_t \sim n \exp(-t/n).$$

Proof. We transform X_t into something close to a martingale. Consider the following random variable

$$H_t = H(X_t, t) = \log X_t + t/n$$

and the stopping time

$$T = \min\{t \ge 0 : X_t < (1/2)\sqrt{n} \log^2 n \text{ or } t = t_f\}.$$

(Note that H_t is chosen so that it is close to a constant along every trajectory of the differential equation $z'(x) = -z(x)$.)

Consider the sequence of random variables $(H_t : 0 \leq t \leq t_f)$. To use the Azuma-Hoeffding inequality, we need to estimate $\mathbb{E}[H_{t+1} - H_t | X_t, t]$ and $|H_{t+1} - H_t|$.

$$
\begin{aligned}
\mathbb{E}[H_{t+1} - H_t | X_t, t] &= \left(1 - \frac{X_t}{n}\right)\frac{1}{n} \\
&\quad + \frac{X_t}{n}\left(\log(X_t - 1) - \log X_t + \frac{1}{n}\right) \\
&= \frac{1}{n} + \frac{X_t}{n}\log\left(1 - \frac{1}{X_t}\right) \\
&= O\left(\frac{1}{nX_t}\right) = O\left(\frac{1}{n^{3/2}\log^2 n}\right),
\end{aligned}
$$

provided $T > t$. Similarly, we find that

$$
\begin{aligned}
|H_{t+1} - H_t| &\leq \left|\log(X_t - 1) - \log X_t + \frac{1}{n}\right| \\
&= O\left(\frac{1}{X_t}\right) + \frac{1}{n} = O\left(\frac{1}{\sqrt{n}\log^2 n}\right),
\end{aligned}
$$

provided $T > t$ again.

Therefore, with the notation $i \wedge T$ denoting $\min\{i, T\}$, we have that

$$
\mathbb{E}[H_{(t+1)\wedge T} - H_{t\wedge T} | X_t, t] = O\left(\frac{1}{n^{3/2}\log^2 n}\right), \text{ and}
$$

$$
|H_{(t+1)\wedge T} - H_{t\wedge T}| = O\left(\frac{1}{\sqrt{n}\log^2 n}\right).
$$

Now we may apply Lemma 6.5.2 to the sequence $(H_{t\wedge T} : 0 \leq t \leq t_f)$, and symmetrically to $(-H_{t\wedge T} : 0 \leq t \leq t_f)$, with $\varepsilon = 1/\log^{1/2} n$, $\beta = O(1/n^{3/2}\log^2 n)$, and $\gamma = O(1/\sqrt{n}\log^2 n)$ to show that a.a.s. for all t at most t_f we have that

$$
|H_{t\wedge T} - H_0| = O(\log^{-1/2} n) = o(1).
$$

As $H_0 = \log n$, this implies, from the definition of H_t, that a.a.s. the desired conclusion holds for every $0 \leq t \leq T$.

To complete the proof we need to show that a.a.s., $T = t_f$. The events asserted by the equation hold a.a.s. up until time T, as shown above. Thus, in particular, a.a.s.

$$X_T \sim n \exp(-T/n) > (1 + o(1))\sqrt{n}\log^2 n,$$

which implies that $T = t_f$ a.a.s. □

6.6 DE Method in Graph Cleaning

The differential equation method is used here to find an upper bound on the number of brushes needed to clean a random d-regular graph using a degree-greedy algorithm. We consider $d = 2$ first, then state some general results, and apply them to the special cases of $3 \leq d \leq 5$ before discussing higher values of d. The results from this section have been published in [7].

2-Regular Graphs

Let $Y = Y_n$ be the total number of cycles in a random 2-regular graph on n vertices. Since exactly two brushes are needed to clean one cycle, we need $2Y_n$ brushes in order to clean a 2-regular graph. We know that the random 2-regular graph is a.a.s. disconnected; by simple calculations we can show that the probability of having a Hamiltonian cycle is asymptotic to $\frac{1}{2}e^{3/4}\sqrt{\pi}n^{-1/2}$; see, for example, [184]. The total number of cycles Y_n is sharply concentrated near $(1/2)\log n$. (It is not difficult to see this by generating the random graph sequentially using the pairing model. The probability of forming a cycle in step i is exactly $1/(2n - 2i + 1)$, so the expected number of cycles is $(1/2)\log n + O(1)$. The variance can be calculated in a similar way.) So we derive that a.a.s. the brush number for a random 2-regular graph is $(1 + o(1))\log n$.

d-Regular Graphs ($d \geq 3$): The General Setting

In this subsection, we assume $d \geq 3$ is fixed with dn even. To obtain an asymptotically almost sure upper bound on the brush number, we study an algorithm that cleans random vertices of minimum degree. This algorithm is called *degree-greedy* because the vertex being cleaned is chosen from those with the lowest degree.

We start with a random d-regular graph $G = (V, E)$ on n vertices. Initially, all vertices are dirty: $D_0 = V$. In step t of the cleaning process, we clean a random vertex α_t, chosen uniformly at random from those vertices with the lowest degree in the induced subgraph $G[D_{t-1}]$, where $D_t = D_{t-1} \setminus \{\alpha_t\}$. In the first step, d brushes are needed to clean a random vertex α_1 (we say that this is "phase zero"). The induced subgraph $G[D_1]$ now has d vertices of degree $d - 1$ and $n - d - 1$ vertices of degree d. Note that α_1 is a.a.s. the only vertex whose degree in $G[D_t]$ is d at the time of cleaning. Indeed, if α_t (where $t \geq 2$) has degree d in $G[D_{t-1}]$, then $G[D_{t-1}]$ consists of some components of G and thus, G is disconnected. It was proven independently in [30, 182] that for constant d, we have that G is disconnected with probability $o(1)$ (this also holds when d is growing with n, as shown in [134]).

In the second step, $d - 2$ brushes are needed to clean a random vertex α_2 of degree $d - 1$. Typically, in the third step, a vertex of degree $d - 1$ is cleaned, and in each subsequent step, a vertex of degree $d - 1$ in $G[D_t]$ is cleaned, until some vertex of degree $d - 2$ is produced in the subgraph induced by the set of dirty vertices. After cleaning the first vertex of degree $d - 2$, we typically return to cleaning vertices of degree $d - 1$, but after some more steps of this type we may clean another vertex of degree $d - 2$. When vertices of degree $d - 1$ become plentiful, vertices of lower degree are more commonly created, and these hiccups occur more often. When vertices of degree $d - 2$ take over the role of vertices of degree $d - 1$, we informally say that the first phase ends and we begin the second phase. In general, in the kth phase, a mixture of vertices of degree $d - k$ and $d - k - 1$ are cleaned.

During the kth phase there are, in theory, two possible endings. It can happen that the vertices of degree $d - k$ become so common that the vertices of degree $d - k - 1$ start to explode (in which case we move to the next phase). It is also possible that the vertices of degree $d - k + 1$ become so rare that those of degree

$d - k$ disappear (in which case the process goes backwards). With various initial conditions, either one could occur. However, the numerical solutions of the differential equations for $d = 4, 5, \ldots, 100$ support the hypothesis that the degree-greedy process we study never goes back. In such cases, the remaining vertices are cleaned for free (that is, after the crucial phases, only $o(n)$ new brushes are required to finish the process). The details of the following instance of the Differential Equation method have been omitted, but can be found in [186].

For $0 \leq i \leq d$, let $Y_i = Y_i(t)$ denote the number of vertices of degree i in $G[D_t]$. (Note that $Y_0(t) = n - t - \sum_{i=1}^{d} Y_i(t)$, so $Y_0(t)$ does not need to be calculated, but it is useful in the discussion.) Let $S(t) = \sum_{\ell=1}^{d} \ell Y_\ell(t)$ and for any statement A, let δ_A denote the Kronecker delta function

$$\delta_A = \begin{cases} 1 & \text{if } A \text{ is true,} \\ 0 & \text{otherwise.} \end{cases}$$

Define

$$f_{i,r}((t-1)/n, Y_1(t-1)/n, Y_2(t-1)/n, \ldots, Y_d(t-1)/n)$$
$$= \mathbb{E}[Y_i(t) - Y_i(t-1) \mid G[D_{t-1}] \text{ and } \deg_{G[D_{t-1}]}(\alpha_t) = r].$$

It is not difficult to see that

$$f_{i,r}((t-1)/n, Y_1(t-1)/n, Y_2(t-1)/n, \ldots, Y_d(t-1)/n)$$
$$= -\delta_{i=r} - r\frac{iY_i(t-1)}{S(t-1)} + r\frac{(i+1)Y_{i+1}(t-1)}{S(t-1)}\delta_{i+1\leq d} \quad (6.9)$$

for $i, r \in [d]$ such that $Y_r(t) > 0$. Indeed, α_t has degree r, hence, the term $-\delta_{i=r}$. When a pair of points in the pairing model is exposed, the probability that the other point is in a bucket of degree i (that is, the bucket contains i unchosen points) is asymptotic to $iY_i(t-1)/S(t-1)$. Thus, $riY_i(t-1)/S(t-1)$ stands for the expected number of the r buckets found adjacent to α_t that have degree i. This contributes negatively to the expected change in Y_i, while buckets of degree $i + 1$ that are reached contribute positively (of course, only if this type of vertex (bucket) exists in the graph; thus, $\delta_{i+1\leq d}$). This explains (6.9).

Suppose that at some step t of phase k, cleaning a vertex of degree $d - k$ creates, in expectation, β_k vertices of degree $d - k - 1$ and cleaning a vertex of degree $d - k - 1$ decreases, in expectation, the number of vertices of degree $d - k - 1$ by τ_k. After cleaning a vertex of degree $d - k$, we expect to then clean (on average) β_k / τ_k vertices of degree $d - k - 1$. Thus, in phase k, the proportion of steps that clean vertices of degree $d - k$ is $1/(1 + \beta_k / \tau_k) = \tau_k/(\beta_k + \tau_k)$. When τ_k falls below zero, vertices of degree $d - k - 1$ begin to build up and do not decrease under repeated cleaning of vertices of this type, and we move to the next phase.

From (6.9) it follows that

$$
\begin{aligned}
\beta_k &= \beta_k(x, y_1, y_2, \ldots, y_d) \\
&= f_{d-k-1,d-k}(x, y_1, y_2, \ldots, y_d) = f_{d-k-1,d-k}(x, y), \text{ and} \\
\tau_k &= \tau_k(x, y_1, y_2, \ldots, y_d) \\
&= -f_{d-k-1,d-k-1}(x, y_1, y_2, \ldots, y_d) = -f_{d-k-1,d-k-1}(x, y),
\end{aligned}
$$

where $x = t/n$ and $y_i(x) = Y_i(t)/n$ for $i \in [d]$. This suggests the following system of differential equations:

$$
\frac{dy_i}{dx} = F(x, y, i, k)
$$

with

$$
F(x, y, i, k) = \begin{cases} \Gamma & \text{for } k \leq d - 2, \\ f_{i,1}(x, y) & \text{for } k = d - 1, \end{cases}
$$

and where $\Gamma = \frac{\tau_k}{\beta_k + \tau_k} f_{i,d-k}(x, y) + \frac{\beta_k}{\beta_k + \tau_k} f_{i,d-k-1}(x, y)$.

At this point we may formally define the interval $[x_{k-1}, x_k]$ to be phase k, where the termination point x_k is defined as the infimum of those $x > x_k$ for which at least one of the following holds: $\tau_k \leq 0$ and $k < d - 1$; $\tau_k + \beta_k = 0$ and $k < d - 1$; $y_{d-k} \leq 0$. Using the final values $y_i(x_k)$ in phase k as initial values for phase $k + 1$, we can repeat the argument inductively, moving from phase to phase, starting from phase 1 with the obvious initial conditions $y_d(0) = 1$ and $y_i(0) = 0$ for $0 \leq i \leq d - 1$.

The general result [186, Theorem 1] studies a deprioritized version of the degree-greedy algorithm, which means that the vertices are chosen for processing in a different way: the vertices are not always chosen from those with minimum degree, but usually from a random mixture of vertices of two different degrees. Once a vertex is chosen, it is treated the same as in the degree-greedy algorithm. The variables Y are defined in an analogous manner. The hypotheses of the theorem are mainly straightforward to verify, but for the full rigorous conclusion to be obtained, several inequalities involving derivatives are required to hold at the terminations of phases. However, in practice, the equations are simply solved numerically in order to find the points x_k, since a fully rigorous bound is not obtained unless one obtains strict inequalities on the values of the solutions. The conclusion is that, for a certain algorithm using a deprioritized mixture of the steps of the degree-greedy algorithm, with variables Y_i defined above, we have that a.a.s.

$$Y_i(t) = ny_i(t/n) + o(n)$$

for $1 \leq i \leq d$ for phases $k = 1, 2, \ldots, m$, where m denotes the smallest k for which either $k = d - 1$, or any of the termination conditions for phase k hold at x_k apart from $x_k = \inf\{x > x_{k-1} : \tau_k \leq 0\}$. We omit all details, pointing the reader to [186] and the general survey [183] about the differential equation method, which is a main tool in proving [186, Theorem 1]. In addition, the theorem gives information on auxiliary variables such as, of importance to our present application, the number of brushes used. Instead of quoting this result precisely, we use it merely as justification for being able to use the above equations as if they applied to the greedy algorithm. (This is no doubt the case, but it is not actually proved in [186]. Instead, we know that they apply in the limit to a sequence of algorithms that use the steps of the degree-greedy algorithm.) The solution to the relevant differential equations for $d = 3$ is shown in Figure 6.3.

In the kth phase, a mixture of vertices of degree $d - k$ and $d - k - 1$ are cleaned. Since $\max\{2\ell - d, 0\}$ brushes are needed to

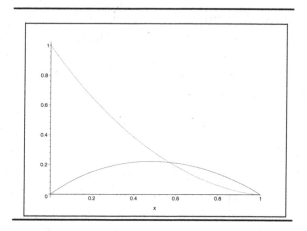

FIGURE 6.3: Solution to the differential equations modeling the brush number in the d-regular case.

clean a vertex of degree ℓ (see (6.1)), we need

$$u_d^k \sim n\left(\max\{d - 2k, 0\} \int_{x_{k-1}}^{x_k} \frac{\tau_k}{\tau_k + \beta_k}\, dx \right.$$

$$\left. + \max\{d - 2k - 2, 0\} \int_{x_{k-1}}^{x_k} \frac{\beta_k}{\tau_k + \beta_k}\, dx \right)$$

brushes in phase k. Thus, the total number of brushes needed to clean a graph using the degree-greedy algorithm is a.a.s. equal to

$$u_d = \sum_{k=1}^{\lfloor (d-1)/2 \rfloor} u_d^k + o(n)$$

$$\sim n\left(\sum_{k=1}^{\lfloor (d-1)/2 \rfloor} ((d - 2k - 2)(x_k - x_{k-1}) \right.$$

$$\left. + 2\int_{x_{k-1}}^{x_k} \frac{\tau_k}{\tau_k + \beta_k}\, dx \right) + \delta_{d\ \text{is odd}} \int_{x_{k-1}}^{x_k} \frac{\beta_k}{\tau_k + \beta_k}\, dx \right).$$

We assume here that the solutions of the differential equations proceed in the way we presume; that is, with no reversion to earlier phases. This discussion implies that only $o(n)$ new brushes are required for the remaining phases.

3-Regular Graphs

Let G be any 3-regular graph on n vertices. The first vertex cleaned must start three brush paths, the last one terminates three brush paths, and all other vertices must start or finish at least one brush path, so the number of brush paths is at least $n/2 + 2$.

The result mentioned above can be shown to result in an upper bound of $n/2 + o(n)$ for the brush number of a random 3-regular (that is, cubic) graph. We do not provide details because of the following stronger result. It is known [172] that a random 3-regular graph a.a.s. has a Hamiltonian cycle. The edges not in a Hamiltonian cycle must form a perfect matching. Such a graph can be cleaned by starting with three brushes at one vertex, and moving along the Hamiltonian cycle with one brush, introducing one new brush for each edge of the perfect matching. Hence, the brush number of a random 3-regular graph with n vertices is a.a.s. $n/2 + 2$. Note that this is also the brush number of any cubic Hamiltonian graph on n vertices.

4-Regular Graphs

For 4-regular graphs, to estimate the brush number one has to carefully analyze phase 1 only: we need two brushes to clean vertices of degree 3, but vertices of degree 2 are cleaned "for free." Note that $y_1(x) = y_2(x) - 0$ throughout phase 1. We have the system of differential equations:

$$\frac{dy_4}{dx} = \frac{-6y_4(x)}{3y_3(x) + 2y_4(x)},$$
$$\frac{dy_3}{dx} = \frac{-3y_3(x) + 4y_4(x)}{3y_3(x) + 2y_4(x)},$$

with the initial conditions $y_4(0) = 1$ and $y_3(0) = 0$. The solution (see Figure 6.4 (a)) to these differential equations is

$$y_4(x) = 5 - 4\sqrt{1 + 3x} + 3x,$$
$$y_3(x) = \frac{4(-3 + 3\sqrt{1 + 3x} - 5x + x\sqrt{1 + 3x})}{2 - \sqrt{1 + 3x}},$$

so $\beta_1 = -3 + 3\sqrt{1 + 3x}$ and $\tau_1 = 3 - 2\sqrt{1 + 3x}$. Thus, phase 1 finishes at time $t_1 = 5n/12$ ($x_1 = 5/12$ is a root of the equation $\tau_1(x) = 0$) and the number of vertices of degree 3 cleaned during this phase is asymptotic to

$$n \int_0^{5/12} \frac{\tau_1}{\tau_1 + \beta_1} dx = n/6.$$

Since we need 2 brushes to clean one such vertex we derive an asymptotically almost sure upper bound of $u_4 \sim n/3$.

The remaining phases can be studied in a similar way, assuring us that no extra brushes are needed. The solutions to the relevant differential equations are shown in Figure 6.4.

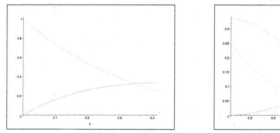 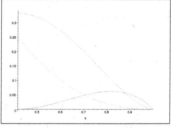

(a) 4-regular graph, phase 1   (b) 4-regular graph, phase 2
(vertices are cleaned "for free")

FIGURE 6.4: Solutions to the differential equations modeling the brush number in the 4-regular case.

5-Regular Graphs

To study the brush number yielded by the degree-greedy algorithm for 5-regular graphs, we cannot consider only phase 1 as before; we need 3 brushes to clean vertices of degree 4 but also 1 brush to clean vertices of degree 3. Thus, two phases must be considered.

In phase 1, $y_1(x) = y_2(x) = y_3(x) = 0$ and we have the system of differential equations:

$$\frac{dy_5}{dx} = \frac{-20y_5(x)}{8y_4(x) + 5y_5(x)},$$

$$\frac{dy_4}{dx} = \frac{-8y_4(x) + 15y_5(x)}{8y_4(x) + 5y_5(x)},$$

with the initial conditions $y_5(0) = 1$ and $y_4(0) = 0$. The numerical solution (see Figure 6.5 (a)) suggests that the phase finishes at time $t_1 = 0.1733n$. The number of brushes needed in this phase is asymptotic to

$$u_5^1 \sim \left(3n \int_0^{t_1/n} \frac{\tau_1}{\tau_1 + \beta_1} dx + n \int_0^{t_1/n} \frac{\beta_1}{\tau_1 + \beta_1} dx \right)$$

$$\sim \left(t_1 + 2n \int_0^{t_1/n} \frac{\tau_1}{\tau_1 + \beta_1} dx \right) \approx 0.3180n.$$

In phase 2, $z_1(x) = z_2(x) = 0$ and we have another system of differential equations:

$$\frac{dz_5}{dx} = \frac{-15z_5(x)}{6z_3(x) + 4z_4(x) + 5z_5(x)},$$

$$\frac{dz_4}{dx} = \frac{-3(4z_4 - 5z_5(x))}{6z_3(x) + 4z_4(x) + 5z_5(x)},$$

$$\frac{dz_3}{dx} = \frac{6z_3(x) + 8z_4(x) - 5z_5(x)}{6z_3(x) + 4z_4(x) + 5z_5(x)},$$

with the initial conditions $z_5(t_1/n) = y_5(t_1/n) = 0.5088$, $z_4(t_1/n) = y_4(t_1/n) = 0.3180$ and $z_3(t_1/n) = 0$. The numerical solution (see Figure 6.5 (b)) suggests that the phase finishes (approximately) at time $t_2 = 0.7257n$. The number of brushes needed in this phase is asymptotic to (the numerical solution)

$$u_5^2 \sim n \int_{t_1/n}^{t_2/n} \frac{\tau_2}{\tau_2 + \beta_2} dx \approx 0.3259n.$$

(Note that there is no $\beta_2/(\tau_2 + \beta_2)$ term this time; each vertex of degree 2 receives 3 extra brushes from already cleaned neighbors and, thus, can be cleaned "for free.") Finally, we derive an asymptotically almost sure upper bound of $u_5 = u_5^1 + u_5^2 \approx 0.6439n$.

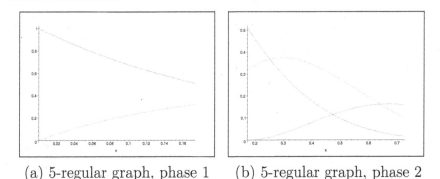

(a) 5-regular graph, phase 1   (b) 5-regular graph, phase 2

FIGURE 6.5: Solutions to the differential equations modeling the brush number in the 5-regular case.

d-Regular Graphs of Higher Order

Note that the lower bound for $d = 4$ will be considerably lower than the lower bound of $n/2 + 2$ for $d = 3$, whereas the upper bound we have been discussing comes from the same degree-greedy algorithm in all cases. However, the upper bound is also sensitive to the parity of d. For the 4-regular case, vertices of degree 2 are processed "for free," so we focus on degree-3 vertices, and when $d = 3$ there are fewer of those processed than degree-2 vertices. But it seems that the parity of d does not greatly affect the value of u_d/n for d big enough (see Figure 6.6 and Table 6.1).

In Figure 6.6, the values of ℓ_d/dn and u_d/dn have been presented for all d-values up to 100, although in Table 6.1 we have only listed the first 30 and a few more values for higher d.

6.7   Game Brush Number

The *brushing game* is a two-player game played on a graph G first introduced in [121]. The game brush number follows in the same spirit as the toppling number [48] studied in Chapter 11, where players with conflicting objectives together make choices

d	ℓ_d/n	u_d/n
3	0.0922	0.500
4	0.220	0.334
5	0.365	0.644
6	0.521	0.684
7	0.686	0.949
8	0.858	1.06
9	1.03	1.31
10	1.21	1.45
11	1.39	1.69
12	1.58	1.85
13	1.77	2.08
14	1.96	2.25
15	2.16	2.49
16	2.35	2.67
17	2.55	2.90
18	2.75	3.08
19	2.95	3.32
20	3.16	3.51
21	3.36	3.74
22	3.56	3.93

d	ℓ_d/n	u_d/n
23	3.77	4.16
24	3.98	4.36
25	4.18	4.59
26	4.39	4.80
27	4.60	5.03
28	4.81	5.23
29	5.02	5.46
30	5.23	5.67
31	5.44	5.90
32	5.66	6.11
99	20.6	21.5
100	20.8	21.7
149	32.1	33.2
150	32.4	33.5
199	43.8	45.1
200	44.1	45.3
249	55.6	57.0
250	55.9	57.3
299	67.5	69.0
300	67.7	69.3

TABLE 6.1: Approximate upper and lower bounds on the brush number.

that produce a feasible solution to some optimization problem. The results of this section derive from [121].

A *configuration* of a graph G is an assignment of some nonnegative integer number of brushes to each vertex of G; we represent a configuration by a map $f : V(G) \to \mathbb{N} \cup \{0\}$. The *brushing game* on a graph G with *initial configuration* f has players *Max* and *Min*. Initially, all vertices of G are deemed *dirty*, and each vertex v contains $f(v)$ brushes.

The players Max and Min alternate turns. At the beginning of each turn, the player whose turn it is adds one brush to any dirty vertex. After this, if some vertex v has at least as many brushes as dirty neighbors, then v *fires*: one brush is added to each dirty neighbor of v, and v itself is marked *clean*. The process of firing v is

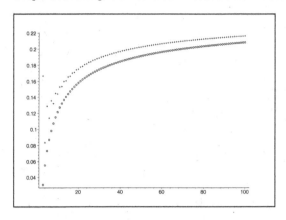

FIGURE 6.6: A graph of u_d/dn and ℓ_d/dn versus d (from 3 to 100).

also referred to as *cleaning v*. Vertices continue to fire, sequentially, until no more vertices may fire, at which point the turn ends. If at this point all vertices of G are clean, then the game ends. When both players play optimally, the number of brushes placed during the game is the *game brush number* of G with initial configuration f, denoted $b_g(G; f)$, if Min takes the first turn and by $\widehat{b}_g(G; f)$ when Max does. If f is identically zero, then we write $b_g(G)$ in place of $b_g(G; f)$ and refer to it as the *game brush number of G*. When Min (respectively, Max) starts the game, we sometimes refer to the process as the *Min-start* (respectively, *Max-start*) game. In the Min-start (respectively, Max-start) game, a *round* of the game consists of one turn by Min (respectively, Max) and the subsequent turn by Max (respectively, Min).

We say that G can be *cleaned by* the configuration f when some list of vertex cleanings, starting from f, results in every vertex of G being cleaned. Given configurations f and g of a graph G, we say that f *dominates* g provided that $f(v) \geq g(v)$ for all $v \in V(G)$. We note the following lemma (proof omitted).

Lemma 6.7.1 ([121]). *Let f and g be configurations of a graph G. If f dominates g, then $b_g(G; f) \leq b_g(G; g)$ and*

$$\widehat{b}_g(G; f) \leq \widehat{b}_g(G; g).$$

Further, we have that $b_g(G; g) - b_g(G; f) \leq 2 \sum_v (f(v) - g(v))$ *and*

$$\widehat{b_g}(G; g) - \widehat{b_g}(G; f) \leq 2 \sum_v (f(v) - g(v)).$$

Our main result of this section is to establish the game brush number of binomial random graphs $\mathbb{G}(n, p)$. First, we consider a fractional variant of the brushing game. The fractional brushing game behaves analogously to the ordinary brushing game, but with two important changes. First, vertices can contain nonintegral numbers of brushes. Second, a vertex v fires when the number of brushes on v is at least p times the number of dirty neighbors, at which time it sends p brushes to each dirty neighbor (where p is a fixed constant between 0 and 1).

Given $p \in (0, 1)$, the *fractional brushing game* on a graph G has players *Max* and *Min*. Throughout the game, each vertex of G contains some nonnegative real number of *brushes*. Initially, all vertices of G are deemed *dirty*, and no vertex contains any brushes.

The players alternate turns. At the beginning of each turn, if the number of brushes on some vertex v is at least p times the number of dirty neighbors, then v *fires*: p brushes are added to each dirty neighbor of v, and v itself is marked *clean*. (The process of firing v is also referred to as *cleaning* v.) Vertices continue to fire, sequentially, until no more vertices may fire. If at this point all vertices of G are clean, then the game ends. Otherwise, the player whose turn it is adds one brush to any dirty vertex, and the turn ends. Min aims to minimize, and Max aims to maximize, the number of turns taken before the game ends. When both players play optimally, the total number of turns taken is the *fractional game brush number* of G with parameter p, denoted $b_{g_p}(G)$, if Min takes the first turn and by $\widehat{b}_{g_p}(G)$ when Max does.

We note the following theorem from [121].

Theorem 6.7.2 ([121]). *If* $p \gg \log n / n$, *then*

$$b_{g_p}(K_n) \sim p b_g(K_n) \sim p n^2 / e.$$

We are now ready to prove the main result of this section.

Theorem 6.7.3 ([121]). *For $p = p(n) \gg \log n/n$ and $G \in$*
$\mathbb{G}(n, p)$, *a.a.s.*

$$b_g(G) \sim p b_g(K_n) \sim p n^2/e.$$

Proof. Theorem 6.7.2 establishes the second equality in the theorem statement, so it suffices to prove the first. Let G be any graph on n vertices. To obtain an upper bound for $b_g(G)$, we provide a strategy for Min that mimics their optimal strategy for the fractional game on K_n, which yields an upper bound on $b_g(G)$ in terms of $b_{g_p}(K_n)$. Max plays their moves in the *real* game on G, but Min interprets them as moves in the *imaginary* fractional game on K_n. Min plays according to an optimal strategy in the imaginary game, then makes the same move in the real game. Our hope is that the two games behave very similarly for $G \in \mathbb{G}(n, p)$.

We introduce an *Oracle* to enforce synchronization between the games. The Oracle can, at any time, clean a vertex in either game by adding extra brushes. Suppose some vertex fires in the real game but not in the imaginary one. When this happens, the Oracle cleans this vertex in the imaginary game; by Lemma 6.7.1, this cannot increase the length of the imaginary game. Now suppose instead that some vertex v fires in the imaginary game but not in the real one. In this case, the Oracle adds brushes to v in the real game until it fires. By Lemma 6.7.1, each brush placed by the Oracle can decrease the length of the real game by at most 2.

To obtain an upper bound on the length of the real game, we must bound the number of brushes added by the Oracle in the real game. Let $\pi = (v_1, v_2, \ldots, v_n)$ be the cleaning sequence produced during the game, that is, the order in which the vertices fire. Consider the state of the game when only vertices $v_1, v_2, \ldots, v_{i-1}$ are clean. In the imaginary game, v_i received $(i-1)p$ brushes from earlier neighbors and needs a total of $(n-i)p$ to fire. In the real game, it received $\deg_\pi^-(v_i) = |N(v_i) \cap \{v_1, v_2, \ldots, v_{i-1}\}|$ brushes and needs $\deg(v_i) - \deg_\pi^-(v_i)$. If the Oracle adds brushes to clean v_i in the real game, then it must be that, compared to the imaginary game, v_i either received fewer brushes or needs more to fire (or both). The Oracle must add at most $D_\pi(v_i)$ brushes, where

$$D_\pi(v_i) = \max\left\{\left((i-1)p-(n-i)p\right)-\left(\deg_\pi^-(v_i)-(\deg(v_i)-\deg_\pi^-(v_i))\right),0\right\}$$

$$= \max\left\{\left(2(i-1)p-(n-1)p\right)-\left(2\deg_\pi^-(v_i)-\deg(v_i)\right),0\right\}$$

$$\leq 2\max\left\{(i-1)p-\deg_\pi^-(v_i),0\right\}+\max\{\deg(v_i)-(n-1)p,0\}.$$

Let $D_1 = 2\sum_{i=1}^n \max\left\{(i-1)p-\deg_\pi^-(v_i),0\right\}$ and $D_2 = \sum_{i=1}^n \max\left\{\deg(v_i)-(n-1)p,0\right\}$. Hence, the length of the game is at most $b_{g_p}(K_n)+2D_\pi(G)$, where $D_\pi(G) = D_1 + D_2$. We do not know, in advance, the cleaning sequence π. However, we still obtain the following upper bound:

$$b_g(G) \leq b_{g_p}(K_n) + 2\max_\pi D_\pi(G).$$

An analogous argument yields a lower bound on $b_g(G)$. We still provide a strategy for Min, but this time, they use an optimal strategy for the real game on G to guide their play in the imaginary (fractional) game on K_n. If a vertex fires in the imaginary game but not in the real one, then the Oracle cleans that vertex by providing extra brushes in the real game; this cannot increase the length of the real game. If a vertex v fires in the real game but not in the imaginary one, then the Oracle adds cleans v by adding brushes in the imaginary game; each brush added can decrease the length of the imaginary game by at most 2. When the Oracle adds brushes in the imaginary game, it is because more brushes were received or fewer were needed (or both) in the real game than in the imaginary game. Using symmetry and the notation introduced above, we obtain that $b_{g_p}(K_n) \leq b_g(G) + 2D_{\bar\pi}(G)$, where $\bar\pi$ is the reverse of π.

Hence, we derive that

$$b_{g_p}(K_n) \leq b_g(G) + 2\max_\pi D_{\bar\pi}(G) = b_g(G) + 2\max_\pi D_\pi(G).$$

It follows that

$$|b_g(G) - b_{g_p}(K_n)| \leq 2\max_\pi D_\pi(G).$$

Now let $G \in \mathbb{G}(n,p)$, let $d = d(n) = p(n-1)$, and let $\omega = \omega(n) = d/\log n$ (note that ω tends to infinity as $n \to \infty$). To complete the proof, it suffices to show that a.a.s. $D_\pi(G) = o(pn^2)$. We can bound the second summation in the definition of $D_\pi(G)$ using the Chernoff bound.

Fix a vertex v_i. It is evident that $\mathbb{E}[\deg(v_i)] = p(n-1) = d$, so it follows immediately from the Chernoff bound that

$$\mathbb{P}\left(|\deg(v_i) - d| \geq \varepsilon d\right) \leq 2\exp\left(-\frac{\varepsilon^2 d}{3}\right) = 2n^{-2},$$

for $\varepsilon = \sqrt{6\log n/d} = \sqrt{6/\omega} = o(1)$. Therefore, by the union bound, a.a.s. for all $i \in \{1, 2, \ldots, n\}$ we have $\deg(v_i) \sim d$. As a consequence, a.a.s.

$$\sum_{i=1}^{n} \max\{\deg(v_i) - (n-1)p, 0\} = \sum_{i=1}^{n} \max\{\deg(v_i) - d, 0\}$$
$$\leq n \cdot o(d) = o(pn^2).$$

Estimating the first summation in the definition of $D_\pi(G)$ is more complicated, since we must consider all possible cleaning sequences π. First, observe that the partial sum containing only the first $n/\omega^{1/5}$ terms can be bounded (deterministically, for any π) as follows:

$$\sum_{i=1}^{n/\omega^{1/5}} \max\{(i-1)p - \deg_\pi^-(v_i), 0\} \leq p \sum_{i=1}^{n/\omega^{1/5}} (i-1)$$
$$= O(pn^2/\omega^{2/5})$$
$$= o(pn^2).$$

Now fix some cleaning sequence π and some i exceeding $n/\omega^{1/5}$. It is straightforward to see that $\mathbb{E}\left[\deg_\pi^-(v_i)\right] = p(i-1)$. Let $\varepsilon = \varepsilon(n) = \sqrt{3}/\omega^{1/5} = o(1)$. We call vertex v_i *bad* if $\mathbb{E}\left[\deg_\pi^-(v_i)\right] - \deg_\pi^-(v_i) > \varepsilon\mathbb{E}\left[\deg_\pi^-(v_i)\right]$. Applying the Chernoff bound again, the probability that v_i is bad is at most

$$\exp\left(-\frac{\varepsilon^2 p(i-1)}{3}\right) \leq \exp\left(-\omega^{-2/5}p(n\omega^{-1/5})\right)$$
$$\leq \exp\left(-\omega^{2/5}\log n\right) = q.$$

Observe that the events $\{v_i$ is bad$\}$ are pairwise independent, so the probability that there are at least $n/\omega^{1/5}$ bad vertices is at most

$$
\begin{aligned}
2^n q^{n/\omega^{1/5}} &= \exp\left(O(n) - \omega^{1/5} n \log n\right) \\
&= \exp\left(-(1 + o(1))\omega^{1/5} n \log n\right) \\
&= o(1/n^n) = o(1/n!).
\end{aligned}
$$

Hence, by the union bound, a.a.s. for all possible cleaning sequences π there are at most $n/\omega^{1/5}$ bad vertices. It follows that a.a.s. the first summation in the definition of $D_\pi(G)$ can be bounded as follows. Let $\Gamma = \sum_{i=1}^{n} \max\left\{(i-1)p - \deg_\pi^-(v_i), 0\right\}$. Then we have that

$$
\begin{aligned}
\Gamma &= o(pn^2) + \sum_{i > n/\omega^{1/5}} \max\left\{(i-1)p - \deg_\pi^-(v_i), 0\right\} \\
&\leq o(pn^2) + n(\varepsilon d) + (n/\omega^{1/5})(np) \\
&= o(pn^2),
\end{aligned}
$$

and the proof follows. $\qquad\square$

Chapter 7

Acquaintance time

Our next graph process is different from previously considered models. The model we consider was recently introduced by Benjamini, Shinkar, and Tsur [20].

Let G be a finite connected graph. We start the process by placing one *agent* on each vertex of G. Every pair of agents sharing an edge is declared to be *acquainted*, and remains so throughout the process. In each round of the process, we choose some matching M in G. The matching M need not be maximal; for example, it may be a single edge. For each edge of M, we swap the agents occupying their endpoints, which may cause more agents to become acquainted. This process may be viewed as a graph searching game with one player, where the player's strategy consists of a sequence of matchings which allow all agents to become acquainted. Some strategies may be better than others, which leads to a graph optimization parameter. The *acquaintance time* of G, denoted by $\mathcal{AC}(G)$, is the smallest number of rounds required for all agents to become acquainted with one another. The problem of determining the acquaintance time is related to problems of routing permutations on graphs via matchings [3], gossiping and broadcasting [109], and target set selection [64, 119, 171].

It is evident that

$$\mathcal{AC}(G) \geq \frac{\binom{|V|}{2}}{|E|} - 1, \tag{7.1}$$

since $|E|$ pairs are acquainted initially, and at most $|E|$ new pairs become acquainted in each round. In [20], it was shown that always $\mathcal{AC}(G) = O\left(\frac{n^2}{\log n / \log \log n}\right)$, where $n = |V|$. Slight progress was made in [120], where it was proved that $\mathcal{AC}(G) = O(n^2 / \log n)$. This general upper bound was recently improved in [10] to $\mathcal{AC}(G) = O(n^{3/2})$, which was conjectured in [20] and is tight up to a multiplicative constant. Indeed, for all functions $f : \mathbb{N} \to \mathbb{N}$ with $1 \leq f(n) \leq n^{3/2}$, there are families $\{G_n : n \geq 3\}$ of graphs with $|V(G_n)| = n$ for all n such that $\mathcal{AC}(G_n) = \Theta(f_n)$.

Let $G \in \mathbb{G}(n, p)$ with $p = p(n) \geq (1 - \varepsilon) \log n / n$ for some $\varepsilon > 0$. Recall that $\mathcal{AC}(G)$ is defined only for connected graphs, and $\log n / n$ is the sharp threshold for connectivity in $\mathbb{G}(n, p)$—see, for example, [35, 115]. Hence, we will not be interested in sparser graphs. It follows immediately from the Chernoff bounds that a.a.s.

$$|E(G)| \sim \binom{n}{2} p.$$

Hence, from the trivial lower bound (7.1) we have that a.a.s. $\mathcal{AC}(G) = \Omega(1/p)$. Despite the fact that no nontrivial upper bound on $\mathcal{AC}(G)$ was known, it was conjectured in [20] that a.a.s.

$$\mathcal{AC}(G) = O(\operatorname{poly} \log(n)/p). \tag{7.2}$$

In [120], a major step toward this conjecture was made by showing that a.a.s. $\mathcal{AC}(G) = O(\log n/p)$, provided that G has a Hamiltonian cycle; that is, when $pn - \log n - \log \log n \to \infty$. The task was finished recently in [80].

It might also be of some interest to study the acquaintance time of other models of random graphs, such as power-law graphs, preferential attachments graphs, or geometric graphs. The latter case was recently considered in [150].

7.1   Dense Binomial Random Graphs

We prove the following result from [120] on dense binomial random graphs and the acquaintance time.

Theorem 7.1.1 ([120]). *Let $\varepsilon > 0$ and $(1 + \varepsilon) \log n / n \leq p = p(n) \leq 1 - \varepsilon$. For $G \in \mathbb{G}(n, p)$, a.a.s. we have that*

$$\mathcal{AC}(G) = O\left(\frac{\log n}{p}\right).$$

Note that whenever G_2 is a subgraph of G_1 on the same vertex set, $\mathcal{AC}(G_1) \leq \mathcal{AC}(G_2)$, since the agents in G_1 have more edges to use. Hence, for any $p \geq 0.99$ (possibly $p \to 1$) and $G_1 \in \mathbb{G}(n,p)$, we have that $\mathcal{AC}(G_1) \leq \mathcal{AC}(G_2)$, where $G_2 \in G(n, 0.99)$. Since a.a.s. $\mathcal{AC}(G_2) = O(\log n)$, we note that a.a.s. $\mathcal{AC}(G_1) = O(\log n)$, and so the condition $p < 1 - \varepsilon$ in the theorem can be eliminated. For denser graphs, this upper bound may not be tight; in particular, for the extreme case $p = 1$, we have that $\mathcal{AC}(G_2) = \mathcal{AC}(K_n) = 0$. Moreover, since the threshold for Hamiltonicity in $\mathbb{G}(n,p)$ is $p = (\log n + \log \log n)/n$ (see, for example, [35]), and for a Hamiltonian graph we have that $\mathcal{AC}(G) = O(n)$ (see Lemma 7.1.3 or [20]), it follows that a.a.s. $\mathcal{AC}(G) = O(n)$, provided that $pn - \log n - \log \log n \to \infty$. Therefore, the desired bound for the acquaintance time holds at the time a random graph becomes Hamiltonian. We obtain the following corollary.

Corollary 7.1.2 ([120]). *If $p = p(n)$ is such that $pn - \log n - \log \log n \to \infty$, then for $G \in \mathbb{G}(n,p)$, a.a.s.*

$$\mathcal{AC}(G) = O\left(\frac{\log n}{p}\right).$$

To prove Theorem 7.1.1 we will use the fact, observed in [20], that for any graph G on n vertices with a Hamiltonian path, we have that $\mathcal{AC}(G) = O(n)$. We need a stronger statement with a different argument. The proof is straightforward and the reader is directed to [120] for more details.

Lemma 7.1.3 ([120]). *Let G be a graph on n vertices. If G has a Hamiltonian path, then there exists a strategy ensuring that within $2n$ rounds every pair of agents gets acquainted (in particular, $\mathcal{AC}(G) = O(n)$) and, moreover, that every agent visits every vertex.*

We are now ready to prove Theorem 7.1.1.

Proof of Theorem 7.1.1. To avoid technical problems with events not being independent, we use a classic technique known as *two-round exposure*. The observation is that a random graph $G \in$

$\mathbb{G}(n,p)$ can be viewed as a union of two independently generated random graphs $G_1 \in \mathbb{G}(n,p_1)$ and $G_2 \in \mathbb{G}(n,p_2)$, provided that $p = p_1 + p_2 - p_1 p_2$ (see, for example, [35, 115] for more information).

Let $p_1 = (1 + \varepsilon/2) \log n/n$ and

$$p_2 = \frac{p - p_1}{1 - p_1} \geq p - p_1 \geq \frac{\varepsilon/2}{1 + \varepsilon} p$$

(recall that $p \geq (1 + \varepsilon) \log n/n$). Fix $G_1 \in \mathbb{G}(n,p_1)$ and $G_2 \in \mathbb{G}(n,p_2)$, with $V(G_1) = V(G_2) = \{v_1, v_2, \ldots, v_n\}$, and view G as the union of G_1 and G_2. It is known that a.a.s. G_1 has a Hamiltonian path (see [35, 115] for additional background). Hence, we may suppose that $P = (v_1, v_2, \ldots, v_n)$ is a Hamiltonian path of G_1 (and thus, also of G).

Now let $k = k(n) = 2.5 \log_{1/(1-p_2)} n$. We partition the path P into many paths, each on k vertices. This partition also divides the agents into $\lceil n/k \rceil$ teams, each team consisting of k agents (except for the "last" team, which may be smaller). Every team performs (independently and simultaneously) the strategy from Lemma 7.1.3. It follows that the length of the full process is at most $2k = 5 \log_{1/(1-p_2)} n$, which is asymptotic to

$$5\frac{\log n}{p_2} \leq 10\frac{(1+\varepsilon)}{\varepsilon}\frac{\log n}{p} = O\left(\frac{\log n}{p}\right)$$

provided that $p = o(1)$; if instead $p = \Omega(1)$, then the number of rounds needed is clearly $O(\log n)$. Note that every pair of agents from the same team gets acquainted.

It remains to show that a.a.s. every pair of agents from different teams gets acquainted. Let us focus on one such pair. It follows from Lemma 7.1.3 that each agent, except those in the "last" team, visits k distinct vertices. Since the agents belong to different teams, at least one belongs to a team of cardinality k, so the two agents occupy at least k distinct pairs of vertices during the process. Considering only those edges in G_2, the probability that the two agents never got acquainted is at most

$$(1 - p_2)^k = o(n^{-2}).$$

Since there are at most $\binom{n}{2}$ pairs of agents, the result holds by the union bound. $\qquad\square$

7.2 Sparse Binomial Random Graphs

In this section, we finish the task by showing that the conjecture (7.2) holds above the threshold for connectivity by considering sparse random graphs.

Theorem 7.2.1 ([80]). *If $p = p(n)$ is such that $pn - \log n \to \infty$, then a.a.s. for $G \in \mathbb{G}(n,p)$ we have that*

$$\mathcal{AC}(G) = O\left(\frac{\log n}{p}\right).$$

The results of [80] prove the conjecture in the strongest possible sense. In particular, we show that it holds right at the time the random graph process creates a connected graph. Before we state the result, let us introduce one more definition.

Recall the random graphs $\mathbb{G}(n, M)$ defined in Chapter 1, where graphs are sampled uniformly from those with n vertices and M edges. Note that in the missing window in which the conjecture is not proved, we have that $p \sim \log n / n$ and so an upper bound of $O(\log n / p)$ is equivalent to $O(n)$. Hence, it is enough to show the following result, which implies Theorem 7.2.1 as will be discussed soon.

Theorem 7.2.2 ([80]). *Let M be a random variable defined as follows:*

$$M = \min\{m : \mathbb{G}(n,m) \text{ is connected}\}.$$

Then for $G \in \mathbb{G}(n, M)$, a.a.s.

$$\mathcal{AC}(G) = O(n).$$

We will need the following lemma proved in [80]. Note that this lemma is also proved for hypergraphs, in a more general setting (see Lemma 7.4.2).

Lemma 7.2.3 ([80]). *For fixed real number $0 < \delta < 1$, there is a positive constant $c = c(\delta)$ such that for $G \in \mathbb{G}(n, c/n)$ we have that*

$$\mathbb{P}(G \text{ has a path of length at least } \delta n) \geq 1 - \exp(-n).$$

Let us start with the following two comments. As we already pointed out, $\mathcal{AC}(G_1) \le \mathcal{AC}(G_2)$ whenever G_2 is a subgraph of G_1 on the same vertex set. Hence, Theorem 7.2.2 implies that for any $m \ge M$ we also have that a.a.s. $\mathcal{AC}(G) = O(n)$, $G \in \mathbb{G}(n, m)$.

Second of all, it is known that the two models ($\mathbb{G}(n, p)$ and $\mathbb{G}(n, m)$) are in many cases asymptotically equivalent, provided $\binom{n}{2}p$ is close to m. For example, Proposition 1.12 in [115] gives us the way to translate results from $\mathbb{G}(n, m)$ to $\mathbb{G}(n, p)$.

Lemma 7.2.4 ([115]). *Let P be an arbitrary property, $p = p(n)$, and $c \in [0, 1]$. If for every sequence $m = m(n)$ such that*

$$m = \binom{n}{2}p + O\left(\sqrt{\binom{n}{2}p(1 - p)}\right)$$

it holds that

$$\mathbb{P}(\mathbb{G}(n, m) \in P) \to c \quad \text{as } n \to \infty,$$

then also

$$\mathbb{P}(\mathbb{G}(n, p) \in P) \to c \quad \text{as } n \to \infty.$$

Using Lemma 7.2.4, Theorem 7.2.2 implies immediately Theorem 7.2.1. Indeed, suppose that $p = p(n) = (\log n + \omega)/n$, where $\omega = \omega(n)$ tends to infinity together with n. One needs to investigate $G \in \mathbb{G}(n, m)$ for $m = \frac{n}{2}(\log n + \omega + o(1))$. It is known that in this range of m, a.a.s. G is connected (that is, a.a.s. $M < m$). Theorem 7.2.2 together with the first observation imply that a.a.s. $\mathcal{AC}(G) = O(n)$, and Theorem 7.2.1 follows from Lemma 7.2.4.

To prove Theorem 7.2.2, we will show that at the time when $G \in \mathbb{G}(n, m)$ becomes connected (that is, at time M), a.a.s. it contains a certain spanning tree T with $\mathcal{AC}(T) = O(n)$. For that, we need to introduce the following useful family of trees. A tree T is *good* if it consists of a path $P = (v_1, v_2, \ldots, v_k)$ (called the *spine*), and some vertices (called *heavy*) that form a set $\{u_i : i \in I \subseteq [k]\}$ that are connected to the spine by a perfect matching (that is, for every $i \in I$, u_i is adjacent to v_i). All other vertices (called *light*)

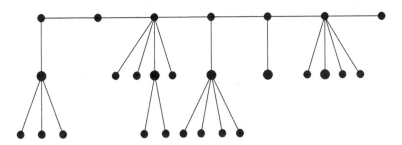

FIGURE 7.1: An example of a good tree.

are adjacent either to v_i for some $i \in [k]$ or to u_i for some $i \in I$ (see Figure 7.1 for an example).

We will use the following result from [20] (Claim 2.1 in that paper).

Lemma 7.2.5 ([20]). *Let $G = (V, E)$ be a tree. Let $S, T \subseteq V$ be two subsets of the vertices of equal cardinality $k = |S| = |T|$, and let $\ell = \max_{v \in S, u \in T} d(v, u)$ be the maximum distance between a vertex in S and a vertex in T. Then there is a strategy of $\ell + 2(k-1)$ matchings that routes all agents from S to T.*

This family of trees is called good for a reason, as shown by the following observation. The proof is omitted here.

Lemma 7.2.6 ([80]). *If T is a good tree on n vertices, then $\mathcal{AC}(T) = O(n)$. In addition, every agent visits every vertex of the spine.*

As we already mentioned, our goal is to show that at the time when $\mathbb{G}(n, m)$ becomes connected, a.a.s. it contains a good spanning tree T. However, it is more straightforward to work with the $\mathbb{G}(n, p)$ model instead of $G(n, m)$. Lemma 7.2.4 provides us with a tool to translate results from $\mathbb{G}(n, m)$ to $\mathbb{G}(n, p)$. The following lemma works the other way round, see, for example, (1.6) in [115].

Lemma 7.2.7 ([115]). *Let P be an arbitrary property, let $m =$*

$m(n)$ be any function such that $m \le n \log n$, and take $p = p(n) = m/\binom{n}{2}$. We then have that

$$\mathbb{P}(\mathbb{G}(n, m) \in P) \le 3\sqrt{n \log n} \cdot \mathbb{P}(\mathbb{G}(n, p) \in P).$$

We will also need the following lemma.

Lemma 7.2.8 ([80]). *If $G \in G(n, \log \log \log n/n)$, then for every set A of cardinality $0.99n$ the following holds. For a set B of cardinality $O(n^{0.03})$ taken uniformly at random from $V \setminus A$, B induces a graph with no edge with probability $1 - o((n \log n)^{-1/2})$.*

Proof. Fix a vertex $v \in V$. The expected degree of vertex v is $(1 + o(1)) \log \log \log n$. It follows from Bernstein's bound (see Theorem 1.6.1), applied with $t = x\mu$ and $x = c \log n / \log \log \log n$ for c large enough, that with probability $o(n^{-2})$, v has degree larger than, say, $2c \log n$. Thus, the union bound over all vertices of G implies that with probability $1 - o((n \log n)^{-1/2})$ all vertices have degrees at most $2c \log n$. Since we aim for such a probability, we may assume that G is a deterministic graph that has this property.

Now, fix any set A of cardinality $0.99n$. Regardless of our choice of A, every vertex of $V \setminus A$ has at most $2c \log n$ neighbors in $V \setminus A$. Take a random set B from $V \setminus A$ of cardinality $O(n^{0.03})$ and fix a vertex $v \in B$. The probability that no neighbor of v is in B is at least

$$\frac{\binom{0.01n - 1 - 2c \log n}{|B| - 1}}{\binom{0.01n - 1}{|B| - 1}} = 1 - O\left(\frac{n^{0.03} \log n}{n}\right) = 1 - O\left(n^{-0.97} \log n\right).$$

Hence, the probability that some neighbor of v is in B is $O\left(n^{-0.97} \log n\right)$. Consequently, the union bound over all vertices of B implies that with probability $O\left(n^{-0.94} \log n\right)$ there is at least one edge in the graph induced by B. The proof of the lemma follows. $\square$

Finally, we are ready to prove the main result of this section.

Proof of Theorem 7.2.2. Let $M_- = \frac{n}{2}(\log n - \log \log n)$ and let $M_+ = \frac{n}{2}(\log n + \log \log n)$; recall that a.a.s. $M_- < M < M_+$.

First, we will show that a.a.s. $\mathbb{G}(n, M_-)$ consists of a good tree T and a small set S of isolated vertices. Next, we will show that between time M_- and M_+ a.a.s. no edge is added between S and light vertices of T. This will finish the proof, since $\mathbb{G}(n, M)$ is connected and so at that point of the process, a.a.s. vertices of S must be adjacent to the spine or heavy vertices of T, which implies that a.a.s. there is a good spanning tree. The result will follow then from Lemma 7.2.6.

As promised, let $p_- = M_-/\binom{n}{2} = (\log n - \log\log n + o(1))/n$ and consider $\mathbb{G}(n, p_-)$ instead of $\mathbb{G}(n, M_-)$. To avoid issues with events not being independent, we again use a classic technique known as *two-round exposure*. Recall that the observation is that a random graph $G \in \mathbb{G}(n, p)$ can be viewed as a union of two independently generated random graphs $G_1 \in \mathbb{G}(n, p_1)$ and $G_2 \in \mathbb{G}(n, p_2)$, provided that $p = p_1 + p_2 - p_1 p_2$.

Let $p_1 = \log\log\log n/n$ and

$$p_2 = \frac{p_- - p_1}{1 - p_1} = \frac{\log n - \log\log n - \log\log\log n + o(1)}{n}.$$

Fix $G_1 \in \mathbb{G}(n, p_1)$ and $G_2 \in \mathbb{G}(n, p_2)$, with $V(G_1) = V(G_2) = V$, and view $G \in \mathbb{G}(n, p_-)$ as the union of G_1 and G_2. It follows from Lemma 7.2.3 that with probability $1 - o((n\log n)^{-1/2})$, G_1 has a long path $P = (v_1, v_2, \ldots, v_k)$ of length $k = 0.99n$ (and thus, G has it also). This path will eventually become the spine of the spanning tree.

Now, we expose edges of G_2 in a very specific order dictated by the following algorithm. Initially, A consists of vertices of the path (that is, $A = \{v_1, v_2, \ldots, v_k\}$), $B = \emptyset$, and $C = V \setminus A$. At each step of the process, we take a vertex v from C and expose edges from v to A. If an edge from v to some v_i is found, we change the label of v to u_i, call it *heavy*, remove v_i from A, and move u_i from C to B. Otherwise (that is, if there is no edge from v to A), we expose edges from v to B. If an edge from v to some heavy vertex u_i is found, we call v *light*, and remove it from C. At the end of this process we are left with a small (with the desired probability, as we will see soon) set C and a good tree T_1. Let $X = |C|$ be the random variable counting vertices not attached to the tree yet.

Since at each step of the process $|A| + |B|$ is precisely $0.99n$, we derive that

$$
\begin{aligned}
\mathbb{E}[X] &= 0.01n\big(1 - p_2\big)^{0.99n} \\
&\sim 0.01 \exp\big(\log n - 0.99(\log n - \log\log n \\
&\quad - \log\log\log n)\big) \\
&= 0.01n^{0.01}\big((\log n)(\log\log n)\big)^{0.99}.
\end{aligned}
$$

Note that the random variable X is a binomial random variable $\mathrm{Bin}(0.01n, (1-p_2)^{0.99n})$. Hence, it follows from the Chernoff bounds that $X \sim \mathbb{E}[X]$ with probability $1 - o((n\log n)^{-1/2})$. Now, let Y be the random variable counting how many vertices not on the path are not heavy. Arguing as before, since at each step of the process $|A| \geq 0.98n$, we have that

$$
\mathbb{E}[Y] \leq 0.01n\big(1 - p_2\big)^{0.98n} \sim 0.01n^{0.02}\big((\log n)(\log\log n)\big)^{0.98}.
$$

Observe that $Y \geq X$ and so $\mathbb{E}[Y]$ is large enough for the Chernoff bounds to be applied again to show that $Y \sim \mathbb{E}[Y]$ with probability $1 - o((n\log n)^{-1/2})$.

Our next goal is to attach almost all vertices of C to T_1 in order to form a larger good tree T_2. Consider a vertex $v \in C$ and a heavy vertex u_i of T_1. An obvious, yet important, property is that when edges emanating from v were exposed in the previous phase, exactly one vertex from v_i, u_i was considered (recall that when u_i is discovered as a heavy vertex, v_i is removed from A). Hence, we may expose these edges and try to attach v to the tree. From the previous argument, since we aim for a statement that holds with probability $1 - o((n\log n)^{-1/2})$, we may assume that the number of heavy vertices is at least $0.01n - n^{0.03}$. For the random variable Z counting vertices still not attached to the path, we derive that

$$
\begin{aligned}
\mathbb{E}[Z] &\sim \mathbb{E}[X]\big(1 - p_2\big)^{0.01n - O(n^{0.03})} \\
&\sim 0.01n^{0.01}\big((\log n)(\log\log n)\big)^{0.99} \\
&\quad \cdot \exp\big(-0.01(\log n - \log\log n - \log\log\log n)\big) \\
&\sim 0.01\big((\log n)(\log\log n)\big),
\end{aligned}
$$

and so $Z \sim \mathbb{E}[Z]$ with probability $1 - o((n \log n)^{-1/2})$ by the Chernoff bounds.

Let us stop for a second and summarize the current situation. We showed that with the desired probability, we have the following structure. There is a good tree T_2 consisting of all but at most $(\log n)(\log \log n)$ vertices that form a set S. T_2 consists of the spine of length $0.99n$, $0.01n - O(n^{0.03})$ heavy vertices and $O(n^{0.03})$ light vertices. Edges between S and the spine and between S and heavy vertices are already exposed and no edge was found. However, edges within S and between S and light vertices are not exposed yet. However, it is straightforward to see that with the desired probability there is no edge there either, and so vertices of S are isolated in G_2. Indeed, the probability that there is no edge in the graph induced by S and light vertices is equal to

$$\left(1 - p_2\right)^{O(n^{0.06})} = \exp\left(- O(n^{0.06} \log n / n)\right) = 1 - o((n \log n)^{-1/2}).$$

Finally, we need to argue that vertices of S are also isolated in G_1. The important observation is that the algorithm we performed that exposed edges in G_2 used only the fact that vertices of $\{v_1, v_2, \ldots, v_{0.99n}\}$ form a long path; no other information about G_1 was ever used. Hence, set S together with light vertices is, from the perspective of the graph G_1, simply a random set of cardinality $O(n^{0.03})$ taken from the set of vertices not on the path. Lemma 7.2.8 implies that with probability $1 - o((n \log n)^{-1/2})$ there is no edge in the graph induced by this set. With the desired property, $\mathbb{G}(n, p)$ consists of a good tree T and a small set S of isolated vertices and so, by Lemma 7.2.7, a.a.s. it is also true in $\mathbb{G}(n, M_-)$.

It remains to show that between time M_- and M_+ a.a.s. no edge is added between S and light vertices of T. Direct computations for the random graph process show that this event holds with probability

$$\binom{\binom{n}{2} - M_- - O(n^{0.06})}{M_+ - M_-} \Big/ \binom{\binom{n}{2} - M_-}{M_+ - M_-}$$

$$= \prod_{i=0}^{M_+ - M_- - 1} \frac{\binom{n}{2} - M_- - O(n^{0.06}) - i}{\binom{n}{2} - M_- - i}$$

$$= \left(1 - \frac{O(n^{0.06})}{n^2}\right)^{M_+ - M_-}$$

$$= 1 - \frac{O(n^{0.06} \log \log n)}{n} \to 1$$

as $n \to \infty$, since $M_+ - M_- = O(n \log \log n)$, and the proof follows.
$\square$

7.3   Is the Upper Bound Tight?

In hopes of improving the trivial lower bound on the acquaintance time of $\mathbb{G}(n, p)$, we consider a variant of the original process. Suppose that each agent has a helicopter and can, on each round, move to any vertex. We retain the requirement that no two agents can occupy a single vertex simultaneously. Hence, in every step of the process, the agents choose some permutation π of the vertices, and the agent occupying vertex v flies directly to vertex $\pi(v)$, regardless of whether there is an edge or even a path between v and $\pi(v)$. Let $\overline{\mathcal{AC}}(G)$ be the counterpart of $\mathcal{AC}(G)$ under this new model; that is, the smallest number of rounds required for all agents to become acquainted with one another. Since helicopters make it easier for agents to become acquainted, we immediately obtain that for every graph G,

$$\overline{\mathcal{AC}}(G) \le \mathcal{AC}(G). \tag{7.3}$$

Note that $\overline{\mathcal{AC}}(G)$ also represents the smallest number of copies of a graph G needed to cover all edges of a complete graph of the same order. Thus, inequality (7.1) can be strengthened to $\overline{\mathcal{AC}}(G) \geq \binom{|V|}{2}/|E| - 1$.

We prove the following lower bound on $\overline{\mathcal{AC}}(G)$ (and hence, on $\mathcal{AC}(G)$). This result also implies that a.a.s. K_n cannot be covered with $o(\log n/p)$ copies of a dense random graph $G \in \mathbb{G}(n,p)$.

Theorem 7.3.1 ([120]). *Let $\varepsilon > 0$, $p = p(n) \geq n^{-1/2+\varepsilon}$ and $p \leq 1 - \varepsilon$. For $G \in \mathbb{G}(n,p)$, a.a.s.*

$$\mathcal{AC}(G) \geq \overline{\mathcal{AC}}(G) \geq \frac{\varepsilon}{2} \log_{1/(1-p)} n = \Omega\left(\frac{\log n}{p}\right).$$

Theorem 7.1.1 and Theorem 7.3.1 together determine the order of growth for the acquaintance time of dense random graphs (in particular, random graphs with average degree at least $n^{1/2+\varepsilon}$ for some $\varepsilon > 0$).

Corollary 7.3.2. *Let $\varepsilon > 0$, $p = p(n) \geq n^{-1/2+\varepsilon}$ and $p \leq 1 - \varepsilon$. For $G \in \mathbb{G}(n,p)$, a.a.s.*

$$\overline{\mathcal{AC}}(G) = \Theta\left(\mathcal{AC}(G)\right) = \Theta\left(\frac{\log n}{p}\right).$$

The behaviors of $\mathcal{AC}(G)$ and $\overline{\mathcal{AC}}(G)$ for sparser random graphs remain undetermined.

The first inequality in the statement of Theorem 7.3.1 is (7.3). It remains to show the desired lower bound for $\overline{\mathcal{AC}}(G)$.

Proof of Theorem 7.3.1. Let $a_1, a_2, \ldots, a_n$ denote the n agents, and let $A = \{a_1, a_2, \ldots, a_n\}$. Take $k = \frac{\varepsilon}{2} \log_{1/(1-p)} n$ and fix k bijections $\pi_i : A \to V(G)$, for $i \in \{0, 1, \ldots k - 1\}$. This corresponds to fixing a $(k-1)$-round strategy for the agents; in particular, agent a_j occupies vertex $\pi_i(a_j)$ in round i. We aim to show that at the end of the process (that is, after $k-1$ rounds) the probability that all agents are acquainted is only $o((1/n!)^k)$. This completes the proof: the number of choices for $\pi_0, \pi_1, \ldots, \pi_{k-1}$ is $(n!)^k$, so by the union bound, a.a.s. no strategy makes all pairs of agents acquainted.

To estimate the probability in question, we consider the following analysis, which iteratively exposes edges of a random graph $G \in \mathbb{G}(n,p)$. For any pair $r = \{a_x, a_y\}$ of agents, we consider all pairs of vertices visited by this pair of agents throughout the process:

$$S(r) = \{\{\pi_i(a_x), \pi_i(a_y)\} : i \in \{0, 1, \ldots k-1\}\}.$$

Note that $1 \le |S(r)| \le k$. Take any pair r_1 of agents and expose the edges of G in $S(r_1)$, one by one. If we expose all of $S(r_1)$ without discovering an edge, then we discard r_1 and proceed. (In fact we have just learned that the pair r_1 never gets acquainted, so we could choose to halt the process immediately. However, to simplify the analysis, we continue normally.) If instead we do discover some edge e of G, then we discard all pairs of agents that ever occupy this edge (that is, we discard all pairs r such that $e \in S(r)$). In either case, we shift our attention to another pair r_2 of agents (chosen arbitrarily from among the pairs not yet discarded). It may happen that some of the pairs of vertices in $S(r_2)$ have already been exposed, but the analysis guarantees that no edge has yet been discovered. Let $T(r_2) \subseteq S(r_2)$ be the set of edges in $S(r_2)$ not yet exposed. As before, we expose these edges one by one, until either we discover an edge or we run out of edges to expose. If an edge is discovered, then we again discard all pairs that ever occupy that edge.

We continue this process until all available pairs of agents have been investigated. Since one pair of agents can force us to discard at most k pairs (including the original pair), the process investigates at least $\binom{n}{2}/k$ pairs of agents. Among these pairs, the probability $\mathbb{P}(r_t)$ that pair r_t gets acquainted is

$$\mathbb{P}(r_t) = 1 - (1-p)^{|T(r_t)|} \le 1 - (1-p)^{|S(r_t)|} \le 1 - (1-p)^k.$$

Hence, the probability that all pairs become acquainted is at most

$$\prod_{t=1}^{\binom{n}{2}/k} \mathbb{P}(r_t) \leq \left(1 - (1-p)^k\right)^{\binom{n}{2}/k} \leq \exp\left(-(1-p)^k \binom{n}{2}/k\right)$$

$$\leq \exp\left(-n^{-\varepsilon/2} \binom{n}{2} 3n^{-1/2+\varepsilon/2}\right) \leq \exp\left(-n^{3/2}\right)$$

$$\leq \exp\left(-n^{1+\varepsilon/2}k\right) = o\left(\exp\left(-kn\log n\right)\right)$$

$$= o\left((1/n!)^k\right),$$

since $k = \Theta(\log n/p) \leq n^{1/2-\varepsilon/2}/3$. As mentioned earlier, it follows that a.a.s. $\overline{\mathcal{AC}}(G) \geq k$, and the proof follows. $\quad\square$

7.4 Hypergraphs

In this section, we consider the acquaintance time of random r-uniform hypergraphs. A *hypergraph* H is an ordered pair $H = (V, E)$, where V is a finite set (the *vertex set*) and E is a family of disjoint subsets of V (the *edge set*). A hypergraph $H = (V, E)$ is *r-uniform* if all edges of H are of cardinality r.

For a given $r \in \mathbb{N} \setminus \{1\}$, the *random r-uniform hypergraph* $H_r(n, p)$ has n labeled vertices from a vertex set $V = [n]$, in which every subset $e \subseteq V$ of cardinality $|e| = r$ is chosen to be an edge of H randomly and independently with probability p. For $r = 2$, this model reduces to the model $\mathbb{G}(n, p)$ of binomial random graphs.

Let us come back to the acquaintance time now. Fix $r \in \mathbb{N} \setminus \{1\}$. For an r-uniform hypergraph $H = (V, E)$, there are a number of ways we can generalize the problem. At each step of the process, we continue choosing some matching M of agents and swapping the agents occupying their endpoints; two agents can be matched if there exists $e \in E$ such that both agents occupy vertices of e. However, this time we have more flexibility in the definition of being acquainted. We can fix $k \in \mathbb{N}$ such that $2 \leq k \leq r$, and every k-tuple of agents sharing an edge is declared to be *k-acquainted*, and remains so throughout the process. The *k-acquaintance time* of H, denoted by $\mathcal{AC}_r^k(H)$, is the smallest number of rounds required for all k-tuples of agents to become acquainted. For any graph G, we have that $\mathcal{AC}_2^2(G) = \mathcal{AC}(G)$.

We show the following result for random r-uniform hypergraphs. Note that $p = \frac{(r-1)!\log n}{n^{r-1}}$ is the sharp threshold for connectivity so the assumption for the function p in the statement of the result is rather mild but could be weakened.

Theorem 7.4.1 ([80]). *Let* $r \in \mathbb{N} \setminus \{1\}$, *let* $k \in \mathbb{N}$ *be such that* $2 \leq k \leq r$, *and let* $\varepsilon > 0$ *be any real number. Suppose that* $p = p(n)$ *is such that* $p \geq (1+\varepsilon)\frac{(r-1)!\log n}{n^{r-1}}$. *For* $H \in H_r(n,p)$, *a.a.s. we have that*

$$\mathcal{AC}_r^k(H) = \Omega\left(\max\left\{\frac{1}{pn^{r-k}}, 1\right\}\right) \quad and$$

$$\mathcal{AC}_r^k(H) = O\left(\max\left\{\frac{\log n}{pn^{r-k}}, 1\right\}\right).$$

We will use the following lemma.

Lemma 7.4.2 ([80]). *For fixed* $r \in \mathbb{N} \setminus \{1\}$ *and real number* $0 < \delta < 1$, *there is a positive constant* $c = c(r, \delta)$ *such that for* $H \in H_r(n,p)$ *with* $p = c/n^{r-1}$

$$\mathbb{P}\left(H \text{ has a loose path of length at least } \delta n\right) \geq 1 - \exp(-n).$$

Proof. We will apply a depth-first search algorithm in order to explore vertices of $H = (V, E) \in H_r(n,p)$ extending some ideas from [125]. Set

$$c = c(r, \delta) = \frac{2(r-1)!\log 4}{\frac{1-\delta}{2(r-1)}\left(\frac{1-\delta}{2}\right)^{r-1}}.$$

(Let us note that we make no effort to optimize c here.) In every step of the process, V is partitioned into the following four sets:

(i) P: ordered set of vertices representing a loose path $(v_1, v_2, \ldots, v_\ell)$ of length ℓ,

(ii) U: set of unexplored vertices that have not been reached by the search yet,

(iii) W and $\tilde{W}$: sets of explored vertices that no longer play a role in the process.

We start the process by assigning to P an arbitrary vertex v from V and setting $U = V \setminus \{v\}$, $W = \tilde{W} = \emptyset$. (Observe that it is vacuously true that $L = (\{v\}, \emptyset)$, a hypergraph consisting of one vertex and no edge, is a loose path.) Suppose that at some point of the process we found a path $P = (v_1, v_2, \ldots, v_\ell)$ of length ℓ. If there exists a set $f = \{u_1, u_2, \ldots, u_{r-1}\} \subseteq U$ of cardinality $r - 1$ such that $f \cup \{v_\ell\} \in E$, then we extend the path by adding the edge $\{v_\ell, u_1, u_2, \ldots, u_{r-1}\}$ to the path; vertices of f are removed from U. (After this operation, $P = (v_1, v_2, \ldots, v_\ell, u_1, u_2, \ldots, u_{r-1})$ has length $\ell + (r - 1)$.) Otherwise, that is, if no such set f can be found in U, then v_ℓ is moved from P to W and, depending on the value of ℓ, we have two cases. If $\ell \neq 1$ (which implies that $\ell \geq r$), then vertices $v_{\ell-1}, v_{\ell-2}, \ldots, v_{\ell-(r-2)}$ are moved from P to $\tilde{W}$. (After this operation, $P = (v_1, v_2, \ldots, v_{\ell-(r-1)})$ has length $\ell - (r - 1)$.) If $\ell = 1$, then we simply take an arbitrary vertex from U and restart the process by taking $P = (v)$ of length 1.

Now, let us assume that H has no loose path of length at least δn. Therefore, in every step of the algorithm $\ell < \delta n$. The process ends with $U = \emptyset$ and $|W| + |\tilde{W}| = |V| - |P| > (1 - \delta)n$. Hence, since $(r - 2)|W| \geq |\tilde{W}|$, we have that $|W| > \frac{1-\delta}{r-1}n$. Observe that in each step of the process either the cardinality of W increases by one, or the cardinality of U decreases by at most $r - 1$ (actually, the cardinality of U decreases either by 1 or $r - 1$). Thus, at some stage we must have that $|W| = \frac{1-\delta}{2(r-1)}n$ and

$$\begin{aligned} |U| &= |V| - |P| - |W| - |\tilde{W}| \\ &> n - \delta n - |W| - (r - 2)|W| = \frac{1 - \delta}{2}n. \end{aligned}$$

Furthermore, note that there is no edge e between U and W such that $|U \cap e| = r - 1$ and $|W \cap e| = 1$.

For fixed sets U and W satisfying $|U| = \frac{1-\delta}{2}n$ and $|W| =$

$\frac{1-\delta}{2(r-1)}n$, let $e(U,W)$ be the number of edges e between U and W in H for which $|U \cap e| = r - 1$ and $|W \cap e| = 1$. Then we have that

$$\mathbb{P}(e(U,W) = 0) = (1-p)^{|W|\binom{|U|}{r-1}}$$
$$\leq \exp\left(-(1+o(1))\frac{c}{n^{r-1}} \cdot \frac{1-\delta}{2(r-1)}n \cdot \left(\frac{1-\delta}{2}n\right)^{r-1}/(r-1)!\right)$$
$$= \exp(-(1+o(1))2n\log 4),$$

and consequently by taking the union over all disjoint subsets $U, W \subseteq V$ satisfying $|U| = \frac{1-\delta}{2}n$ and $|W| = \frac{1-\delta}{2(r-1)}n$, we obtain

$$\mathbb{P}\left(H \text{ has no loose path of length at least } \delta n\right)$$

$$\leq \mathbb{P}\left(\bigcup_{U,W} e(U,W) = 0\right)$$
$$\leq 2^n 2^n \exp(-(1+o(1))2n\log 4)$$
$$= \exp(-(1+o(1))n\log 4) \leq \exp(-n),$$

as required. $\qquad\qquad\qquad\qquad\qquad\qquad\qquad\qquad\qquad\qquad\qquad\square$

To prove Theorem 7.4.1 we will also need the following result about Hamiltonian paths in $H_r(n,p)$, which follows from a series of papers [93, 77, 78] devoted to loose Hamiltonicity of random hypergraphs. Observe that if an r-uniform hypergraph of order n contains a loose Hamiltonian path (that is, a loose path that covers each vertex of H), then necessarily $r - 1$ must divide $n - 1$.

Lemma 7.4.3 ([93, 77, 78]). *For a fixed integer $r \geq 3$, let $p = p(n)$ be such that $pn^{r-1}/\log n$ tends to infinity together with n and $H \in H_r(n,p)$. Then a.a.s. H has a loose Hamiltonian path provided that $r - 1$ divides $n - 1$.*

Now, we come back to the acquaintance time. First, we mention that the trivial bound for graphs (7.1) can easily be generalized to r-uniform hypergraphs and any $2 \leq k \leq r$:

$$\mathcal{AC}_r^k(H) \geq \frac{\binom{|V|}{k}}{|E|\binom{r}{k}} - 1.$$

For $H = (V, E) \in H_r(n, p)$ with $p > (1 + \varepsilon)\frac{(r-1)! \log n}{n^{r-1}}$ for some $\varepsilon > 0$, we derive immediately that a.a.s.

$$\mathcal{AC}_r^k(H) = \Omega\left(\frac{1}{pn^{r-k}}\right),$$

since the expected number of edges in H is $\binom{n}{r}p = \Omega(n \log n)$ and so the concentration follows from the Chernoff bounds. Hence, the lower bound in Theorem 7.4.1 holds.

Now we prove the upper bound. For this, we will split Theorem 7.4.1 into two parts and then, independently, prove each of them.

Theorem 7.4.1a. *Let* $r \in \mathbb{N} \setminus \{1\}$, *let* $k \in \mathbb{N}$ *be such that* $2 \leq k \leq r$, *and let* $\varepsilon > 0$ *be any real number. Suppose that* $p = p(n) = (1 + \varepsilon)\frac{(r-1)! \log n}{n^{r-1}}$. *For* $H \in H_r(n, p)$, *a.a.s.*

$$\mathcal{AC}_r^k(H) = O(n^{k-1}) = O\left(\frac{\log n}{pn^{r-k}}\right).$$

Theorem 7.4.1b. *Let* $r \in \mathbb{N} \setminus \{1\}$, *let* $k \in \mathbb{N}$ *be such that* $2 \leq k \leq r$, *and let* $\omega = \omega(n)$ *be any function tending to infinity together with* n. *Suppose that* $p = p(n) = \omega\frac{\log n}{n^{r-1}}$. *For* $H \in H_r(n, p)$, *a.a.s.*

$$\mathcal{AC}_r^k(H) = O\left(\max\left\{\frac{\log n}{pn^{r-k}}, 1\right\}\right).$$

Note that when, say, $p \geq 2k(r - k)!\frac{\log n}{n^{r-k}}$, the expected number of k-tuples that do not become acquainted initially (that is, those that are not contained in any hyperedge) is equal to

$$\binom{n}{k}(1 - p)^{\binom{n-k}{r-k}} \leq \exp\left(k \log n - (1 + o(1))2k(r - k)!\right.$$
$$\left. \cdot \frac{\log n}{n^{r-k}} \cdot \frac{n^{r-k}}{(r - k)!}\right)$$
$$= o(1),$$

and so, by Markov's inequality, a.a.s. every k-tuple gets acquainted immediately and $\mathcal{AC}_r^k(H) = 0$. This is also the reason for taking the largest of the two values in the statement of the result.

For a given hypergraph $H = (V, E)$, sometimes it will be convenient to think about its *underlying graph* $\tilde{H} = (V, \tilde{E})$ with $\{u, v\} \in \tilde{E}$ if and only if there exists $e \in E$ such that $\{u, v\} \subseteq e$. Two agents are matched in $\tilde{H}$ if and only if they are matched in H. Therefore, we may always assume that agents walk on the underlying graph of H.

Let $H = (V, E)$ be an r-uniform hypergraph. A *1-factor* of H is a set $F \subseteq E$ such that every vertex of V belongs to exactly one edge in F. A *1-factorization* of H is a partition of E into 1-factors; that is, $E = F_1 \cup F_2 \cup \cdots \cup F_\ell$, where each F_i is a 1-factor and $F_i \cap F_j = \emptyset$ for all $1 \leq i < j \leq \ell$. Denote by K_n^r the *complete r-uniform hypergraph of order* n (that is, each r-tuple is present). We will use the following well-known result of Baranyai [19].

Lemma 7.4.4 ([19]). *If r divides n, then K_n^r is 1-factorizable.*

The result on factorization is helpful to deal with loose paths. The proof of the next lemma is omitted.

Lemma 7.4.5 ([80]). *Let $r \in \mathbb{N} \setminus \{1\}$ and let $k \in \mathbb{N}$ be such that $2 \leq k \leq r$. For a loose r-uniform path P_n on n vertices,*

$$\mathcal{AC}_r^k(P_n) = O(n^{k-1}).$$

Further, every agent visits every vertex of the path.

It remains to prove the Theorems 7.4.1a and Theorems 7.4.1b.

Proof of Theorem 7.4.1a. As in the proof of Theorem 7.2.2, we use the two-round exposure technique: $H \in H_r(n, p)$ can be viewed as the union of two independently generated random hypergraphs $H_1 \in H_r(n, p_1)$ and $H_2 \in H_r(n, p_2)$, with $p_1 = \frac{\varepsilon}{2} \cdot \frac{(r-1)! \log n}{n^{r-1}}$ and

$$p_2 = \frac{p - p_1}{1 - p_1} \geq p - p_1 = \left(1 + \frac{\varepsilon}{2}\right) \frac{(r-1)! \log n}{n^{r-1}}.$$

It follows from Lemma 7.2.3 that a.a.s. H_1 has a long loose path

P of length δn (and therefore, H has it, also), where $\delta \in (0,1)$ can be made arbitrarily close to 1, as needed.

Now, we will use H_2 to show that a.a.s. every vertex not on the path P belongs to at least one edge of H_2 (and so of H also) that intersects with P. Indeed, for a given vertex v not on the path P, the number of r-tuples containing v and at least one vertex from P is equal to

$$
\begin{aligned}
s &= \binom{n-1}{r-1} - \binom{(1-\delta)n-1}{r-1} \\
&\sim \left(\frac{n^{r-1}}{(r-1)!} - \frac{((1-\delta)n)^{r-1}}{(r-1)!} \right) \\
&= \frac{n^{r-1}}{(r-1)!} \left(1 - (1-\delta)^{r-1} \right),
\end{aligned}
$$

and so the probability that none of them occurs as an edge in H_2 is

$$
\begin{aligned}
(1-p_2)^s &\leq \exp\left(-(1+o(1))\left(1+\frac{\varepsilon}{2}\right) \frac{(r-1)!\log n}{n^{r-1}} \right. \\
&\quad \left. \cdot \frac{n^{r-1}}{(r-1)!} \left(1-(1-\delta)^{r-1}\right) \right) \\
&= o(n^{-1}),
\end{aligned}
$$

after taking δ close enough to 1. The claim holds by the union bound.

The rest of the proof is straightforward. We partition all agents into $(k+1)$ groups, $A_1, A_2, \ldots, A_{k+1}$, each of them consisting of at least $\lfloor n/(k+1) \rfloor$ agents. Since δ can be arbitrarily close to 1, we may assume that $\lceil n/(k+1) \rceil > (1-\delta)n$. For every $i \in [k+1]$, we route *all* agents from all groups but A_i and some agents from A_i to P, and after $O(n^{k-1})$ rounds they get acquainted by Lemma 7.4.5. Since each k-tuple of agents intersects at most k groups, it must become acquainted eventually, and the proof is complete. □

Proof of Theorem 7.4.1b. First, we assume that $\omega = \omega(n) \leq n^{k-1}$ and $r-1$ divides $n-1$. We consider two independently generated random hypergraphs $H_1 \in H_r(n,p_1)$ and $H_2 \in H_r(n,p_2)$, with

$p_1 = \frac{\omega}{2} \cdot \frac{\log n}{n^{r-1}}$ and $p_2 > p - p_1 = p_1$. It follows from Lemma 7.4.3 that a.a.s. H_1 has a Hamiltonian path $P = (v_1, v_2, \ldots, v_n)$.

Let C be a large constant that will be determined soon. We cut P into $(\omega/C)^{1/(k-1)}$ loose sub-paths, each of length $\ell = (C/\omega)^{1/(k-1)}n$; there are $r - 2$ vertices between each pair of consecutive sub-paths that are not assigned to any sub-path for a total of at most $r\ell = r \cdot (\omega/C)^{1/(k-1)} \leq rnC^{-1/(k-1)}$ vertices, since $\omega \leq n^{k-1}$. This can be made smaller than, say, $\lceil n/(k+1) \rceil$, provided that C is large enough. We will call agents occupying these vertices *passive*; agents occupying sub-paths will be called *active*. Active agents will be partitioned into units, with respect to the sub-path they occupy. Every unit performs (independently and simultaneously) the strategy from Lemma 7.4.5 on their own sub-path. After $s = O(Cn^{k-1}/\omega)$ rounds, each k-tuple of agents from the same unit will be k-acquainted.

Now, we will show that a.a.s. every k-tuple of active agents, not all from the same unit, becomes acquainted. Let us focus on one such k-tuple, say f. If one starts from a random order in the strategy from Lemma 7.4.5, then it is evident that with probability $1 - o(n^k)$ agents from f visit at least $s/2 = Cn^{k-1}/(2\omega)$ different k-tuples of vertices. Hence, by considering vertices that are not on paths on which agents from f walk, we derive that with probability $1 - o(n^k)$, the number of distinct r-tuples visited by agents from f is at least

$$t = \frac{s}{2}\binom{n - k\ell}{r - k} \geq \frac{Cn^{k-1}}{2\omega} \cdot \frac{(0.9n)^{r-k}}{(r-k)!} = \frac{C(0.9)^{r-k}}{2(r-k)!} \cdot \frac{n^{r-1}}{\omega}.$$

Considering only those edges in H_2, the probability that agents from f never got acquainted is at most

$$(1 - p_2)^t \leq \exp\left(-\frac{\omega \log n}{n^{r-1}} \cdot \frac{C(0.9)^{r-k}}{2(r-k)!} \cdot \frac{n^{r-1}}{\omega}\right)$$

$$= \exp\left(-\frac{C(0.9)^{r-k}}{2(r-k)!} \log n\right) = o(n^{-k}),$$

for C sufficiently large. Since there are at most $\binom{n}{k} = O(n^k)$ k-tuples of agents, a.a.s. all active agents got acquainted.

It remains to deal with passive agents, which can be done the same way as in the proof of Theorem 7.4.1a. We partition all agents into $(k+1)$ groups, and shuffle them so that all become acquainted. In addition, if $r - 1$ does not divide $n - 1$, we may take N to be the largest integer such that $N \le n$ and $r - 1$ divides $N - 1$, and deal with a loose path on N vertices. It is straightforward to see that a.a.s. each of the remaining $O(1)$ vertices belong to at least one hyperedge intersecting with P. We may treat these vertices as passive ones and proceed as before.

Finally, suppose that $\omega = \omega(n) > n^{k-1}$. Since $\mathcal{AC}_r^k(H_r(n, p))$ is a decreasing function of p,

$$\mathcal{AC}_r^k(H_r(n, p)) \le \mathcal{AC}_r^k(H_r(n, (\log n)/n^{r-k})) = O(1),$$

by the previous case. □

7.5 Random Geometric Graphs

We finish the chapter by considering the acquaintance time of random geometric graphs. The model for random geometric graphs differs from the one used earlier, where we identify points in the unit sphere $[0, 1]^2$ (rather than in $[0, \sqrt{n}]^2$). For notational convenience (and following Penrose [155]) we let

$$\mathcal{X}_n = \{X_1, X_2, \ldots, X_n\}, \tag{7.4}$$

where $X_1, X_2, \ldots \in \mathbb{R}^2$ is an infinite set of random points, i.i.d. on the unit square.

The results of this section derive from [150]. Note that the acquaintance time for random percolated geometric graphs was also considered in [150], but we do not discuss those results here.

The main theorem of this section is the following.

Theorem 7.5.1 ([150]). *If $(r_n)_n$ is such that $\pi n r_n^2 - \log n \to \infty$, then $\mathcal{AC}(G(n, r_n)) = \Theta(r_n^{-2})$ a.a.s.*

We prove only the upper bound of 7.5.1. Before we proceed, we need a few lemmas on the acquaintance time.

Theorem 7.5.2 ([10]). *For every connected graph G we have $\mathcal{AC}(G) \leq 20 \cdot \Delta(G) \cdot |V(G)|$, where $\Delta(G)$ is the maximum degree of G.*

For graphs G and H, the graph $G[H]$ has vertex set $V(G) \times V(H)$ with an edge between (u, v) and (u', v') if either 1) $u = u'$ and $vv' \in E(H)$, or 2) $uu' \in E(G)$. In particular, $G[K_s]$ is the graph we derive by replacing each vertex of G by an s-clique and adding all edges between the cliques corresponding to adjacent vertices of G.

Lemma 7.5.3 ([150]). *We have that $\mathcal{AC}(G[K_s]) \leq \mathcal{AC}(G)$ for all connected G and all $s \in \mathbb{N}$.*

The final lemma we state here is analogous to Lemma 7.1.3.

Lemma 7.5.4 ([150]). *If G is a connected graph, then there is a strategy such that all agents get acquainted, and return to their initial vertices in $2 \cdot \mathcal{AC}(G)$ rounds.*

We are now ready to prove the main theorem of the section.

Proof of Theorem 7.5.1. We first consider the dense case. Let $\mathcal{D}(m)$ be the *dissection* of $[0, 1]^2$ into squares of side length $s(m) = 1/m$, with $m \in \mathbb{N}$. We will call the squares of this dissection *cells*, and given cell $c \in \mathcal{D}(m)$, we will denote by $V(c)$ the set of points of $\mathcal{X}_n$ that fall in c.

For $(r_n)_n$, an arbitrary sequence of numbers with $0 < r_n < \sqrt{2}$, let us define $m_n = \lceil 1000/r_n \rceil$. Then we have $1/m_n \leq r_n/1000$ and $1/m_n = \Omega(r_n)$. Further, $1/m_n \sim r_n/1000$ if $r_n \to 0$. Let $\mathcal{D}_n = \mathcal{D}(m_n)$ and let $\mu_n = n/m_n^2$ denote the expectation $\mathbb{E}|V(c)|$. The proof of the following lemma is omitted.

Lemma 7.5.5 ([150]). *There is a constant $K > 0$ such that if $(r_n)_n$ is such that $\pi n r_n^2 \geq K \log n$, then $0.9 \cdot \mu_n \leq |V(c)| \leq 1.1 \cdot \mu_n$ for all $c \in \mathcal{D}_n$, a.a.s.*

Now let $V \subseteq [0,1]^2, 0 < r < \sqrt{2}$ be such that the conclusion of Lemma 7.5.5 holds, but otherwise, arbitrary. It suffices to show that $G = G(V,r)$ satisfies $\mathcal{AC}(G) = O(r^{-2})$.

For each $c \in \mathcal{D}_n$ we partition $V(c)$ into three parts $V_1(c), V_2(c), V_3(c)$, each of cardinality at most $0.4 \cdot \mu_n$. For each pair $1 \le i < j \le 3$ and each cell $c \in \Gamma_n$, let $W_{ij}(c) \subseteq V(c)$ be a set of cardinality exactly $t = \lfloor 0.9\mu_n \rfloor$ such that $V_i(c) \cup V_j(c) \subseteq W_{ij}(c)$. Set $W_{ij} = \bigcup_{c \in \Gamma_n} W_{ij}(c)$. Let $G_{ij} = G[W_{ij}]$ denote the subgraph induced by W_{ij}. We now observe that, since points in touching cells of the dissection $\mathcal{D}_n$ have distance at most r, the graph G_{ij} has a spanning subgraph that is isomorphic to $H[K_t]$ where H denotes the $m_n \times m_n$-grid. It follows from Theorem 7.5.2 and Lemmas 7.5.3 and 7.5.4 that we can acquaint all agents on vertices of W_{ij} with each other, and return them to their starting positions in $O(m_n^2) = O(r_n^{-2})$ rounds.

By repeating this procedure for each of W_{12}, W_{13}, W_{23}, we acquaint all agents with each other in $O(r_n^{-2})$ rounds, and the proof in the dense case follows.

Before we prove the sparse case (which is more involved), we need to recall some definitions and results from [27]. Consider any geometric graph $G = (V,r)$, where $V = \{x_1, x_2, \ldots, x_n\} \subset [0,1]^2$.

Let $m \in \mathbb{N}$ be such that $s(m) = 1/m \le r/1000$ and consider $\mathcal{D} = \mathcal{D}(m)$. Given $T > 0$ and $V \subseteq [0,1]^2$, we call a cell $c \in \mathcal{D}$ *good* with respect to T, V if $|c \cap V| \ge T$ and *bad* otherwise. When the choice of T and V is clear from the context we will just speak of good and bad. Let $\Gamma = \Gamma(V, m, T, r)$ denote the graph whose vertices are the good cells of $\mathcal{D}(m)$, with an edge $cc' \in E(\Gamma)$ if and only if the lower left corners of c, c' have distance at most $r - s\sqrt{2}$.

We will usually just write Γ when the choice of V, m, T, r is clear from the context. Let us denote the components of Γ by $\Gamma_1, \Gamma_2, \ldots$ where Γ_i has at least as many cells as Γ_{i+1} (ties are broken arbitrarily). For convenience we will also write $\Gamma_{\max} = \Gamma_1$. We will often be a bit sloppy and identify Γ_i with the union of its cells, and speak of $\operatorname{diam}(\Gamma_i)$ and the distance between Γ_i and Γ_j and so forth.

A point $v \in V$ is *safe* if there is a good cell $c \in \Gamma_{\max}$ such that $|B(v; r) \cap V \cap c| \ge T$. Otherwise, if there is a good cell $c \in \Gamma_i$,

$i \geq 2$, such that $|B(v; r) \cap V \cap c| \geq T$, we say that v is *risky*. Otherwise we call v *dangerous*.

For $i \geq 2$ we let Γ_i^+ denote the set of all points of V in cells of Γ_i, together with all risky points v that satisfy $|B(v; r) \cap V \cap c| \geq T$ for at least one $c \in \Gamma_i$. The following is a list of desirable properties that we would like V and $\Gamma(V, m, T, r)$ to have:

(S1) The component $\Gamma_{\max}$ contains more than $0.99 \cdot |\mathcal{D}|$ cells.

(S2) For all $i \geq 2$, $\mathrm{diam}(\Gamma_i^+) < r/100$.

(S3) If $u, v \in V$ are dangerous, then either $\|u - v\| < r/100$ or $\|u - v\| > r \cdot 10^{10}$.

(S4) For all $i > j \geq 2$ the distance between Γ_i^+ and Γ_j^+ is at least $r \cdot 10^{10}$.

(S5) If $v \in V$ is dangerous and $i \geq 2$, then the distance between v and Γ_i^+ is at least $r \cdot 10^{10}$.

Finally, we introduce some terminology for sets of dangerous and risky points. Suppose that $V \subseteq [0, 1]^2$ and m, T, r are such that **(S1)-(S5)** above hold. Dangerous points come in groups of points of diameter $< r/100$ that are far apart. We formally define a *dangerous cluster* (with respect to V, m, T, r) to be an inclusion-wise maximal subset of V with the property that $\mathrm{diam}(A) < r \cdot 10^{10}$ and all elements of A are dangerous.

A set $A \subseteq V$ is an *obstruction* (with respect to V, m, T, r) if it is either a dangerous cluster or Γ_i^+ for some $i \geq 2$. We call A an *s-obstruction* if $|A| = s$. By **(S3)-(S5)**, obstructions are pairwise separated by distance $r \cdot 10^{10}$. A consequence is that a vertex in a good cell is adjacent in G to at most one obstruction. A point $v \in V$ is *crucial* for A if $A \subseteq N(v)$, and v is *safe*.

We are interested in the following choice of m for our dissection. For $n \in \mathbb{N}$ and $\eta > 0$ a constant, define

$$m_n = \left\lceil \sqrt{\frac{n}{\eta^2 \log n}} \right\rceil. \tag{7.5}$$

We need three additional lemmas (with proofs omitted).

Lemma 7.5.6 ([150]). *For every sufficiently small $\eta > 0$, there exists a $\delta = \delta(\eta) > 0$ such that the following holds. Let m_n be given by (7.5), let $\mathcal{X}_n$ be as in (7.4), let $T_n \leq \delta \log n$ and let r_n be such that $\pi n r_n^2 = \log n + o(\log n)$. Then **(S1)-(S5)** hold for $\Gamma(\mathcal{X}_n, m_n, T_n, r_n)$ a.a.s.*

Lemma 7.5.7 ([150]). *For every sufficiently small $\eta > 0$, there exists a $\delta = \delta(\eta) > 0$ such that the following holds. Let $(m_n)_n$ be given by (7.5), let $T_n \leq \delta \log n$ and let $V_n = \mathcal{X}_n$ with $\mathcal{X}_n$ as in (7.4), and let $(r_n)_n$ be a sequence of positive numbers such that $\pi r_n^2 - \log n \to \infty$.*

Then a.a.s., it holds that for every $s \geq 2$, every s-obstruction has at least $s - 100$ crucial vertices.

Lemma 7.5.8 ([150]). *If $\eta > 0$ is fixed, $(m_n)_n$ is as given by (7.5) and $V_n = \mathcal{X}_n$, then there is a constant C such that, a.a.s., every cell contains at most $C \log n$ points.*

We now prove the upper bound for a sequence r_n such that $\pi n r_n^2 = \log n + \omega(1)$ and $\pi n r_n^2 \leq K \log n$ for a large constant K. For this range of r_n, it suffices to consider the case when $1 \ll \pi n r_n^2 - \log n \ll \log n$, and prove that in that case a.a.s. $\mathcal{AC}(\mathbb{G}(n, r_n)) = O(n / \log n)$.

If $\pi n r_n^2 = \log n + \omega(1)$, then we have that a.a.s.

$$\mathcal{AC}(G) \geq \frac{\binom{n}{2}}{|E(G)|} - 1 = \Omega(r_n^{-2}). \tag{7.6}$$

By (7.6) we already have the asymptotically almost sure lower bound $\mathcal{AC}(\mathbb{G}(n, r_n)) = \Omega(r_n^{-2}) = \Omega(n / \log n)$ for all sequences $(r_n)_n$ satisfying $\pi n r_n^2 \leq K \log n$. We therefore, pick such a sequence r_n and assume that η, δ have been chosen in such a way that the conclusions of Lemmas 7.5.6 and 7.5.7 hold a.a.s. By [156], in this range a.a.s. $\mathbb{G}(n, r_n)$ is connected.

Let $V \subseteq [0, 1]^2$ be an arbitrary set of points, and $r, m, \eta, \delta > 0$ be arbitrary numbers such that $G(V, r)$ is connected and the conclusions of Lemmas 7.5.6, 7.5.7 and 7.5.8 are satisfied. It suffices to show that every such graph $G = G(V, r)$ has acquaintance time $O(n / \log n)$, as we will now show.

Claim 7.5.9. *There is a constant c_1 such that every obstruction consists of at most $c_1 \log n$ points.*

Proof. Every obstruction O has diameter $r/100$ and hence, there are only $O(1)$ cells that contain points of O. Since each cell contains $O(\log n)$ points, we are done. $\qquad\square$

Each point v that is safe, but not in a cell of $\Gamma_{\max}$ has at least T neighbors in some cell $c \in \Gamma_{\max}$. We arbitrarily "assign" v to such a cell.

Claim 7.5.10. *There is a constant $c_2 > 1$ such that the following holds. For every obstruction O there is a cell $c \in \Gamma_{\max}$ such that at least $|O|/c_2$ vertices that are crucial for O have been assigned to c.*

Proof. Since G is connected, every obstruction has at least one crucial vertex. (It is adjacent to at least one vertex $v \in V \setminus O$ and this v cannot be dangerous or risky.) This shows that, by choosing c_2 sufficiently large, the claim holds whenever $|O| < 1000$. Let us therefore, assume $|O| \geq 1000$. Since O has diameter $< r/100$, there is a constant D such that the crucial vertices are all assigned to one of the D cells within range $2r$ of O. Since $|O| > 1000$ there are at least $|O| - 100 > |O|/2$ crucial vertices for O, and hence, at least $|O|/2D$ of these crucial vertices are assigned to the same cell. $\qquad\square$

For each obstruction O, we now assign all its vertices to a cell $c \in \Gamma_{\max}$ such that at least $|O|/c_2$ vertices that are crucial for O have been assigned to c. For a cell $c \in \Gamma$ let $V(c)$ denote the set of points that fell in c. For each $c \in \Gamma_{\max}$, let $A(c)$ denote the union of $V(c)$ with all vertices that have been assigned to c.

Claim 7.5.11. *There exists a constant c_3 such that $|A(c)| \leq c_3 \log n$ for all $c \in \Gamma_{\max}$.*

Proof. Note that if a vertex v has been assigned to c, it must lie within distance $2r$ of c. Hence, there are only $O(1)$ cells in which $A(c)$ is contained, and since every cell contains $O(\log n)$ points, we are done. $\qquad\square$

We now partition $A(c)$ into sets $A_1(c), A_2(c), \ldots, A_L(c)$ each of size at most $T/100$, with $L = \lceil c_3/(100\delta) \rceil$. The following observation will be key to our strategy.

Claim 7.5.12. *There is a constant c_4 such that the following holds. For every $c \in \Gamma_{\max}$ and every $A' \subseteq A(c)$ with $|A'| \leq T/50$ there is a sequence of at most c_4 moves that results in the agents on vertices A' being placed on vertices of $V(c)$ (and uses only edges of $G[A(c)]$).*

Proof. We first move all agents on vertices of $A' \setminus V(c)$ on safe vertices not in $V(c)$ onto vertices of $V(c) \setminus A'$ in one round. (To see that this can be done, note that $|A'| < T/50$ and each safe vertex of A' has at least T neighbors in $V(c)$.) Let $W \subseteq V(c)$ be the set of vertices now occupied by agents that were originally on A'.

If A' also contains (part of) some obstruction O, then we partition $O \cap A'$ into $O(1)$ sets $O_1, O_2, \ldots, O_K$ of cardinality at most $|O|/c_2$ where $K \leq \lceil 1/c_2 \rceil$ (and hence, is a constant). We first move the agents on vertices O_1 onto crucial vertices assigned to c, and then on vertices of $V(c) \setminus W$, in two rounds. (Note this is possible since $|A'| \leq T/50$ and each crucial vertex is adjacent to at least T vertices of $V(c)$.) Similarly, supposing that the agents on $O_1, O_2, \ldots, O_{i-1}$ have already been moved onto vertices of $V(c)$, we can move the agents on vertices of O_i onto vertices of $V(c) \setminus W$ not occupied by agents from $O_1, O_2, \ldots, O_{i-1}$ in two rounds.

We thus, have moved all agents on vertices of A' onto vertices of $V(c)$ in a constant number of rounds, as required. $\square$

We are now ready to describe the overall strategy. We write $A_i = \bigcup_{c \in \Gamma_{\max}} A_i(c)$. For each pair of indices $1 \leq i < j \leq L$ we do the following.

First, we move all agents of $A_i(c) \cup A_j(c)$ onto vertices of $V(c)$ (in constantly many moves, simultaneously for all cells $c \in \Gamma_{\max}$). Next, we select a set $B(c) \subseteq V(c)$ for each $c \in \Gamma_{\max}$ with $|B(c)| = T$ and all agents that were on $A_i(c) \cup A_j(c)$ originally are now on vertices of $B(c)$. By a result of [27], if G is a connected geometric graph, G has a spanning tree of maximum degree at most five. Hence, the largest component $\Gamma_{\max}$ of the cells-graph has a spanning tree H of maximum degree at most five. Thus, by Theorem 7.5.2, we have $\mathcal{AC}(\Gamma_{\max}) \leq O(\Delta(H) \cdot |V(H)|) = O(|\Gamma|) = O(n/\log n)$. Now note that the graph spanned by $\bigcup_{c \in \Gamma_{\max}} B(c)$ contains a spanning subgraph isomorphic to $\Gamma_{\max}[K_T]$. Using Lemma 7.5.3 and 7.5.4, we can thus, acquaint all vertices of $\bigcup_{c \in \Gamma_{\max}} B(c)$ with each other and return them to their starting vertices in $O(n/\log n)$ rounds. Hence, in particular, we have acquainted the agents of $A_i \cup A_j$ and returned them to their starting positions in $O(n/\log n)$ moves. Once we have repeated this procedure for each of the $\binom{L}{2} = O(1)$ pairs of indices, all agents will be acquainted, still in $O(n/\log n)$ rounds. $\qquad\square$

Chapter 8

Firefighting

Along the way, we have captured robbers, cleaned graphs and become acquainted. Now it is time to fight some fires!

The Firefighter graph process was first introduced by Hartnell [108], and it may be viewed as a simplified deterministic model of the spread of contagion (such as fire, diseases, and computer viruses) in a network. In *Firefighter*, the process proceeds over discrete time-steps. Vertices are either *burning* or not. There are k-many *firefighters* who are attempting to control the fire, where k is a positive integer; the original and more common version is to have $k = 1$, although we will consider the general case in this chapter. If a vertex is visited by a firefighter, then it can never burn in any subsequent round, and is called *protected*. The fire begins at some vertex in the first round, and the firefighter chooses some vertex to save. The firefighter can visit any vertex in a given round, but cannot protect a vertex on fire. Observe that the Firefighters, unlike cops in Cops and Robbers, may move to any vertices of the graph, not only to vertices neighboring their present position. The fire spreads to all nonprotected neighbors at the beginning of each time-step; such vertices are called *burned*. Once a vertex has been protected, its state cannot change; that is, it can never be on fire. The process stops when the fire can no longer spread. A vertex is *saved* if it is not burned at the end of the process. See Figure 8.1 for an example of the Firefighter process.

Firefighter has been a considered in several familiar graph classes such as multi-dimensional grids (such as Cartesian, strong and other grids) and trees, and been studied both from a structural and algorithmic points of view. For the latter, it was shown in [86] that the Firefighter process (for all k) is **NP**-complete even when restricted to the family of trees of maximum degree three. For a

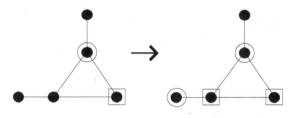

FIGURE 8.1: The fire, represented by squares, starts on the bottom right and moves left until fully blocked. Protected vertices are surrounded by circles.

survey, the reader is directed to [87]. See also the final chapter of [52].

We introduce a quantitative measure of how robust a graph is to the spread of fire in Firefighter. The *surviving rate* $\rho(G)$ of a graph G was introduced by Cai and Wang [62] and defined as the expected proportion of vertices that can be saved when a fire breaks out at a vertex of G chosen uniformly at random. Exact values for the surviving rate have been determined for paths and cycles [62] and bounds have been determined for trees [61, 62], planar graphs [62], K_4-free minor graphs [88], and outerplanar graphs [61]. For sparse graphs the Firefighter can more readily block the fire, and so we would expect their surviving rates should be relatively large. Finbow, Wang, and Wang [88] showed using the discharging method that any graph G with $n \geq 2$ vertices and at most $\left(\frac{4}{3} - \varepsilon\right) n$ edges has the property that $\rho(G) \geq \frac{6}{5}\varepsilon$, where $0 < \varepsilon < \frac{5}{24}$ is fixed. For more on the discharging method, see [72].

In [163], the above bound for the surviving rate was improved to show that graphs with size at most $\left(\frac{15}{11} - \varepsilon\right) n$ have surviving rate $\rho(G) \geq \frac{1}{60}\varepsilon$, where $0 < \varepsilon < \frac{1}{2}$ is fixed. We note that a construction of a random graph was given in [163] to show that no further improvement is possible; that is, $\frac{15}{11}$ is the threshold. Those results will form a theme in the chapter, which we will revisit several times. We survey the results from [163] in the next section. In Section 8.3, we consider an extension to general $k > 1$ discussed in [165]. The final section focuses on a variant of Firefighter, called

Constrained Firefighter, where fire may spread only to a prescribed number of neighbors.

8.1 Tool: Expander Mixing Lemma

The results in the next section refer to the probability space of random d-regular graphs with uniform probability distribution introduced in Chapter 1. Recall that this space is denoted $\mathcal{G}_{n,d}$, and asymptotics are for $n \to \infty$ with $d \geq 2$ fixed, and n even if d is odd.

To prove Theorem 8.2.2 in the next section, we will exploit spectral expansion properties of random d-regular graphs. The adjacency matrix $A = A(G)$ of a given d-regular graph G with n vertices is an $n \times n$ real and symmetric matrix. Thus, the matrix A has n real eigenvalues which we denote by $\lambda_1 \geq \lambda_2 \geq \cdots \geq \lambda_n$.

We are interested in the following: $\lambda = \lambda(G) = \max(|\lambda_2|, |\lambda_n|)$. For additional background, see the general survey [112] about expanders, or Chapter 9 of [8]. The value of λ for random d-regular graphs has been studied extensively. The celebrated result of Friedman [92] is important in this context.

Lemma 8.1.1 ([92]). *For every fixed $\varepsilon > 0$ and for $G \in \mathcal{G}_{n,d}$, a.a.s.*

$$\lambda(G) \leq 2\sqrt{d-1} + \varepsilon.$$

The number of edges $|E(S,T)|$ between two sets S and T in a random d-regular graph on n vertices is expected to be close to $d|S||T|/n$. A small λ implies in turn that this deviation is small. The following useful bound is proved in [2] (see also [8]).

Lemma 8.1.2 (Expander Mixing Lemma). *If G is a d-regular graph with n vertices and set $\lambda = \lambda(G)$, then for all $S,T \subseteq V$*

$$\left| |E(S,T)| - \frac{d|S||T|}{n} \right| \leq \lambda\sqrt{|S||T|}.$$

We will apply a stronger lower estimate than the one in 8.1.2 for $|E(S, V \setminus S)|$, first proved in [5] (see also [8]).

$$|E(S, V \setminus S)| \geq \frac{(d - \lambda)|S||V \setminus S|}{n} \tag{8.1}$$

for all $S \subseteq V$.

8.2   The $k = 1$ Case

It was shown in [163] that any graph G with average degree strictly smaller than $30/11$ has the surviving rate bounded away from zero. The proof is involved and so it is omitted here.

Theorem 8.2.1 ([163]). *If a graph G has $n \geq 2$ vertices and $m \leq (\frac{15}{11} - \varepsilon)n$ edges, for some $0 < \varepsilon < \frac{1}{2}$, then $\rho(G) \geq \frac{\varepsilon}{60}$.*

In contrast to Theorem 8.2.1, there exist dense graphs with large surviving rates. For example, consider a disjoint union of a large number of cliques. A deterministic family of sparse random graphs with the surviving rate tending to zero as the number of vertices n goes to infinity was given in [163]. Therefore, Theorem 8.2.1 is tight and the constant $\frac{15}{11}$ cannot be improved.

We next prove the following theorem that implies a weaker statement—the constant $\frac{15}{11} \approx 1.36$ cannot be replaced by anything larger than $3/2 = 1.5$. The results and proofs for the remainder of this section are derived from [163].

Theorem 8.2.2 ([163]). *If $G \in \mathcal{G}_{n,3}$, then a.a.s.*

$$\rho(G) = O(\log n/n) = o(1).$$

We need the following lemma about cycles in random graphs. The proof is omitted; however, let us mention that the proof of Lemma 6.3.5 can be easily adjusted.

Lemma 8.2.3 ([163]). *Let $K \geq 3$, $d \geq 3$ be fixed integers and $G \in \mathcal{G}_{n,d}$. A.a.s. the number of vertices that belong to a cycle of length at most K is at most $\log \log n$.*

Proof of Theorem 8.2.2. Let $U \subseteq V(G)$ be the set of vertices that do not belong to a cycle of length at most 30. Since $|U| \geq n - \log \log n$, a.a.s. by Lemma 8.2.3 applied with $K = 30$, it is enough to show that a.a.s. $\mathrm{sn}(G, u) = O(\log n)$ for all $u \in U$. Once this is proven, then a.a.s. we have that

$$
\begin{aligned}
\rho(G) &= \frac{1}{n^2} \sum_{v \in V} \mathrm{sn}(G, v) \\
&= \frac{1}{n^2} \sum_{v \in U} \mathrm{sn}(G, v) + \frac{1}{n^2} \sum_{v \in V \setminus U} \mathrm{sn}(G, v) \\
&= \frac{n - O(\log \log n)}{n^2} \cdot O(\log n) + \frac{O(\log \log n)}{n^2} \cdot O(n) \\
&= O(\log n / n).
\end{aligned}
$$

Let $u \in U$ and let s_t denote the number of vertices burning at the end of time-step t. Since u does not belong to a short cycle, to minimize s_t during a few first steps of the process, $t \subset \{1, 2, \ldots, 15\}$, the firefighter should use a *greedy strategy*; that is, protect a vertex adjacent to the vertex on fire. If the greedy strategy is used, then at the end of time-step $t \in \{1, 2, \ldots, 15\}$ there are at least ρ_t vertices that capture fire in this time-step, where ρ_t's satisfy the following recursion: $\rho_1 = 2$ (during the first round, the firefighter protects one vertex but 2 other vertices capture fire) and $\rho_t = 2\rho_{t-1} - 1$ for $t \geq 2$ (there are $2\rho_{t-1}$ vertices adjacent to the fire, one of them gets protected).

We derive that $\rho_t = 2^{t-1} + 1$ and, in particular, $s_{15} \geq \rho_{15} \geq 16000$. Note that from (8.1) and Lemma 8.1.1 it follows that we may assume that

$$
|E(S, V \setminus S)| \geq \frac{(d - \lambda)|S||V \setminus S|}{n} \geq 0.08|S|,
$$

for all sets S of cardinality at most $n/2$. Therefore, if vertices from set S ($|S| \leq n/2$) are on fire, then at least $(0.02)|S|$ vertices are not on fire (including perhaps protected vertices) but are adjacent to at least one vertex on fire. Thus, at least

$$0.02s_{15} - 16 \geq 0.01s_{15} + 160 - 16 \geq 0.01s_{15}$$

new vertices are going to capture fire at the next round and so $s_{16} \geq 1.01s_{15}$. We may apply induction now to show that $s_{t+1} \geq 1.01s_t$, provided that $s_t \leq n/2$. Hence, for any $t \geq 15$ with $s_t \leq n/2$, we have that

$$\begin{aligned} s_{t+1} &\geq 1.02s_t - (t+1) \\ &\geq 1.01s_t + 0.01 \cdot 16000 \cdot 1.01^{t-15} - (t+1) \\ &\geq 1.01s_t. \end{aligned}$$

The above inequality implies that at least $n/2$ vertices are on fire at time $t_0 \leq \log_{1.01} n$.

For $t > t_0$, we consider the $r_t = n - s_t$ many vertices that are not burning at time t. By using (8.1) and Lemma 8.1.1, we obtain that at least $0.02r_t$ nonburning vertices are adjacent to some vertex on fire at time t. Thus, we derive that $r_{t+1} = 0.99r_t$ provided that $r_t \geq 100(t+1)$. Therefore, we derive that

$$r_{t+1} \leq 0.98r_t + (t+1) \leq 0.99r_t.$$

The above inequality implies that at least $n - O(\log n)$ vertices are on fire at time $T \leq \log_{1.01} n + \log_{1/0.99} n = O(\log n)$. $\square$

We consider next a graph G with the maximum degree at most 3 and average degree at most $3 - \varepsilon$, for some $\varepsilon > 0$. This means that a positive fraction of all vertices (at least $\varepsilon n/3$ vertices) must have degree at most two. If the fire breaks out at such a vertex (which happens with probability at least $\varepsilon/3$), then we can use a greedy algorithm to protect an arbitrary vertex adjacent to the fire at any step of the process. By doing this, at most one new vertex catches fire at any time-step of the process, and in this way we can save at least $1/3$ of the vertices. It follows that $\rho(G) \geq \varepsilon/9 > 0$. Therefore, in order to construct a sparser graph that burns fast, perhaps surprisingly, we need to introduce vertices of higher degree.

As already mentioned, a random graph construction from [163] shows that the constant $\frac{15}{11}$ cannot be improved. Fix $K \in \mathbb{N}$ that is congruent to 3 (mod 5), and $n \in \mathbb{N}$. Begin with n disjoint paths of length K on the vertex set X, and the set Y consisting of $2n(K+2)/5$ isolated vertices. Place random edges between X and Y such that every vertex in X has degree 4, and every vertex in Y has degree 5. We may derive multiple edges from this process, but we restrict our probability space to simple graphs. It can be shown that we generate a simple graph with probability asymptotic to some constant $c > 0$, so the pairing model can be used again to study an asymptotic behavior of this random graph. No edge between vertices in Y is added, so Y forms an independent set. Finally, subdivide each edge (both in the random part as well as in the n paths we started with) to obtain a graph $G(K, n)$. The random graph $G(K, n)$ is defined for any value of K (congruent to 3 (mod 5) but for our purpose we take, say, $K = K(n) \sim \log\log\log n$ such that it is tending to infinity together with n but not so fast.

In the random graph $G(K, n)$, we have a.a.s. $(1 + o(1))Kn$ vertices of degree 4 and $(\frac{2}{5} + o(1))Kn$ vertices of degree 5. Thus, the number of edges in the graph G before subdividing edges is

$$\frac{1}{2} \sum_{v \in V} \deg_G(v) \sim \frac{4 \cdot Kn + 5 \cdot \frac{2}{5}Kn}{2}$$

$$\sim 3Kn,$$

so this is also the number of vertices of degree 2 after subdividing. The average degree of $G(K, n)$ is approximated as the following.

$$\mathrm{Av}(G(K, n)) = \frac{\sum_{v \in V} \deg_{G(K,n)}(v)}{|V(G)|}$$

$$\sim \frac{4 \cdot Kn + 5 \cdot \frac{2}{5}Kn + 2 \cdot 3Kn}{Kn + \frac{2}{5}Kn + 3Kn}$$

$$\sim \frac{30}{11}.$$

We have the following result.

Theorem 8.2.4 ([163]). *If $K = K(n) \sim \log\log\log n$ and $K \equiv 3$ (mod 5), then a.a.s.*

$$\rho(G(K, n)) = o(1).$$

Although we omit the proof of the theorem, we sketch some ideas. Suppose that some vertices of degree 5 are adjacent to two vertices on the same path (before edges are subdivided). If this happens, then we call the path *bad*; otherwise, the path is *good*. We can show that a.a.s. $o(n)$ paths are bad; in other words, a.a.s. almost all paths are good. Since our goal is to show the result that holds a.a.s., we may assume that the graph has this property.

If the fire breaks out on a vertex $v \in V = V(G(K,n))$, then the firefighter must try to avoid hitting vertices of degree 5. Indeed, the same argument as for random 3-regular graphs can be used to show that if this happens, then the fire will start spreading exponentially fast and there is no hope to stop this process. It follows that $\mathrm{sn}(G,v) = o(Kn)$ if $\deg(v) = 5$. If v is on (or adjacent to) a bad path, then there is a chance to save a large fraction of the graph. However, since almost all paths are good, this does not affect an asymptotic behavior of the surviving rate. Suppose then that v is on (or adjacent to) a good path. If $\deg(v) = 4$, then the firefighter must start with protecting two vertices corresponding to the random part, but they have to give up after the next two rounds. The fire reaches a vertex of degree 5, and $\mathrm{sn}(G,v) = o(Kn)$. See Figure 8.2.

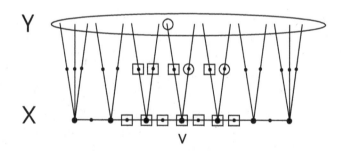

FIGURE 8.2: The fire starts on a degree 4 vertex. The circles are protected vertices and squares are the fire.

If $\deg(v) = 2$, then the firefighter can keep pushing fire along the path (see Figure 8.3) but, since at the end of the path there is a vertex adjacent to 3 random edges, it is impossible to avoid vertices of degree 5. Once the fire reaches a vertex of degree 5, it spreads quickly and the result holds.

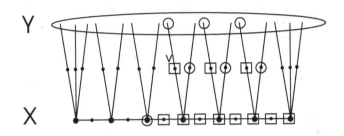

FIGURE 8.3: The fire starts on a degree 2 vertex.

8.3 The $k > 1$ Case

In this section, we consider k-Firefighter for an integer $k > 1$. Let $\mathrm{sn}_k(G, v)$ and $\rho_k(G)$ denote the corresponding parameters of $\mathrm{sn}(G)$ and $\rho(G)$, respectively, when the firefighters protect k vertices at each time-step. We say that a family of graphs $\mathcal{G}$ is k-flammable if for any $\varepsilon > 0$ there exists $G \in \mathcal{G}$ such that $\rho_k(G) < \varepsilon$. The situation $k \geq 2$ is simpler than the $k = 1$ case. The main result we provide in this section is that $\tau_k = k+2-\frac{1}{k+2}$ is the threshold in this case; that is, all graphs with average degree strictly less than τ_k are not k-flammable, but the family of graphs with average degree at least τ_k is k-flammable. Results and proofs for this section are derived from [165].

Theorem 8.3.1 ([165]). *Let $k \in \mathbb{N}$, $\tau_k = k+2-\frac{1}{k+2}$, and $\varepsilon > 0$. If the graph G has $n \geq 2$ vertices and average degree at most $\tau_k - \varepsilon$, then $\rho_k(G) \geq \frac{2\varepsilon}{5\tau_k}$.*

To demonstrate that this result is sharp, consider the family $\mathbb{G}(n, d, d+2)$ of bipartite $(d, d+2)$-regular graphs ($n \in \mathbb{N}$, $d \geq 3$).

This family consists of all bipartite graphs with two parts X and Y such that $|X| = (d+2)n$ and $|Y| = dn$. Each vertex in X has degree d, whereas vertices in Y have degree $d+2$. Our next result refers to the probability space of random $(d, d+2)$-regular graphs with uniform distribution. We show for $G \in \mathbb{G}(n, k+1, k+3)$ that $\rho_k(G) = o(1)$ a.a.s., which implies that this family is k-flammable. Since the average degree of G is

$$\frac{(k+1)n(k+3) + (k+3)n(k+1)}{(k+1)n + (k+3)n} = k+2 - \frac{1}{k+2},$$

Theorem 8.3.1 is sharp.

Theorem 8.3.2 ([165]). *If $k \geq 2$ and $G \in \mathbb{G}(n, k+1, k+3)$, then a.a.s.*

$$\rho_k(G) = \Theta(\log n/n) = o(1).$$

The main tool in the proof of Theorem 8.3.1 is the discharging method commonly used in graph coloring problems, and that was also used in [88] in the case $k = 1$.

Proof of Theorem 8.3.1. Let $k \in \mathbb{N}$, $\tau_k = k+2 - \frac{1}{k+2}$, and $\varepsilon > 0$. Let G be a graph on $n \geq 2$ vertices, m edges, and with $\frac{2m}{n} \leq \tau_k - \varepsilon$.

Let $V_1 \subseteq V$ be the set of vertices of degree at most k. Let $V_2 \subseteq V \setminus V_1$ be the set of vertices of degree $k+1$ that are adjacent to at least one vertex of degree at most $k+1$. Finally, let $V_3 \subseteq V \setminus (V_1 \cup V_2)$ be the set of vertices of degree $k+1$ that are adjacent to at least one vertex of degree $k+2$ with at least two neighbors of degree $k+1$. A main note here is that the fire can be managed when it starts at any vertex in $V_1 \cup V_2 \cup V_3$. If the fire starts at $v \in V_1$, the firefighter can protect all neighbors of v saving at least half of the vertices. If the fire starts at $v \in V_2$, then the firefighter protects all neighbors of v except the vertex u of degree at most $k+1$. The fire spreads to u in the next round but it is stopped there. At least a $\frac{n-2}{n} \geq \frac{(2+k)-2}{2+k} \geq \frac{1}{2}$ fraction of vertices are saved. Analogously, when the fire starts at $v \in V_3$, the firefighter can direct the fire to vertex of degree $k+2$, then to another vertex of degree $k+1$, and finish the job there saving at least a $\frac{n-3}{n} \geq \frac{(3+k)-3}{3+k} \geq \frac{2}{5}$ fraction of the graph.

It remains to show that if G is sparse (that is, when $\frac{2m}{n} \leq \tau_k - \varepsilon$), then $V_1 \cup V_2 \cup V_3$ contains a positive fraction of all vertices. To show this, we use the discharging method as referenced before the proof.

The method is as follows. To each vertex $v \in V \setminus (V_1 \cup V_2 \cup V_3)$, we assign an initial weight of $\omega(v) = \deg(v) \geq k+1$. The weights then spread to neighbors. Every vertex of degree at least $k+2$ gives $\frac{1}{k+2}$ to each of its neighbors of degree $k+1$ that is in $v \in V \setminus (V_1 \cup V_2 \cup V_3)$. Let $\omega'(v)$ be a new weight after this discharging operation. We note that

$$\sum_{v \in V \setminus (V_1 \cup V_2 \cup V_3)} \omega(v) = \sum_{v \in V \setminus (V_1 \cup V_2 \cup V_3)} \omega'(v).$$

For each vertex $v \in V \setminus (V_1 \cup V_2 \cup V_3)$ of degree $k+1$ we have that

$$\omega'(v) = \omega(v) + (k+1)\frac{1}{k+2} = k+2 - \frac{1}{k+2} = \tau_k,$$

since all neighbors of v are of degree at least $k+2$; otherwise, v would be in V_2. For each vertex $v \in V \setminus (V_1 \cup V_2 \cup V_3)$ of degree $k+2$ we have that

$$\omega'(v) \geq \omega(v) - \frac{1}{k+2} = k+2 - \frac{1}{k+2} = \tau_k,$$

since at most one neighbor of v from $V \setminus (V_1 \cup V_2 \cup V_3)$ has degree $k+1$; otherwise, all neighbors of v of degree $k+1$ would be in $V_2 \cup V_3$ and so v would not receive weight.

In the final step, every vertex $v \in V \setminus (V_1 \cup V_2 \cup V_3)$ of degree at least $k+3$ must have

$$
\begin{aligned}
\omega'(v) &\geq \omega(v) - \deg(v)\frac{1}{k+2} \\
&\geq \frac{(k+3)(k+1)}{k+2} = k+2 - \frac{1}{k+2} \\
&= \tau_k.
\end{aligned}
$$

From this and the fact that G is sparse, we may derive that

$$(\tau_k - \varepsilon)\, n \;\geq\; 2m = \sum_{v \in V} \deg(v)$$

$$\geq \sum_{v \in V \setminus (V_1 \cup V_2 \cup V_3)} \deg(v)$$

$$= \sum_{v \in V \setminus (V_1 \cup V_2 \cup V_3)} \omega'(v)$$

$$\geq \tau_k \left(n - |V_1| - |V_2| - |V_3| \right).$$

Hence, we derive the inequality

$$|V_1| + |V_2| + |V_3| \geq \frac{\varepsilon n}{\tau_k},$$

which implies that with probability at least ε/τ_k the fire starts on $V_1 \cup V_2 \cup V_3$ (since a fire breaks out at a random vertex of G). If this is the case, then we showed that at least a 2/5 fraction of vertices can be saved, so $\rho_k(G) \geq (2/5)(\varepsilon/\tau_k)$. The proof of the theorem now follows. $\square$

We next investigate properties of random $(d, d + 2)$-regular graphs, where each vertex has degree d or $d + 2$. Instead of working directly in the uniform probability space of random $(d, d + 2)$-regular graphs on $(2d+2)n$ vertices $\mathcal{G}(n, d, d+2)$, we use a *pairing model*, which is described next (analogous to the one for random regular graphs; see Subsection 6.3 for a detailed discussion). In our present case, consider $d(d + 2)n$ points, which we label as the set P_X. This set is partitioned into $(d + 2)n$ labeled buckets $x_1, x_2, \ldots, x_{(d+2)n}$ of d points each. Another $d(d+2)n$ points (forming P_Y) are partitioned into dn labeled buckets $y_1, y_2, \ldots, y_{dn}$, this time each consisting of $d + 2$ points. A *pairing* of these points is a perfect matching between P_X and P_Y into $d(d + 2)n$ pairs. Given a pairing P, we may construct a multigraph $G(P)$, with loops allowed, as follows: the vertices are the buckets $x_1, x_2, \ldots, x_{(d+2)n}$ and $y_1, y_2, \ldots, y_{dn}$; a pair $\{x, y\}$ in P ($x \in P_X$, $y \in P_Y$) corresponds to an edge $x_i y_j$ in $G(P)$ if x and y are contained in the buckets x_i and y_j, respectively.

The probability of a random pairing corresponding to a given simple graph G is independent of the graph. Hence, the restriction of the probability space of random pairings to simple graphs is precisely $\mathcal{G}(n, d, d + 2)$. Further, it can be shown (see Lemma 8.3.3) that a random pairing generates a simple graph with probability asymptotic to $e^{-(d^2-1)/2}$ depending on d but not on n. We conclude the important fact that an event holding a.a.s. over the probability space of random pairings also holds a.a.s. over the corresponding space $\mathcal{G}(n, d, d + 2)$. For this reason, asymptotic results over random pairings suffice, rather than working directly with the uniform space $\mathcal{G}(n, d, d+2)$. Unlike in the uniform space, in this model the pairs may be chosen sequentially so that the next pair is chosen uniformly at random over the remaining unchosen points.

As in the pairing model, we have the following lemma (whose proof is omitted) that derives the probability that $G(P)$ is simple.

Lemma 8.3.3 ([165]). *If P is a random pairing, then $G(P)$ is simple with probability tending to $e^{-(d^2-1)/2}$ as $n \to \infty$.*

We next observe that a.a.s. almost all vertices have the property that the neighborhood of the vertex induces a tree. This is, of course, not a good property for the firefighter, and it will be used to establish an upper bound for the threshold we investigate. For $d \geq 3$, a cycle is called *short* if it has length at most $L = \log_{d^2-1} \log n$. Since the proof is analogous to the one for random d-regular graphs, it is omitted here.

Lemma 8.3.4 ([165]). *If $d \geq 3$ and $G \in \mathbb{G}(n, d, d+2)$, then a.a.s. the number of vertices that belong to a short cycle is at most $\log n$.*

Now, we will move to a more technical lemma showing that a.a.s. $\mathbb{G}(n, d, d + 2)$ has good expansion properties. We start with investigating subsets of X and subsets of Y only (using Lemma 8.3.5), and then generalize it to any subset of $V = X \cup Y$ (Lemma 8.3.6). Let $N(K)$ denote the set of vertices in $V \setminus K$ that have at least one neighbor in K.

Lemma 8.3.5 ([165]). *Let $d \geq 3$, $\varepsilon = 0.237$, and $G = (X, Y, E) \in \mathbb{G}(n, d, d+2)$. The following properties hold a.a.s.*

(a) *For every* $K \subseteq Y$ *with* $1 \leq k = |K| \leq \frac{1}{2}|Y| = \frac{dn}{2}$, *we have that*

$$|N(K)| \geq k\frac{d+2}{d}(1+\varepsilon).$$

(b) *For every* $K \subseteq X$ *with* $1 \leq k = |K| \leq \frac{1}{2}|X| = \frac{(d+2)n}{2}$, *we have that*

$$|N(K)| \geq k\frac{d}{d+2}(1+\varepsilon).$$

Lemma 8.3.5 implies that for every $K \subseteq X$ (or $K \subseteq Y$) with at most half of the vertices from X (or Y), respectively, $N(K)$ contains substantially more points (in the pairing model) compared to the number of points that are associated with K. Note also that the choice of ε is optimized for the best possible outcome, and is obtained numerically.

Lemma 8.3.6 ([165]). *If* $d \geq 3$, $\varepsilon' = 0.088$, *and* $G = (X, Y, E) \in \mathbb{G}(n, d, d+2)$, *then a.a.s. for every* $K \subseteq V = X \cup Y$ *with* $1 \leq k = |K| \leq \frac{1}{2}|V| = \frac{1}{2}(|X| + |Y|) = (d+1)n$, *we have that*

$$|N(K)| \geq \varepsilon'k.$$

With these results available, we are ready to prove Theorem 8.3.2.

Proof of Theorem 8.3.2. Let $k \geq 2$, $\varepsilon' = 0.088$ as in Lemma 8.3.6, and $G = (X, Y, E) \in \mathbb{G}(n, k+1, k+3)$. Let U be the set of vertices that do not belong to a cycle of length at most $L = \log_{k^2+2k} \log n$. Since we would like this property to hold a.a.s., we assume without loss of generality that the properties described in Lemma 8.3.4 and 8.3.6 hold.

It follows from Lemma 8.3.4 with $d = k+1$ that

$$|U| \geq |V| - \log n = (2d+2)n - \log n.$$

Since

$$\rho_k(G) = \frac{1}{|V|^2} \sum_{v \in V} \mathrm{sn}_k(G, v)$$

$$= \frac{1}{|V|^2} \sum_{v \in U} \mathrm{sn}_k(G, v) + \frac{1}{|V|^2} \sum_{v \in V \setminus U} \mathrm{sn}_k(G, v)$$

$$= (1 + o(1)) \frac{1}{|U|} \sum_{v \in U} \frac{\mathrm{sn}_k(G, v)}{|V|} + O\left(\frac{|V \setminus U|}{|V|}\right)$$

$$= (1 + o(1)) \frac{1}{|U|} \sum_{v \in U} \frac{\mathrm{sn}_k(G, v)}{|V|} + O\left(\frac{\log n}{n}\right),$$

it is enough to show that $\mathrm{sn}_k(G, v) = \Theta(\log n)$ for every $v \in U$. Since G is a $(k + 1, k + 3)$-regular graph, it takes at least $\frac{1}{2} \log_{k^2+2k} n - O(1)$ steps to discover $n/2$ vertices. The firefighter can deterministically save $\Omega(\log n)$ vertices until the process is completed. Hence, it remains to show that $\mathrm{sn}_k(G, v) = O(\log n)$ for every $v \in U$.

Let $v \in U$ and let s_t denote the number of vertices that capture fire at time t. It is clear that in order to minimize s_t during the first few steps of the process when the game is played on a tree, the firefighter should use a greedy strategy and protect any vertex adjacent to the fire. Suppose that $v \in U \cap X$; that is, $\deg(v) = k+1$. It is straightforward to see that $s_1 = 1$ (initial vertex v is on fire), $s_2 = 1$ (v has $k + 1$ neighbors but only one catches fire, since k of them are protected), $s_3 = (k + 2) - k = 2$ (the neighbor of v on fire has $k + 2$ new neighbors but k of them will be saved), and $s_4 = 2k - k = k$ (two vertices on fire have $2k$ neighbors but k of them are saved). We obtain the following recurrence relation: $s_2 = 1$ and for every $r \in \mathbb{N}$,

$$s_{2r+2} = s_{2r+1} k - k = (s_{2r}(k + 2) - k)k - k.$$

After solving this relation we find that for $r \in \mathbb{N}$

$$s_{2r} = \frac{k - 1}{k(k + 2)(k^2 + 2k - 1)} \left(k(k + 2)\right)^r + \frac{k(k + 1)}{k^2 + 2k - 1}.$$

Hence, at time $T = 2\lfloor L/4 \rfloor = \frac{1}{2} \log_{k^2+2k} \log n + O(1)$ we find that

$s_T = \Omega(\sqrt{\log n})$. It is evident that the same bound holds for $v \in U \cap Y$; that is, when $\deg(v) = k + 3$.

From that point on, the number of vertices on fire is large (relative to the number of firefighters introduced) so that the fire will be spreading very fast. Let $q_t = \sum_{r=1}^{t} s_t$ be the number of vertices on fire at time t; hence, we have that $q_T \geq s_T = \Omega(\sqrt{\log n})$. We claim that for every $t \geq T$ we have that $q_t \geq n/2$ or $q_t \geq q_T(1+\varepsilon'/2)^{t-T}$, and the proof is by induction. The statement holds for $t = T$. For the inductive step, suppose that the statement holds for $t \geq T$. If $q_t \geq n/2$, then $q_{t+1} \geq q_t \geq n/2$ as well. Suppose then that $q_t < n/2$. It follows from Lemma 8.3.6 that at least $\varepsilon' q_t$ vertices are not on fire (including perhaps some protected vertices) but are adjacent to vertices on fire. Thus, by the inductive hypothesis, at least

$$\varepsilon' q_t - t = (1 - o(1))\varepsilon' q_t > \frac{\varepsilon'}{2} q_t$$

new vertices are going to capture fire at the next round, so

$$q_{t+1} \geq q_t(1 + \varepsilon'/2) \geq q_T(1 + \varepsilon'/2)^{t+1-T},$$

and the claim holds. Note that the claim implies that at time $\hat{T} \leq \log_{1+\varepsilon'/2} n = O(\log n)$ at least half of the vertices are on fire.

For $t \geq \hat{T}$, it is easier to focus on P_t, the set of vertices that are not burning at time t (including t vertices protected by the firefighter till this point of the process). Let $p_t = |P_t| = n - q_t$. Note that for every $t \geq \hat{T}$, we have $p_t \leq p_{\hat{T}} < n/2$. We will prove that if $p_t \geq \frac{2(k+3)}{\varepsilon'} t$, then

$$p_{t+1} \leq p_t \left(1 - \frac{\varepsilon'}{2(k+3)} \right),$$

and this will finish the proof. If this holds, then the inequality $p_t \geq \frac{2(k+3)}{\varepsilon'} t$ must be false for

$$
\begin{aligned}
t \geq \bar{T} &= \hat{T} + \log_{1/\left(1 - \frac{\varepsilon'}{2(k+3)}\right)} n \\
&= \log_{1+\varepsilon'/2} n + \log_{1/\left(1 - \frac{\varepsilon'}{2(k+3)}\right)} n \\
&= O(\log n).
\end{aligned}
$$

Therefore, at most $p_{\bar{T}} < \frac{2(k+3)}{\varepsilon'}\bar{T} = O(\log n)$ vertices can be saved.

We will prove the claim by induction. Suppose that $\frac{2(k+3)}{\varepsilon'}t \leq p_t < n/2$. Using Lemma 8.3.6 for the last time, we derive that $|N(P_t)| \geq \varepsilon' p_t$; that is, at least $\varepsilon' p_t$ of burning vertices are adjacent to some vertex from P_t. Since the maximum degree of G is $k+3$, this implies that at least $\varepsilon' p_t/(k+3)$ vertices of P_t are adjacent to the fire. Hence, at least

$$\frac{\varepsilon' p_t}{k+3} - t \geq \frac{\varepsilon' p_t}{2(k+3)}$$

new vertices will capture the fire in the next round. $\qquad\square$

8.4 Fighting Constrained Fires in Graphs

We close out the chapter by considering a variant of firefighting, called *k-Constrained Firefighter*, first introduced in [49]. In this process, k is a positive integer and in each round, the fire chooses at most k neighboring vertices to burn. In k-Firefighter, a move for the fire is to spread to at most k unprotected vertices. Unlike the usual process of Firefighter, this results in a two-player game, where an optimal strategy for the fire aims to burn as many vertices as possible, and an optimal strategy for the firefighter aims to save as many vertices as possible.

The choice of strategy for the fire is important in k-Constrained Firefighter. Consider the graph G in Figure 8.4. Suppose the fire breaks out at x and the firefighter protects a. In Firefighter, in the second round the fire spreads to both y and b. The firefighter can then only save two vertices of G. However, in 1-Constrained Firefighter, if the fire chooses to burn y in the second round, then the firefighter can save all of K_m by protecting b. If the fire burns b rather than y in the second round, then the firefighter can save 3 vertices in the 5-vertex clique. For graphs like cliques or paths, however, the game does not depend on how the fire spreads.

We now define the surviving rate of graphs in the game of k-Constrained Firefighter, which is analogous to the definition in

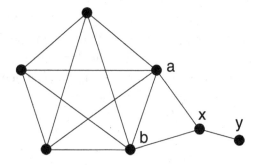

FIGURE 8.4: In the graph G, the choice of vertices burned by the fire affects how many vertices are saved.

Firefighter. For a vertex v in G, define $\mathrm{sn}_{k,C}(G, v)$ to be the number of vertices that can be saved if a fire breaks out at v. For a graph G, define its *k-Constrained surviving rate* to be

$$\rho(G, k) = \frac{1}{n^2} \sum_{u \in V(G)} \mathrm{sn}_{k,C}(G, u).$$

For example, for a complete graph we have that

$$\rho(K_n, k) = \frac{\lceil (n-1)/(k+1) \rceil}{n} \geq \frac{1}{k+1} \left(1 - \frac{1}{n} \right).$$

For a path, it is an exercise to show that

$$\rho(P_n, k) = \rho(P_n) = \frac{2}{n} \cdot \frac{n-1}{n} + \frac{n-2}{n} \cdot \frac{n-2}{n} = 1 - \frac{2}{n} + \frac{2}{n^2}.$$

We show that random regular graphs have low k-Constrained surviving rates for all values of k by using analogous methods (such as spectral techniques) from previous sections in this chapter. Our discussion here stems from [49]. As we will show in Theorem 8.4.2, a.a.s. random regular graphs have a k-surviving rate at most

$$\frac{1 + 2d^{-1/2}(\sqrt{k} + O(1))}{k+1} = \frac{(1 + O(d^{-1/2}))}{k+1},$$

where d is the degree of regularity. This tends to $\frac{1}{k+1}$ as d tends to infinity, which is the smallest possible surviving rate in k-Firefighter (see (8.2)).

As adding edges does not increase the k-surviving rate, it follows that cliques have the smallest surviving rates. Hence, for a graph with n vertices we have that

$$\rho(G, k) \geq \frac{\lceil (n-1)/(k+1) \rceil}{n} \geq \frac{1}{k+1}\left(1 - \frac{1}{n}\right). \qquad (8.2)$$

We now reference a useful upper bound for the k-surviving rate of a connected graph as a function of k and its order.

Theorem 8.4.1 ([49]). *For a connected graph G on n vertices,*

$$\begin{aligned} \rho(G, k) &\leq 1 - \frac{2}{n} + \frac{1}{n^2} + \frac{1}{n^2}\left\lceil \frac{n-1}{k+1} \right\rceil. \qquad (8.3) \\ &\leq 1 - \frac{1}{n}\left(2 - \frac{1}{k+1}\right) + O\left(\frac{1}{n^2}\right). \end{aligned}$$

Note that the bound in (8.3) is sharp as equality holds for a star on n vertices.

Random d Regular Graphs Are Flammable

We now consider k-Constrained Firefighter played on random d-regular graphs $\mathcal{G}_{n,d}$. We once again use the *pairing model* of random regular graphs defined earlier. The following theorem from [49] gives an asymptotic upper bound for the k-Constrained surviving rate of random d-regular graphs for all values of d and k.

Theorem 8.4.2 ([49]). *Let $d \geq 3$, $k \geq 1$, and fix $\varepsilon > 0$. If $\lambda = 2\sqrt{d-1} + \varepsilon$, then for $G \in \mathcal{G}_{n,d}$ we obtain that a.a.s.*

$$\begin{aligned} \rho(G, k) &\leq \frac{(1 + o(1))}{k+1}\left(1 + \frac{\lambda}{d}\left(\sqrt{k} + \frac{d}{d-\lambda}\right)\right) \\ &= \frac{(1 + O(d^{-1/2}))}{k+1}. \end{aligned}$$

Before we prove Theorem 8.4.2 we need a lemma (stated without proof), which is analogous to Lemma 8.3.4. For $d \geq 3$, a cycle is called *short* if it has length at most $L = \log_{d-1} \log_{d-1} n$.

Lemma 8.4.3 ([49]). *If $d \geq 3$ and $G \in \mathcal{G}_{n,d}$, then a.a.s. the number of vertices that belong to a short cycle is at most $\log^2 n$.*

Proof of Theorem 8.4.2. Consider the vertex set U consisting of vertices from G that do not belong to a short cycle of length at most L. By Lemma 8.4.3, since $|U| \geq n - \log^2 n$ a.a.s., we have that a.a.s.

$$
\begin{aligned}
\rho(G, k) &= \frac{1}{n^2} \sum_{v \in V(G)} \mathrm{sn}_{k,C}(G, v) \\
&= \frac{1}{n^2} \sum_{v \notin U} \mathrm{sn}_{k,C}(G, v) + \frac{1}{n^2} \sum_{v \in U} \mathrm{sn}_{k,C}(G, v) \\
&= o(1) + \frac{1}{n|U|} \sum_{v \in U} \mathrm{sn}_{k,C}(G, v).
\end{aligned}
$$

Therefore, it is enough to show that a.a.s. for all $u \in U$

$$
\mathrm{sn}_{k,C}(G, u) \leq (1 + o(1)) \frac{n}{k+1} \left(1 + \frac{\lambda}{d} \left(\sqrt{k} + \frac{d}{d - \lambda} \right) \right).
$$

Let $u \in U$ and let B_t, F_t denote the set of vertices burned or protected in round t, respectively. We will investigate three successive phases for the fire spreading from u.

In the first phase, consisting of rounds up to some constant t_0, the fire spreads to all nonprotected vertices adjacent to the fire but fewer than k new vertices capture fire in each round. We will use the fact that G is locally a tree to bound t_0. In the second phase, defined as those rounds up to a certain $t_1 = t_1(n)$, the fire spreads to k neighbors, but there are still no short cycles. The subgraph induced by B_{t_1} is a tree and there will be $(1 + o(1))kt_1$ burned vertices. In the third stage after round t_1, there are short cycles that clearly help the firefighter; however, many vertices are already burned, which makes it impossible for the firefighter to save too many vertices.

In the first phase, in order to minimize $|B_t|$, the firefighter uses a greedy strategy (as we have seen several times in the chapter). However, if $k \geq d$, then the fire cannot use the whole power in this phase and can only spread to less than k vertices in each round.

This property can occur only for a constant number of initial steps. In the first round the fire spreads to at least $d - 1$ vertices, then to at least $\left((d-1)^2 - 1\right)$ new ones, and so on. In round t during this phase, there are at least

$$\left((d-1)^t \quad -(d-1)^{t-2} - (d-1)^{t-3} - \cdots - (d-1) - 1\right)$$
$$= (d-1)^t - \frac{(d-1)^{t-1}-1}{d-2}$$
$$= \frac{(d-1)^{t-1}(d^2-3d+1)+1}{d-2}$$

new vertices on fire. We therefore, have that

$$t_0 \leq \left\lceil \log_{d-1}\left(\frac{k(d-2)-1}{d^2-3d+1}\right) + 1 \right\rceil,$$

so it is a constant that does not depend on n.

In the second phase, after round t_0, the fire is free to spread to k new vertices in each round. After $t_1 = \lfloor \frac{1}{2}L \rfloor = \lfloor \frac{1}{2}\log_{d-1}\log_{d-1}n \rfloor$ rounds, there are

$$|B_{t_1}| = O(kt_0) + k(t_1 - t_0) = (1 + o(1))kt_1$$

vertices on fire that form a tree.

We now consider the third stage after round t_1. It is straightforward to see that the fire will be spreading up to round $\hat{T}$ when $E(B_{\hat{T}}, V \setminus B_{\hat{T}}) = E(B_{\hat{T}}, F_{\hat{T}})$; that is, there is no vertex adjacent to the fire that is not protected. Observe that if for every $t_0 < t \leq T$, then we have that $|E(B_t, V \setminus B_t)| \geq |E(B_t, F_t)| + dk$, the fire spreads with the full speed; that is, $|B_{t+1}| = |B_t| + k$ during this stage of the process. (Since there are at least dk edges from burned to non-protected vertices, there must be at least k nonprotected vertices adjacent to the fire.) Hence, $|B_t| \sim kt$ during this time period. Note that from (8.1) it follows that for every round t

$$|E(B_t, V \setminus B_t)| \geq \frac{(d-\lambda)|B_t||V \setminus B_t|}{n}$$
$$\sim \frac{(d-\lambda)kt(n-kt)}{n}.$$

Since F_T can receive at most $d|F_T| = dT$ edges, T can be taken to be arbitrarily large as long as T satisfies the inequality

$$(1 + o(1))\frac{(d - \lambda)kT(n - kT)}{n} \geq dT + dk \sim dT.$$

If T is maximized, $T + 1$ does not satisfy the inequality, which implies that

$$1 - \frac{k(T + 1)}{n} \leq (1 + o(1))\left(\frac{d}{d - \lambda}\right)\frac{1}{k}.$$

Therefore, we have that

$$T \geq (1 + o(1))\frac{n}{k}\left(1 - \left(\frac{d}{d - \lambda}\right)\frac{1}{k}\right). \qquad (8.4)$$

A lower bound for T can be derived by estimating the number of edges between B_t and F_t using Lemma 8.1.2, and using inequality (8.4):

$$|E(B_t, F_t)| \leq \frac{d|B_t||F_t|}{n} + \lambda|F_t|\sqrt{\frac{|B_t|}{|F_t|}}$$

$$\sim \left(\frac{dkt^2}{n} + \lambda t\sqrt{k}\right).$$

We therefore, have the milder condition for T

$$\frac{(d - \lambda)kT(n - kT)}{n} \geq (1 + o(1))\left(\frac{dkT^2}{n} + \lambda T\sqrt{k}\right),$$

and so we have that

$$1 - \frac{(k + 1)T}{n} \geq (1 + o(1))\frac{\lambda}{d}\left(1 + \frac{1}{\sqrt{k}} - \frac{kT}{n}\right).$$

Via (8.4), we obtain that T can be arbitrarily large, provided that T satisfies the following inequality

$$1 - \frac{(k + 1)T}{n} \geq (1 + o(1))\frac{\lambda}{d}\left(\frac{1}{\sqrt{k}} + \left(\frac{d}{d - \lambda}\right)\frac{1}{k}\right).$$

If T is taken such that $T + 1$ does not have the desired property, we have that the third phase lasts until time T and

$$T \geq (1 + o(1)) \frac{n}{k+1} \left(1 - \frac{\lambda}{d} \left(\frac{1}{\sqrt{k}} + \left(\frac{d}{d - \lambda} \right) \frac{1}{k} \right) \right).$$

Hence, we have that the number of vertices saved $\mathrm{sn}_{k,\mathrm{C}}(G, u)$ is at most

$$n - (1 + o(1)) kT \leq (1 + o(1)) \frac{n}{k+1} \left(1 + \frac{\lambda}{d} \left(\sqrt{k} + \frac{d}{d - \lambda} \right) \right),$$

and the proof follows from Lemma 8.1.1. $\qquad\qquad\square$

Chapter 9

Acquisition Number

We consider a graph searching process that is a simplified model for the spread of gossip or information in a network. For this, we consider the notion of acquisition in a graph defined as follows.

Suppose each vertex of a graph G begins with a weight of 1 (this can be thought of as one bit of information starting at that vertex). A *total acquisition move* is a transfer of all the weight from a vertex v onto a vertex u, provided that immediately prior to the move, the weight on u is at least the weight on v. For example, a vertex of weight 3 can receive a transfer of weights of either 1, 2, or 3, but no higher weights. Suppose we continue with acquisition moves until no further moves may be made (which occurs as the graphs we consider are finite). Such a maximal sequence of moves is referred to as an *acquisition protocol* and the collection of vertices that retain positive weight after an acquisition protocol is called a *residual set*. Every residual set is necessarily independent (otherwise, weights may continue to transfer).

In a wheel or star, the maximum degree vertex u can be transferred weights from all remaining vertices resulting in a residual consisting solely of u. Hence, in both these cases, the acquisition number is 1. Figure 9.1 gives another example.

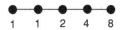

FIGURE 9.1: In this configuration of weights, the right end-vertex is the sole member of a residual set.

As with most of the games and processes we have studied in this book, we are interested in optimizing (in this case, minimizing) the cardinality of a residual set. The corresponding graph parameter is the *total acquisition number of G*, denoted $a_t(G)$. The parameter $a_t(G)$ was introduced by Lampert and Slater [128] and studied in [176, 130]. In [128], it is shown that $a_t(G) \leq \lfloor \frac{n+1}{3} \rfloor$ for every connected graph G on n vertices; further, they proved that this bound is tight. Slater and Wang [176], via a reduction to the three-dimensional matching problem, show that it is **NP**-complete to determine whether $a_t(G) = 1$ for general graphs G. In [130], upper bounds on the acquisition number of trees are shown in terms of the diameter and the order of the graph. They show further that for graphs of order n that are diameter two, $a_t(G) \leq 32 \log n \log \log n$.

In the present chapter, we focus on the behavior of the acquisition number of random graph models, including the binomial random graph in Section 9.1 and random geometric graphs in Section 9.2. We finish by randomizing the process by considering the acquisition number of the path, where vertices have random weights determined by a sequence of independent Poisson variables with unit mean.

9.1   Binomial Random Graphs

In this section, we study the total acquisition number of the binomial random graph $\mathbb{G}(n, p)$. The results and proofs from this section derive from the paper [15]. The main theorem we present is the following.

Theorem 9.1.1 ([15]). *Fix $\varepsilon > 0$. If*

$$p = p(n) \geq \frac{1 + \varepsilon}{\log 2} \cdot \frac{\log n}{n},$$

then a.a.s. $a_t(\mathbb{G}(n, p)) = 1$.

Before we tackle Theorem 9.1.1, we need the following elementary lemma, whose short proof we include for completeness.

Lemma 9.1.2 ([15]). *If vertex v is to acquire weight w during the acquisition process, then v has degree at least $\log_2 w$.*

Proof. The vertex v can only ever acquire weight $1+2+\ldots+2^{d(v)-1}$, in addition to the 1 it receives at the beginning of the process. Hence, this gives a total of $2^{d(v)}$. □

We note the following, complementary result proven in [15].

Theorem 9.1.3 ([15]). *Suppose that* $p = \frac{c+o(1)}{\log 2} \cdot \frac{\log n}{n}$ *for some fixed* $c \in (0, 1)$. *If* $0 < \varepsilon < \min\{c, 1-c\}$, *then a.a.s.*

$$n^{1-c-\varepsilon} \leq a_t(\mathbb{G}(n, p)) \leq n^{1-c+\varepsilon}.$$

Theorem 9.1.3 implies that at the connectivity threshold of the random graph (that is, where $p = \frac{\log n}{n}$), the total acquisition number is already of polynomial size; namely, it is at least, say, $n^{0.3}$. Theorems 9.1.1 and 9.1.3 together imply that $p = \frac{\log_2 n}{n}$ is the sharp threshold for the property $a_t(G) = 1$.

Our next goal is to define a property of rooted trees that ensures the root can acquire all the weight on the tree, given that each vertex begins with weight 1. Let T be a tree rooted at r. We say T has the *cut-off property* if the following holds: for each vertex v with children $v_1, v_2, \ldots, v_k$, and denoting by T_i the subtree rooted at v_i, there exists an i' (which may depend on v) such that $|T_i| = 2^{i-1}$ for $i \leq i'$, and $|T_i| \leq 2^{i'}$ for $i > i'$. In this case, the vertices v_i for $i \leq i'$ are called *exact*. A vertex with an exact ancestor is called *tight* and otherwise, vertices are *loose*.

Lemma 9.1.4 ([15]). *If T is a tree rooted at r that has the cut-off property, then $a_t(T) = 1$. In particular, vertex r can acquire all the weight.*

Proof. We proceed by induction on the depth of T. The depth 0 case is trivial. For the induction step, let r have children $v_1, v_2, \ldots, v_k$ and let T_i be the subtree of T rooted at the v_i. Then the T_i inherit the cut-off property, and all have depth strictly less than the depth of T, and so by induction, all the weight from subtree T_i can be transferred to the root v_i. By the cut-off property, r may acquire the weight of each child v_i, going in order of increasing index. □

We now give a recursive construction of a tree with the cut-off property. For any $\rho, m, \sigma \in \mathbb{N}$, and positive integer sequence c_{m-k-1}, construct the rooted tree T_ρ by the following process.

(i) In the first step, we introduce a *root* vertex $r = \langle\rangle$ with *weight* $w(r) = \rho$, and of *level* $k = 0$.

(ii) At level k, for a vertex $\langle i_1, i_2, \ldots, i_k \rangle$ with weight $w = w(\langle i_1, i_2, \ldots i_k \rangle) > 1$, we iterate the following.

   (a) If $1 < w \leq c_{m-k-1}$, then add $w - 1$ end-vertices to the vertex $\langle i_1, i_2, \ldots, i_k \rangle$ each with weight 1.

   (b) If $w > c_{m-k-1}$, then attach $c = c_{m-k-1}$ children to vertex $\langle i_1, i_2, \ldots, i_k \rangle$, labeled

$$\langle i_1, i_2, \ldots, i_k, 1 \rangle, \ldots, \langle i_1, i_2, \ldots, i_k, c \rangle.$$

Let i' be the smallest integer $i \geq 0$ such that

$$\frac{w - 2^i - \sigma}{c - i - \sigma} \leq 2^i + \sigma.$$

Assign weights to the children via the following procedure.

$$w(\langle i_1, i_2, \ldots, i_{k+1} \rangle) = \begin{cases} 2^{i_{k+1}-1} & \text{if } i_{k+1} \leq i' \\ 1 & \text{if } i' < i_{k+1} \leq i'+\sigma \quad (9.1) \\ \frac{w-2^{i'}-\sigma}{c-i'-\sigma} & otherwise. \end{cases}$$

Here, we assume that $\frac{w-2^{i'}-\sigma}{c-i'-\sigma}$ is an integer and that $i' + \sigma < c$ such that i' is well-defined.

See Figure 9.2 for a depiction of the tree T_ρ.

Observe that the tree T_ρ has the cut-off property. The quantity $w(v)$ represents the number of vertices that will end up in the subtree rooted at v. The sequence c provides a threshold for the recursive part of the definition to occur. Hence, if the weight on v is at most c, then the subtree appears in the form of end-vertices.

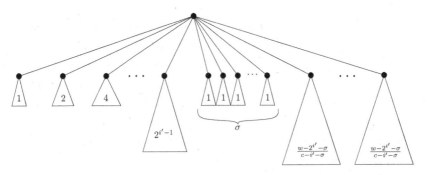

FIGURE 9.2: The recursive tree T_ρ with the cut-off property.

If the weight on v exceeds c, then v will have exactly c children and the weight is distributed according to (9.1).

We give a definition that will aid in the selection of parameters of T_ρ. Let $d = \frac{1+\varepsilon}{\log 2} \log n$ and let σ, β, α be constants. Let

$$
c_j^* = \begin{cases} \beta \log n & \text{if } j \le \alpha \frac{\log n}{\log \log n}, \\ \frac{1+\frac{\varepsilon}{2}}{\log 2} \log n & \textit{otherwise.} \end{cases}
$$

and let c be a sequence such that $c_j^* \le c_j \le c_j^*(1 + o(1))$. Define the function

$$
i^*(x) = \begin{cases} 0 & \text{if } x \le \sigma \\ \lceil \log_2(x - \sigma) \rceil & \text{otherwise.} \end{cases}
$$

Define sequences $\rho_1, \rho_2, \ldots$ and $b_1, b_2, \ldots$ recursively by setting $\rho_1 = 2$ and

$$
\rho_{j+1} = \sigma + 2^{i^*(\rho_j)} + \left(c_j - i^*(\rho_j) - \sigma \right) \cdot \rho_j,
$$

$b_1 = 1$ and

$$
b_{j+1} = \left(c_j - i^*(\rho_j) - \sigma \right) \cdot b_j.
$$

The sequences ρ, b depend on the choice of the sequence c, where we assume that the constants $\varepsilon, \sigma, \beta, \alpha$ are fixed. The purpose of this recursive sequence is to calculate (for a given sequence c and depth m) the weight of the root; in fact, ρ_j is the weight of each loose vertex at level j (counted from the bottom) so the weight of the whole tree is ρ_m. The purpose of i' was to make sure that the total weight of subtrees rooted at exact vertices together with the weight of the root is at least the weight of each subtree rooted at nonexact children. It is straightforward to see that i^* has the same purpose and so these values are always the same. We fix $\rho_1 = 2$, indicating that every loose vertex at the level directly above the bottom has precisely one end-vertex. The parameter b_j counts how many such end-vertices we have in the tree rooted at loose vertices at level j.

Let ρ^* be the sequence obtained by setting $c = c^*$ as defined above. Let m be the largest integer such that $\rho_m^* \leq \frac{8}{5}n$. Note that for every $j \geq 2$, $\rho_j \geq \rho_2 = \Omega(\log n)$ and so

$$\rho_{j+1} = \left(c_j - \log_2 \rho_j + O(1)\right) \cdot \rho_j \leq c_j \rho_j. \tag{9.2}$$

It follows that $\rho_m^* = \Omega\left(\frac{n}{\log n}\right)$, since ρ grows by at most a log factor each time.

Throughout, we keep $m = m(n)$ as defined above (9.2) and consider the sequences ρ, b only up to the terms ρ_m, b_m. We will consider sequences c with terms that might be larger than those of c^*. As a result, $\rho_j \geq \rho_j^*$ for all j. However, we will only consider sequences c such that the corresponding sequence ρ has $\rho_m = n^{1+o(1)}$.

Lemma 9.1.5 ([15]). *Let c be any sequence such that $c_j^* \leq c_j \leq c_j^*(1+o(1))$ and the ρ-sequence corresponding to c has $\rho_m = n^{1+o(1)}$. If $\alpha < \frac{\beta \log 2}{2}$, then*

$$\rho_j = \exp\{(j-1)\log\log n + O(j)\}$$

for all $2 \leq j \leq m$.

Proof. It is straightforward to see that $\rho_2 = \Theta(\log n)$ and it follows immediately from (9.2) that for every $2 \leq j < m$ we have that

$$\frac{\rho_{j+1}}{\rho_j} \leq c_j \leq 2c_j^* \leq 2\max\left(\beta, \frac{1+\frac{\varepsilon}{2}}{\log 2}\right)\log n.$$

Hence, $\rho_j \leq \exp\{(j-1)\log\log n + O(j)\}$ for all $j \leq m$ and so the upper bound holds. In particular, as long as $j \leq \alpha \frac{\log n}{\log \log n}$ we have that

$$\rho_j \leq \rho_{\alpha \frac{\log n}{\log \log n}} \leq n^{\alpha \cdot (1 + o(1))}. \tag{9.3}$$

For the lower bound, we use (9.2) and (9.3) to derive that for every j such that $2 \leq j \leq \alpha \frac{\log n}{\log \log n}$, we have that

$$\frac{\rho_{j+1}}{\rho_j} \geq \left(\beta(1 + o(1)) - \frac{\alpha}{\log 2}(1 + o(1)) \right) \log n \geq \frac{\beta}{2} \log n.$$

Note that this follows by our choice of α. As for every $j \leq m$ we have that $\rho_j \leq \rho_m = n^{1 + o(1)}$, for every j such that $\alpha \frac{\log n}{\log \log n} < j < m$ we derive that

$$\frac{\rho_{j+1}}{\rho_j} \geq \left(\frac{1 + \frac{\varepsilon}{2} + o(1)}{\log 2} - \frac{1 + o(1)}{\log 2} \right) \log n \geq \frac{\varepsilon}{4 \log 2} \log n.$$

We now obtain the following, which completes the proof: $\rho_j \geq \exp\{(j - 1)\log \log n + O(j)\}$ for all $j \leq m$. □

We assume that $\alpha < \frac{\beta \log 2}{2}$. We immediately obtain the following corollary.

Corollary 9.1.6 ([15]). $m \sim \frac{\log n}{\log \log n}$.

We assumed earlier that i' was well-defined; that is, that the condition $i' + \sigma < c$ holds. Because of the relationship between the two definitions, this condition is equivalent to the condition $c_j - i^*(\rho_j) - \sigma \geq 1$. In the next lemma, we show that the same condition for α as in Lemma 9.1.5 is enough to guarantee that i' is well-defined. The following is a useful property that we will use in the next few arguments.

Lemma 9.1.7 ([15]). *The quantity i' exists. Further, the following stronger property holds: for every j such that $1 \leq j \leq m$ we have*

$$c_j - \log_2 \rho_j = \Omega(\log n).$$

Proof. To show that i' exists, we will show that the equivalent condition $c_j - i^*(\rho_j) - \sigma \geq 1$ holds. We actually prove the stronger statement that $c_j - \log_2 \rho_j = \Omega(\log n)$ for every $1 \leq j \leq m$. For $j \leq \alpha \frac{\log n}{\log \log n}$ we have that

$$
\begin{aligned}
c_j - \log_2 \rho_j &\geq c_j - \log_2 \rho_{\alpha \frac{\log n}{\log \log n}} \\
&\sim \left(\beta - \frac{\alpha}{\log 2} \right) \log n \\
&= \Omega(\log n).
\end{aligned}
$$

If $j > \alpha \frac{\log n}{\log \log n}$, then

$$
\begin{aligned}
c_j - \log_2 \rho_j &\geq c_j - \log_2 \rho_m \\
&\sim \left(\frac{1 + \frac{\varepsilon}{2}}{\log 2} - \frac{1}{\log 2} \right) \log n \\
&= \Omega(\log n),
\end{aligned}
$$

and the proof follows. $\square$

We next show that we can adjust a sequence c^* to derive another sequence c with $\rho_m \sim 8n/5$. We omit the technical proof.

Lemma 9.1.8 ([15]). *There exists a sequence of integers c_j with $c_j^* \leq c_j \leq c_j^*(1 + o(1))$ and $\rho_m \sim 8n/5$.*

Define the tree T'_{ρ_j} to be the tree T_{ρ_j} with each end-vertex in the bottom level being removed if it has a loose parent. Call the parents that lose their children *bereft*. Note that by induction and definition of ρ_j, b_j, and T'_{ρ_j}, we see that T'_{ρ_j} has $\rho_j - b_j$ many vertices, b_j of which are bereft. It is not difficult to see that by construction, T'_{ρ_j} has the cut-off property. Further, if we form another tree T''_{ρ_j} by reattaching at most one end-vertex to each bereft parent of T'_{ρ_j}, then T''_{ρ_j} still has the cut-off property.

Our next goal is to show that almost all vertices of T'_{ρ_j} are bereft. Since each bereft vertex has exactly one child in T_{ρ_j}, we have that $|T_{\rho_j}| \sim 2|T'_{\rho_j}|$. In particular, $|T'_{\rho_j}| \sim 4n/5$. The proof of the following lemma is omitted.

Lemma 9.1.9 ([15]). *For all $1 \leq j \leq m$ we have $\frac{b_j}{\rho_j - b_j} \to 1$ as $n \to \infty$.*

Now, we are ready to show that $T'_{\rho m}$ can be embedded into the random graph $\mathbb{G}(n, p)$.

Lemma 9.1.10 ([15]). *If $\beta < \frac{1}{10 \log 2}$ and $0 < \alpha < \frac{\beta \log 2}{2}$, then a.a.s. $\mathbb{G}(n, p)$ (with $p = \frac{1+\epsilon}{\log 2} \cdot \frac{\log n}{n}$) contains a copy of $T'_{\rho m}$.*

Proof. We will embed $T'_{\rho m}$ in $\mathbb{G}(n, p)$. Select any vertex (arbitrarily) that will serve as a root of the tree. The embedding is done greedily and from the top down, and at each step we reveal the neighborhood of one vertex. We group vertices in the same level (that is, distance from the root) consecutively. The embedding will be determined iteratively as we reveal the random graph. We will not put a vertex of $\mathbb{G}(n, p)$ into our partial embedding until we have exposed all of its children.

We say that a vertex in level k is *bad* if its neighborhood (into the unexposed vertices) contains less than c_{m-k-1}. We will show that a.a.s. the root is not bad and no vertex has more than σ bad children. Any bad children will be put into the partial embedding as end-vertices, and the other vertices will be arbitrarily assigned (to non-end-vertices first and then to end-vertices, if the number of bad children is smaller than σ).

The tree $T'_{\rho m}$ has at most

$$\left(\frac{1 + \frac{\varepsilon}{2}}{\log 2} \log n\right)^{m - \alpha \frac{\log n}{\log \log n}} = n^{(1-\alpha)(1+o(1))} - o(n)$$

vertices total in levels 0 through $m - \alpha \frac{\log n}{\log \log n} - 1$. Thus, the expected degree (into the unexposed vertices) of each vertex exposed in such a level k is

$$(1 + o(1))d \sim \frac{1+\varepsilon}{\log 2} \log n > c_k + \frac{\varepsilon}{3 \log 2} \log n \qquad (9.4)$$

and so it follows from the Chernoff bound that the probability that a fixed vertex is bad is polynomially small, that is, at most $n^{-\Theta(\varepsilon^2)}$. For levels k farther to the bottom, note that the number of vertices that are not embedded yet is always at least $\frac{1+o(1)}{5} n$ and so the expected degree of each exposed vertex in layer k is at least

$$\frac{1 + o(1)}{5} d \sim \frac{1+\varepsilon}{5 \log 2} \log n > c_k + \frac{1}{10 \log 2} \log n,$$

again yielding that the probability that a fixed vertex is bad is polynomially small (this time the exponent is a universal constant, not a function of ε).

Therefore, if $\sigma = \Theta(1/\varepsilon^2)$ is a large enough constant, then a.a.s. each vertex has at most σ bad children. This proves that our embedding procedure is successful a.a.s. □

Let B be the set of bereft vertices in $\mathbb{G}(n,p)$ and R be the set of remaining, unexposed vertices that are not embedded in the tree yet. Note that $|B| = b_m \sim 4n/5$ and $|R| = n - (1+o(1))b_m \sim n/5$. An important property is that no edge between B and R is exposed at this point, so the next lemma shows that a.a.s. set B dominates set R but in such a way that at most one vertex of T is assigned to each bereft vertex.

These observations provide the proof of our main result, Theorem 9.1.1. The proof of the following theorem uses Hall's theorem for bipartite graphs (see Theorem 5.1.4) and is omitted.

Lemma 9.1.11 ([15]). *A.a.s. there is a matching from R to B that saturates R.*

9.2 Random Geometric Graphs

In this section, we consider the acquisition number of random geometric graphs, $\mathbb{G}(\mathcal{X}_n, r_n)$. The results and proofs of this section are based on those in [114]. The main result we present is the following.

Theorem 9.2.1 ([114]). *If $r = r_n$ is any positive real number, then a.a.s. $a_t(\mathbb{G}(n,r)) = \Theta(f_n)$, where we have that*

$$f_n = \begin{cases} n & \text{if } r < 1, \\ \frac{n}{(r \log r)^2} & \text{if } r \geq 1 \text{ and } r \log r \leq \sqrt{n}, \\ 1 & \text{if } r \log r > \sqrt{n}. \end{cases}$$

We prove only the lower bound of Theorem 9.2.1. The proof begins with the following simple but useful observation.

Lemma 9.2.2 ([114]). *Let G be a graph. If $v \in V(G)$ is to acquire weight w (at any time during the process of moving weights around), then $deg(v) \geq \log w$. Additionally, all vertices that contributed to the weight of w (at this point of the process) are at graph distance at most $\log w$ from v.*

Proof. During each total acquisition move, when weight is shifted onto v from some neighboring vertex, the weight of v can at most double. Thus, v can only ever acquire $1 + 2 + \ldots + 2^{deg(v)-1}$, in addition to the 1 it starts with, and so v can acquire at most weight $2^{deg(v)}$. To see the second part, suppose that some vertex u_0 moved the initial weight of 1 it started with to v through the path $(u_0, u_1, \ldots, u_{k-1}, u_k = v)$. It is straightforward to see that after u_{i-1} transfers its weight onto u_i, u_i has weight at least 2^i. Thus, if u_0 contributed to the weight of w, then u_0 must be at graph distance at most $\log w$ from v. □

First, we concentrate on dense graphs for which we show a stronger result that no vertex can acquire large weight a.a.s.

Theorem 9.2.3 ([114]). *Suppose that*

$$r = r_n \geq c\sqrt{\log n}/\log\log n$$

for some sufficiently large $c \in \mathbb{R}$, and consider any acquisition protocol on $\mathbb{G}(n,r)$. Then a.a.s. each vertex in the residual set acquires $O((r\log r)^2)$ weight. As a result, a.a.s.

$$a_t(\mathbb{G}(n,r)) = \Omega\left(\frac{n}{(r\log r)^2}\right).$$

Proof. Let $\ell = 2\log r + 2\log\log r + \log(8\pi)$. Suppose for a contradiction that at some point of the process some vertex v acquires weight $w \geq 2^\ell = 8\pi(r\log r)^2$. Since one total acquisition move corresponding to transferring all the weight from some neighbor of v onto v, increases the weight on v by a factor of at most 2, we may assume that $w < 2^{\ell+1}$. It follows from Lemma 9.2.2 that all vertices contributing to the weight of w are at graph distance at most $\ell+1$ from v (and so at Euclidean distance at most $(\ell+1)r$).

The desired contradiction will be obtained if no vertex has at least 2^ℓ vertices (including the vertex itself) at Euclidean distance at most $(\ell + 1)r$.

The remaining part is a simple consequence of the Chernoff bounds and the union bound over all vertices. For a given vertex v, the number of vertices at Euclidean distance at most $(\ell + 1)r$ is a random variable Y that is stochastically bounded from above by the random variable

$$X \sim \text{Bin}(n - 1, \pi(\ell + 1)^2 r^2/n)$$

with $\mathbb{E}[X] \sim \pi\ell^2 r^2 \sim 4\pi(r \log r)^2$. Note that $Y = X$ if v is at distance at least $(\ell + 1)r$ from the boundary; otherwise, $Y \leq X$. It follows from the Chernoff bounds that

$$
\begin{aligned}
\mathbb{P}(Y \geq 2^\ell) &\leq \mathbb{P}\left(X \geq (2 + o(1))\mathbb{E}[X]\right) \\
&\leq \exp\left(-(1/3 + o(1))\mathbb{E}[X]\right) \\
&\leq \exp\left(-(4\pi/3 + o(1))(r \log r)^2\right) \\
&\leq \exp\left(-(\pi c^2/3 + o(1))\log n\right) \\
&= o(1/n),
\end{aligned}
$$

provided that c is large enough. The conclusion follows from the union bound over all n vertices of $\mathbb{G}(n, r)$. $\qquad\square$

Tool: De-Poissonization

For sparser graphs we will make use of a technique known as de-Poissonization, which has many applications in the field of geometric probability (see [155], for example).

Consider the following related model of a random geometric graph. Let $V = V'$, where V' is a set obtained as a homogeneous Poisson point process of intensity 1 in $[0, \sqrt{n}]^2$. Hence, V' consists of N points in the square $[0, \sqrt{n}]^2$ chosen independently and uniformly at random, where N is a Poisson random variable of mean n (which was introduced in Chapter 1). Exactly as we did for the model $\mathbb{G}(n, r)$, again identifying vertices with their spatial locations, we connect by an edge u and v in V' if the Euclidean distance between them is at most r. We denote this new model by $\mathrm{Po}(n, r)$.

The main advantage of defining V' as a Poisson point process is motivated by the following two properties: the number of vertices of V' that lie in any region $A \subseteq [0, \sqrt{n}]^2$ of area a has a Poisson distribution with mean a, and the number of vertices of V' in disjoint regions of $[0, \sqrt{n}]^2$ are independently distributed. By conditioning $\mathrm{Po}(n, r)$ upon the event $N = n$, we recover the original distribution of $\mathbb{G}(n, r)$. Therefore, since $\Pr(N = n) = \Theta(1/\sqrt{n})$, any event holding in $\mathrm{Po}(n, r)$ with probability at least $1 - o(f_n)$ must hold in $\mathbb{G}(n, r)$ with probability at least $1 - o(f_n \sqrt{n})$.

Back to the Acquisition Number

We now return to the lower bound. For sparser graphs we cannot guarantee that no vertex acquires large weight a.a.s. but a lower bound of the same order holds.

Theorem 9.2.4 ([114]). *If $r = r_n \geq c$ for some sufficiently large $c \in \mathbb{R}$, then a.a.s. we have that*

$$a_t\left(\mathbb{G}(n, r)\right) = \Omega\left(\frac{n}{(r \log r)^2}\right).$$

Proof. Since Theorem 9.2.3 applies to dense graphs, we may assume here that $r = O(\sqrt{\log n}/\log \log n)$ (in particular, $r \log r = o(\sqrt{n})$). Tessellate $[0, \sqrt{n}]^2$ into $\lfloor \sqrt{n}/(20r \log r) \rfloor^2$ squares, each one of side length $(20 + o(1))r \log r$. Consider the unit circle centered on the center of each square and call it the *center circle*. We say that a given square is *dangerous* if the corresponding center circle contains at least one vertex and the total number of vertices contained in the square is less than $1200(r \log r)^2$.

Consider any acquisition protocol. First, we show that at least one vertex from each dangerous square must belong to the residual set. Let u_0 be a vertex inside the corresponding center circle. Suppose to the contrary that the square has no vertex in the residual set. In particular, this implies that u_0 moved the initial weight of 1 it started with onto some vertex outside the square through some path $(u_0, u_1, \ldots, u_k)$. Note that the Euclidean distance between u_0 and the border of the square (and so also u_k) is at least $(20 + o(1))r \log r / 2 - 1 \geq 9r \log r$, provided that c is large enough, and so $k \geq 9 \log r$.

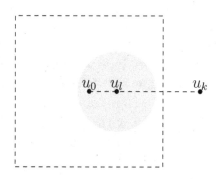

FIGURE 9.3: Residual sets contain at least one vertex from each dangerous square.

Consider the vertex u_ℓ on this path, where $\ell = \lfloor 4 \log r \rfloor \geq 3 \log r$, provided c is large enough; see Figure 9.3. Right after $u_{\ell-1}$ transferred all the weight onto u_ℓ, u_ℓ had weight at least $2^\ell \geq r^3 > 1200(r \log r)^2$, provided c is large enough. As argued in the proof of the previous theorem, at some point of the process u_ℓ must have acquired weight w satisfying $2^\ell \leq w < 2^{\ell+1}$. Lemma 9.2.2 implies that all vertices contributing to the weight of w are at Euclidean distance at most $(\ell+1)r$ from v and so inside the square (as always, provided c is large enough). However, dangerous squares contain less than $1200(r \log r)^2$ vertices, and so we derive a contradiction. The desired claim holds.

Showing that a.a.s. a positive fraction of the squares is dangerous is straightforward. In $\text{Po}(n, r)$, the probability that the center circle contains no vertex is $\exp(-\pi) \leq 1/3$. The number of vertices falling into the square is a Poisson random variable X with expectation $\mu \sim 400(r \log r)^2$. By the Chernoff bounds applied with $\varepsilon = e - 1$, we find that

$$\mathbb{P}(X \geq e\mu) \leq \left(\frac{e^{e-1}}{(1 + (e-1))^e} \right)^\mu = \exp(-\mu).$$

Hence, we obtain that

$$\mathbb{P}(X \geq 1200(r \log r)^2) \leq \mathbb{P}(X \geq e\mu) \leq \exp(-\mu) \leq 1/3,$$

provided c is large enough. Hence, the expected number of dangerous squares is at least $(1/3)(1/400 + o(1))n/(r \log r)^2 \gg \log n \to \infty$. By the Chernoff bounds, with probability at least $1 - o(n^{-1/2})$, the number of dangerous squares in $\text{Po}(n, r)$ is at least $(1/2500)n/(r \log r)^2$. By the de-Poissonization argument mentioned before this proof, the number of dangerous squares in $\mathbb{G}(n, r)$ is a.a.s. also at least $(1/2500)n/(r \log r)^2$. $\square$

The only range of $r = r_n$ not covered by the two theorems is when $r < c$ for c as in Theorem 9.2.4. However, in such a situation a.a.s. there are $\Omega(n)$ isolated vertices which clearly remain in the residual set. If r is such that $r \log r > \sqrt{n}$, then the trivial lower bound $\Omega(1)$ applies. The lower bound in the main theorem holds for the whole range of r.

9.3 Randomly Weighted Path

We finish the chapter with a discussion of the total acquisition number of a path where weights are distributed by a random process. The results and proofs in this section are derived from [102]. For the P_n with unit weights, we begin our discussion with the following elementary lemma. As the proof is short, it is included here.

Lemma 9.3.1 ([102]). *The sequence $\{a_t(P_n)\}_{n=1}^{\infty}$ is subadditive.*

Proof. Consider the graph P_{n+m}, with $m, n \in \mathbb{Z}$. If P_{n+m} is subdivided into P_n and P_m and distinct acquisition protocols are run on each, then the sum of the resulting residual sets is $a_t(P_m) + a_t(P_n)$. Now, since $a_t(P_{n+m})$ is the cardinality of the minimal residual set, the fact that it is possible to obtain a set of cardinality $a_t(P_n) + a_t(P_m)$ gives the bound

$$a_t(P_{n+m}) \leq a_t(P_n) + a_t(P_m),$$

and the proof follows.   □

The following lemma now follows directly by using the above and Fekete's lemma [84] on limits of subadditive sequences.

Corollary 9.3.2 ([102]). *For $\{a_t(P_i)\}_{i=1}^{\infty}$, the limit $\lim\limits_{n\to\infty} \frac{a_t(P_n)}{n}$ exists and is equal to $\inf \frac{a_t(P_n)}{n}$.*

Although it is clear that this limit equals $1/4$, the applications of subadditivity will be more evident in the next section where we use random weights.

Poisson Distribution

In this subsection, we consider the total acquisition number of P_n when each vertex begins with weight Poi(1); that is, the vertices have random weights determined by a sequence of independent Poisson variables with unit mean. We denote this specific configuration as P_n^{Poi}. In general, the uppercase letter A will be used for the acquisition number when it is viewed as a random variable. We can begin by proving that the limit

$$\lim_{n\to\infty} \frac{E(A_t(P_n^{\text{Poi}}))}{n}$$

exists, and provide upper and lower bounds for $E(A_t(P_n^{\text{Poi}}))$.

We may determine whether a given weighting of $P_n = (v_1, v_2, \ldots, v_n)$ can be used to transfer the total weight onto one vertex in the following way. Starting from v_1, we transfer its weight to the right as much as possible, ending at v_k for some $k \leq n$. Then independently (and using the initial weighting), we start from v_n and transfer its weight to the left as much as possible, ending at v_ℓ for some $\ell \geq 1$. It follows that if $k \geq \ell - 1$, then our task is possible; otherwise, it is not.

Determining $a_t(P_n)$ for a given weighting can be done as follows. Suppose that weights on the subpath $(v_1, v_2, \ldots, v_k)$ can be moved to one vertex. If weights on the subpath $(v_1, v_2, \ldots, v_k, v_{k+1})$ can also be moved to one vertex, then this is at least as good a strategy as moving only weights from $(v_1, v_2, \ldots, v_k)$ and then applying the best strategy for the remaining path. Therefore, finding

$a_t(P_n)$ for any weighting can be done with a straightforward greedy algorithm.

Finally, we derive from the above remarks that $E(A_t(P_n^{\text{Poi}}))$ is an increasing function of n.

We now present observations that are counterparts of Lemma 9.3.1 and 9.3.2, respectively.

Lemma 9.3.3 ([102]). *The sequence* $\{E(A_t(P_n^{Poi}))\}_{n=1}^{\infty}$ *is subadditive.*

Corollary 9.3.4 ([102]). *The limit* $\lim\limits_{n\to\infty} \frac{E(A_t(P_n^{Poi}))}{n}$ *exists and is equal to* $\inf \frac{E(A_t(P_n^{Poi}))}{n}$.

Finally, we move to the main result of this section.

Theorem 9.3.5 ([102]). *The expected acquisition number of* P_n^{Poi} *is bounded as*

$$0.242n \leq E(A_t(P_n^{Poi})) \leq 0.375n.$$

Proof. Define an *island* as the set of vertices to the left of the first zero weight; to the right of the last zero weight; or in between any two successive zero weights. Islands consist of a possibly empty set of nonzero numbers. The island cardinality therefore, has a geometric distribution with success probability $1/e$ and expected cardinality $e - 1$, yielding an expected number of $n/e + c$, where $0 \leq c \leq 1$, for the random number Λ of islands. The expected total acquisition number may be calculated using the conditional probability expression

$$
\begin{aligned}
E(A_t(P_n^{\text{Poi}})) &= \sum_{j=1}^{\Lambda} E\left[a_t(P_{|\Lambda_j|})\right] \\
&= \left(\frac{n}{e} + c\right) E[a_t(P_{|\Lambda_1|})] \\
&= \left(\frac{n}{e} + c\right) \sum_{j=0}^{n} E(A_t(P_j^{\text{Poi}})) P(|\Lambda_1| = j),
\end{aligned}
$$

where the second equality follows from Wald's Lemma [179] on expectations of sums of random variables. However, calculating $E\left[A_t(P_j^{\text{Poi}})\right]$ is difficult for arbitrary j. The probability that an island is of cardinality j equals $\left(1 - \frac{1}{e}\right)^j \left(\frac{1}{e}\right)$. Thus, there is a roughly 84% probability that an island has cardinality three or less, and a reasonable lower bound can be obtained by restricting the calculation to those cases. It is evident that $E\left[A_t(P_0^{\text{Poi}})\right] = 0$ and that $E\left[A_t(P_1^{\text{Poi}})\right] = E\left[A_t(P_2^{\text{Poi}})\right] = 1$. For P_3, however, it is possible to have $A_t(P_3) = 1$ or $A_t(P_3) = 2$. Because $A_t(P_3) = A_t(a - b - c)$ requires that $w(a) > w(b) \geq 1$ and $w(c) > w(b) \geq 1$, we condition on each of $w(a), w(b)$, and $w(c)$ being nonnegative, and, setting $W(b) = k$, we obtain that

$$
\begin{aligned}
P\left[A_t(P_3) = 2\right] &= \sum_{k \geq 1} \frac{e^{-1}}{1 - e^{-1}} \frac{1}{k!} \left(\sum_{j \geq k+1} \frac{e^{-1}}{(1 - e^{-1})} \frac{1}{j!} \right)^2 \\
&= \frac{1}{(e-1)^3} \sum_{k \geq 1} \frac{1}{k!} \left(\sum_{j \geq k+1} \frac{1}{j!} \right)^2 \\
&= \frac{e^2}{(e-1)^3} \sum_{k=1}^{\infty} \frac{1}{k!} \left(1 - \frac{\Gamma(k+1, 1)}{\Gamma(k+1)} \right)^2 \\
&\approx 0.10648.
\end{aligned}
$$

Thus, $E[A_t(P_3)]$ can be calculated as follows:

$$
\begin{aligned}
E[A_t(P_3)] &= P\left[A_t(P_3) = 1\right](1) + P\left[A_t(P_3) = 2\right](2) \\
&\approx 1.10648.
\end{aligned}
$$

Using (1) and the monotonicity of $E(A_t(P_n^{\text{Poi}}))$, we can now calculate a lower bound for $E(A_t(P_n^{\text{Poi}}))$ as

$$
\begin{aligned}
E(A_t(P_n^{\text{Poi}})) &\geq \left(\frac{n}{e} + c\right) \left(\left(1 - \frac{1}{e}\right) \frac{1}{e} + \left(1 - \frac{1}{e}\right)^2 \frac{1}{e} \right. \\
&\qquad \left. + 1.106 \left(1 - \frac{1}{e}\right)^3 \right) \\
&\geq 0.242n.
\end{aligned}
$$

To obtain an upper bound, we use the fact that $a_t(P_j) \leq \frac{j+1}{2}$

for every j. By using the above conditional probability expression, we derive an upper bound for $E(A_t(P_n^{\text{Poi}}))$ as

$$E(A_t(P_n^{\text{Poi}})) = \left(\frac{n}{e}+c\right)\left(\sum_{j=1}^{3} E(A_t(P_j))P(|\Lambda_1|=j)\right.$$
$$\left. + \sum_{j\geq 4} E(A_t(P_j))P(|\Lambda_1|=j)\right)$$

$$\leq \left(\frac{n}{e}+c\right)\sum_{j=1}^{3} E(A_t(P_j))P(|\Lambda_1|=j)$$
$$+ \left(\frac{n}{e}+c\right)\sum_{j\geq 4}\frac{j+1}{2}P(|\Lambda_1|=j).$$

It follows that

$$E(A_t(P_n^{\text{Poi}})) \leq 0.178n + \frac{n}{e}\frac{1-\frac{1}{e}}{2e}\sum_{j=4}^{\infty} j\left(1-\frac{1}{e}\right)^{j-1}+0.029n$$

$$= 0.207n + \frac{n}{e}\frac{1-\frac{1}{e}}{2e}\frac{(e-1)^3(3+e)}{e^2}$$

$$\approx 0.375n,$$

which gives us the desired bounds for $E(A_t(P_n^{\text{Poi}}))$. $\qquad\square$

With the support of a computer, it is straightforward to find $a_t(P_j)$ for a given initial weighting. By considering all possible k^j configurations of weights at most $k = k(j)$, we can prove a bound on $E(A_t(P_j))$ from both below and above. In [102], all paths on at most 21 vertices were considered (with different values of $k = k(j)$ for various j) to obtain the following bounds.

Corollary 9.3.6 ([102]). *The expected acquisition number of P_n^{Poi} is bounded as*

$$0.29523n \leq E(A_t(P_n^{Poi})) \leq 0.29576n.$$

Based on computer simulations completed in [102] on paths of length $n = 100,000,000,000$, it was conjectured there that $E(A_t(P_n^{\text{Poi}})) \approx 0.295531n$.

Chapter 10

Temporal parameters

Implicit in all of the graph searching games and related dynamic graph processes that we have considered is the notion of time or length of the game or process. For example, in Cops and Robbers, we may consider how many rounds it takes for the cops to capture the robber (assuming both sides play optimally, as described more explicitly in Section 10.1 below). For this, we may play with $c(G)$-many cops and consider the capture time of the graph, or instead, play with more cops than actually needed to capture the robber. The latter case is the new setting of Overprescribed Cops and Robbers, and is described in detail in Section 10.2. The length of the game is also considered, along with a new dynamic process of graph burning, and a drunk robber, who moves at random; see Sections 10.3 and 10.5, 10.6, respectively.

10.1 Capture Time

The *length* of a Cops and Robbers game played on G is the number of rounds it takes (not including the initial round) to capture the robber on G. We say that a play of the game with $c(G)$ cops is *optimal* if its length is the minimum over all possible strategies for the cops, assuming the robber is trying to evade capture for as long as possible. There may be several optimal plays possible (for example, on the path P_4 with four vertices, the cop may start on either of the two vertices in the center), but the length of an optimal game is an invariant of G. If k cops play on a graph with $k \geq c(G)$, we denote this invariant $\text{capt}_k(G)$, which we call the *k-capture time* of G. In the case $k = c(G)$, we write $\mathrm{capt}(G)$ and refer to this as the *capture time* of G.

We focus on the case $k = c(G)$ in the present section, and $k > c(G)$ in the next. There is small but growing literature on the capture time for graphs with a higher cop number. In [45], the authors proved that if G is cop-win of order $n \geq 5$, then $\mathrm{capt}(G) \leq n - 3$. By considering small order cop-win graphs, the bound was improved to $\mathrm{capt}(G) \leq n-4$ for $n \geq 7$ in [99]. Examples were given of planar cop-win graphs in both [45, 99] which prove that the bound of $n - 4$ is optimal. Beyond the case $k = 1$, a recent paper by Mehrabian [142] investigates the capture time of Cartesian grids. It was shown in [142] that if G is the Cartesian product of two trees, then $\mathrm{capt}(G) = \lfloor \mathrm{diam}(G)/2 \rfloor$. In particular, the capture time of an $m \times n$ Cartesian grid is $\lfloor \frac{m+n}{2} \rfloor - 1$.

The determination of the capture time of higher-dimensional grids was left open in [142]. Perhaps the simplest such higher-dimensional grid is the hypercube. In this section we derive the asymptotic order of the capture time of the hypercube. It was established in [137] that the cop number of a Cartesian product of m trees is $\lceil \frac{m+1}{2} \rceil$; in particular, $c(Q_n) = \lceil \frac{n+1}{2} \rceil$. We note that the cop numbers of other graph products were investigated in [151].

The following result was established recently in [44]. We present the sketch of the proof of the lower bound, which uses random walks. Hence, we have another example of a deterministic parameter analyzed using the probabilistic method. For the upper bound, the reader is directed to [44].

Theorem 10.1.1 ([44]). *For n an integer, we have that*

$$\mathrm{capt}(Q_n) = \Theta(n \log n).$$

To prove the lower bound of Theorem 10.1.1, we will show that $\mathrm{capt}(Q_n) \geq (1 - o(1))\frac{1}{2}n \log n$. Our proof of this bound involves two strategies, one for the cops and one for the robber. On each turn, a robber using the *random strategy* moves to an adjacent vertex chosen uniformly at random; that is, the robber performs a random walk on Q_n. A random walk by the robber appears to be optimal or near-optimal with regards to maximizing the capture time for them.

A cop using the *zombie strategy* moves one step closer to the robber on each turn; when there are multiple ways to do this, the cop chooses between them arbitrarily. For both strategies, the players may choose their starting locations intelligently. The zombie strategy is in some sense the optimal response to the random strategy. The proof is omitted here and can be found in [44].

Lemma 10.1.2 ([44]). *Fix T as a positive integer. Consider a cop C attempting to capture a robber R using the random strategy on the cube Q_n, and suppose R has the next move. For $d \leq n$, define*

$$p(d) = \mathbb{P}\left(C \text{ catches } R \text{ in } < T \text{ rounds} | C \text{ and } R \text{ are distance } d\right),$$

and let

$$p_d = \max p(d),$$

where the maximum is taken over all possible strategies for C. Then the following inequalities hold for $1 \leq k \leq n/2$: $p_{2k-2} \geq p_{2k}$ and $p_{2k} \geq p_{2k-1}$.

Lemma 10.1.2 shows that, against a random robber, the cop should keep the distance between the players even and, subject to that, minimize the distance. We have, therefore, that the following cop strategy is optimal against a random robber: if the robber is an even distance away, pass; otherwise, play the zombie strategy. The lemma shows that this strategy is optimal in that it maximizes the probability of capturing the robber quickly.

A single cop using the zombie strategy will eventually capture a random robber. In this scenario, the capturing process behaves analogously to the *coupon collector process* (which was discussed in more detail in Chapter 6). There is an urn with m distinct labeled balls in it. Draw a ball (or *coupon*) at random, observe its label, and put the ball back in the urn. We repeat this experiment until we see all m coupons. If all but i coupons have already been collected (out of m total), then the probability of obtaining a new coupon on the next turn is i/m, so the number of turns needed to collect a new coupon is geometrically distributed with probability of success i/m. As we will show, similar behavior arises when a cop using the zombie strategy plays against a random robber on Q_n.

The approach is to bound the probability that the actual capture time is much less than its expectation.

Lemma 10.1.3 ([44]). *Consider the coupon collector process with m coupons in total, of which all but m_0 have already been collected. Let X be a random variable denoting the amount of time needed to collect the remaining m_0 coupons. For $\varepsilon > 0$, we have that*

$$\mathbb{P}\left(X < (1-\varepsilon)(m-1)\log m\right) \leq \exp(-m^{-1+\varepsilon}m_0).$$

Proof. Let $T = (1-\varepsilon)(m-1)\log m$. For the ith uncollected coupon, let X_i be an indicator random variable for the event that this type of coupon is collected within T rounds. It is straightforward to see that

$$
\begin{aligned}
\mathbb{P}\left(X_i = 1\right) &= 1 - \left(1 - \frac{1}{m}\right)^T \leq 1 - \exp(-T/(m-1)) \\
&= 1 - \exp(-(1-\varepsilon)\log m) = 1 - m^{-1+\varepsilon}.
\end{aligned}
$$

Fix $I \subseteq [m_0]$ and $j \in [m_0] \setminus I$. Observe that

$$\mathbb{P}\left(X_i = 1 \text{ for all } i \in I \mid X_j = 1\right) \leq \mathbb{P}\left(X_i = 1 \text{ for all } i \in I\right).$$

Thus,

$$
\begin{aligned}
\mathbb{P}\left(X_i = 1 \text{ for all } i \in I \cup \{j\}\right) &\leq \mathbb{P}\left(X_i = 1 \text{ for all } i \in I\right) \\
&\quad \cdot \mathbb{P}(X_j = 1).
\end{aligned}
$$

By induction, we have that

$$
\begin{aligned}
\mathbb{P}\left(X_i = 1 \text{ for all } i \in [m_0]\right) &\leq \prod_{i=1}^{m_0} \mathbb{P}(X_i = 1) \\
&\leq (1 - m^{-1+\varepsilon})^{m_0} \\
&\leq \exp(-m^{-1+\varepsilon}m_0),
\end{aligned}
$$

where the last inequality follows because $(1+x) \leq \exp(x)$ for all x. $\square$

We are now ready to prove our lower bound on $\mathrm{capt}(Q_n)$.

Theorem 10.1.4 ([44]). *In Q_n, a robber can escape capture against n^c cops for at least $(1 - o(1))\frac{1}{2}n \log n$ turns, where $c > 0$ is a constant.*

Proof. Let $T = \frac{1}{2}(n - 1) \log n$ and

$$\varepsilon = \log((4c + 1) \log n) / \log n = o(1).$$

Fix a cop strategy. We claim that for sufficiently large n, a random robber evades capture for at least $(1-\varepsilon)T$ turns with positive probability; it would then follow that some particular set of random choices and, therefore, some particular deterministic strategy, allows the robber to survive this long.

For the initial placement, the robber aims to choose a starting vertex that is relatively far from all cops. The number of vertices of Q_n within distance k of any given cop is $\sum_{i=0}^{k} \binom{n}{i}$, so the number of vertices within distance k of any of the n^c cops is at most

$$
\begin{aligned}
n^c \sum_{i=0}^{k} \binom{n}{i} &\leq n^c(k + 1)\binom{n}{k} \\
&\leq 2n^c k \left(\frac{ne}{k}\right)^k \\
&\leq n^{c+1}\left((4e)^{1/4}\right)^n \\
&< n^{c+1}1.85^n,
\end{aligned}
$$

when $k = n/4$. For sufficiently large n we have $n^c \sum_{i=0}^{k} \binom{n}{i} < 2^n$, so the robber can choose an initial position at least distance $n/4+1$ from every cop.

To understand the effectiveness of the robber's strategy, we use the following approach. First, we focus on a single cop. We determine the probability that this cop captures the robber within $(1-\varepsilon)T$ turns. We then apply the union bound to obtain an upper bound on the probability that any of the cops can capture the robber within $(1 - \varepsilon)T$ turns. If this probability is less than 1, then with positive probability the robber evades capture.

Suppose cop C' is at distance $2k$ from the robber (on the robber's turn). By Lemma 10.1.2, we may assume C' uses a zombie strategy. On any given round, the robber moves toward C' with probability $2k/n$. Thus, after the robber's move and the response of C', the distance between the two is $2k - 2$ with probability $2k/n$ and $2k$ otherwise. Letting X_i denote the number of turns for which the cop remains at distance $2i$ from the robber, we see that X_i is geometrically distributed with probability of success $2i/n$; the number of turns needed for C' to capture the robber is at most $\sum_{i=1}^{k} X_i$. Hence, this process is equivalent to the coupon collector process with $n/2$ coupons, if we suppose that all but k coupons have already been collected. Since the robber begins at least distance $n/4 + 1$ from C', we may (again by Lemma 10.1.2) suppose $k = \lfloor n/8 \rfloor$. Now the expected length of the process is

$$\sum_{i=1}^{\lfloor n/8 \rfloor} \mathbb{E}[X_i] = \sum_{i=1}^{\lfloor n/8 \rfloor} \frac{n}{2i} = \frac{n}{2} \sum_{i=1}^{\lfloor n/8 \rfloor} \frac{1}{i},$$

which tends to $\frac{n}{2}(\log n - \log 8 + \gamma)$, where $\gamma \approx 0.557$ denotes the Euler-Mascheroni constant.

By Lemma 10.1.3, the probability that C' captures the robber in under $(1 - \varepsilon)T$ turns is at most

$$\exp(-(n/2)^{-1+\varepsilon} \lfloor n/8 \rfloor) = O(\exp(-(n/2)^{\varepsilon}/4)).$$

By the union bound, the probability that any of the cops captures the robber in under $(1 - \varepsilon)T$ turns is thus, at most $O(n^c \exp(-(n/2)^{\varepsilon}/4))$; since

$$\frac{(n/2)^{\varepsilon}}{4} = \frac{1}{4} \exp\left(\varepsilon\left(\log n + O(1)\right)\right)$$

$$\sim \frac{1}{4}(4c + 1)\log n$$

$$> \left(c + \frac{1}{5}\right)\log n,$$

this probability tends to zero as n tends to infinity. Hence, with probability approaching 1 as $n \to \infty$, the robber escapes capture for at least $(1 - \varepsilon)T$ turns. $\qquad\square$

10.2 Overprescribed Cops and Robbers

When playing Cops and Robbers with exactly $c(G)$-many cops in a graph G, the length of the game (assuming optimal play by both sets of players) is longest. However, if we add more cops, then the length of the game shortens. We may continue adding cops all the way up to $\gamma(G)$-many, in which case the game lasts only one turn. Hence, it is natural to consider *Overprescribed Cops and Robbers*: as more cops are added, the capture time (the minimum length of the game assuming optimal play) monotonically decreases. We call this effect on the length of the game *temporal speed up*, and we have a sequence of game lengths that are a function of the number of cops. In the present section, we provide results on temporal speed up in hypercubes and random graphs. For the case of hypercubes, probabilistic methods are key to understanding the dynamics of Overprescribed Cops and Robbers. The results of this section are taken from [54].

Hypercubes

As described in Section 10.1, the cop number of Q_n, the hypercube on 2^n vertices, is $\left\lceil \frac{n+1}{2} \right\rceil$, with the associated capture time $\Theta(n \log n)$. On the other hand, the domination number of Q_n is $(1 + o(1))\frac{2^n}{n}$, with the associated capture time 1. We investigate the capture time for the asymptotic number of cops between the cop number and the domination number in the hypercube.

To analyze temporal speed up in hypercubes, we utilize coverings by certain special subgraphs. A *retraction* is a homomorphism that is the identity on its image. A *retract* of a graph G is an induced subgraph H which is the image of some retraction. Retracts play an important role in the game of Cops and Robbers as first noted in [152], who proved that for a retract H of G, $c(H) \le c(G)$. In particular, if G is cop-win, then so is H, naturally leading to the recursive characterization of cop-win graphs as dismantlable (see Theorem 2.0.1). We first consider the following theorem.

Theorem 10.2.1 ([54]). *If $V(G) = V_1 \cup \cdots \cup V_t$, where $G[V_i]$*

is a retract of G for every i and $k = \sum_{i \in [t]} k_i$, then $\operatorname{capt}_k(G) \leq \max_{i \in [t]} \operatorname{capt}_{k_i}(G[V_i])$. Note that if $k_i < c(G[V_i])$, then we say that $\operatorname{capt}_{k_i}(G[V_i]) = \infty$.

Proof. For each i, we assign a team of k_i cops to $G[V_i]$, which we refer to as the *territory* of those cops. Each team of cops plays their optimal strategy on their territory to capture the image of the robber under the retract to $G[V_i]$. After $\max_i \operatorname{capt}_{k_i}(G[V_i])$ turns, every team of cops has caught their projection of the robber via the corresponding retraction. Now, one team of cops will capture the robber. $\square$

We now state the upper bound that works well for a large number of cops.

Theorem 10.2.2 ([54]). *If $k = k(n) \in \mathbb{N}$ is such that*

$$k \geq 36 \cdot 2^n \, \frac{(2d+1)_{d+1}}{(n-d)_{d+1}}, \tag{10.1}$$

for some $d \leq cn - 2$, where $c = 1/2 - \sqrt{2}/4 \approx 0.1464$, then for n sufficiently large,

$$\operatorname{capt}_k(Q_n) \leq 2d+1.$$

In particular, the desired upper bound for the capture time holds provided that

$$k \geq 36 \cdot 2^n n \left(\frac{3d}{n(1-c)} \right)^{d+1}.$$

Proof of Theorem 10.2.2. The cops select a set of vertices of cardinality k uniformly at random, and then they begin the game on this set. Suppose that the robber starts the game on vertex v. We show that a.a.s., regardless of where they start, after $d+1$ (cops') moves, the cops can completely occupy $N_d = N_d(v)$. As the first move belongs to the cops, the robber will not be able to escape from the ball $N_{\leq d}(v)$ around their initial position; they will be trapped there.

We show that with probability at least $1 - 2^{-2n+2}$, there exists a matching saturating N_d between vertices of N_d and cops initially occupying $N_{2d+1} = N_{2d+1}(v)$. To do it, we are going to use Hall's theorem for matchings in bipartite graphs. A neighbor in N_{2d+1} of a vertex $w \in N_d$ (in this auxiliary bipartite graph) is a vertex in N_{2d+1} that contains a cop and is at distance exactly $d+1$ from w. For a given $S \subseteq N_d$ with $s = |S| \geq 1$, we wish to find t_s, a lower bound for the number of vertices in N_{2d+1} at distance $d+1$ from some vertex in S. As each vertex in N_d has $\binom{n-d}{d+1}$ vertices in N_{2d+1} that are at distance $d+1$, and each vertex in N_{2d+1} has $\binom{2d+1}{d+1}$ vertices in N_d that are at distance $d+1$, we find that

$$t_s \geq \frac{\binom{n-d}{d+1}}{\binom{2d+1}{d+1}} s = \frac{(n-d)_{d+1}}{(2d+1)_{d+1}} s.$$

Let X be the random variable counting how many of these vertices initially contain cops. Using the assumption for k, we find that $\mathbb{E}X \geq t_s k/2^n \geq 36ns$, and it follows from the Chernoff bounds (applied to X, which is a random variable following a hypergeometric distribution; for more details, see Chapter 1) that

$$
\begin{aligned}
\mathbb{P}\left(X < s\right) &\leq \mathbb{P}\left(X \leq \mathbb{E}X/2\right) \\
&\leq 2\exp(-\mathbb{E}X/12) \\
&\leq 2\exp(-3ns).
\end{aligned}
$$

Taking a union bound over all $\left(\binom{n}{d}\right) \leq \binom{2^n}{s} \leq 2^{ns}$ choices for sets S of cardinality s, we conclude that with probability at least $1 - 2^{-2ns+1}$, Hall's condition holds for all sets of cardinality s. Summing the failure probability over $1 \leq s \leq \binom{n}{d}$, we find that the desired condition holds for all sets with probability at least $1 - 2^{-2n+2}$. Finally, by taking a further union bound over all 2^n choices for v, the initial vertex the robber starts on, we conclude that a.a.s., regardless of where the robber initially starts, the desired matching can be found. We may assume then that this is the case.

Let us suppose that the robber starts at vertex v. We give a strategy for the cops for the remainder of the game. The cops in $N_{2d+1}(v)$ move to destinations in $N_d(v)$ according to the matching guaranteed above (moving along any shortest path), thereby occupying every vertex of $N_d(v)$, taking exactly $d+1$ steps. As we already mentioned, the robber is now trapped in the ball around

v. In the next d steps, the cops move toward v by covering at each step one full layer $N_i(v)$. Note that for any i with $1 \leq i \leq d-1$ (and in particular $i < cn < n/2$), there exists a matching between $N_i(v)$ and $N_{i+1}(v)$ saturating $N_i(v)$. Indeed, we notice that for every $S \subseteq N_i(v)$, $|N(S) \cap N_{i+1}(v)| \geq \frac{n-i}{i+1}|S| \geq |S|$. Hence, Hall's condition holds for the bipartite graph induced by layers $N_i(v)$ and $N_{i+1}(v)$. The robber is captured after another d steps, and the proof follows. $\square$

For lower bounds, we present a result that works well for a small number of cops.

Theorem 10.2.3 ([54]). *Fix any constants $0 < \alpha < \alpha' < 1$, and suppose that $c(Q_n) \leq k = k(n) \leq e^{n^\alpha}$. Then we have that*

$$\text{capt}_k(Q_n) \geq \frac{1-\alpha'}{2}(n-1)\log n.$$

Proof. The robber performs a random walk on Q_n. Following the proof of Theorem 10.1.1 with $T = (1/2)(n-1)\log n$ and $\epsilon = \alpha'$, the probability that any of the cops captures the robber in under $(1-\epsilon)T$ rounds is at most $k \exp(-(n/2)^{\alpha'}/4) = o(1)$. $\square$

The next result works well for a large number of cops.

Theorem 10.2.4 ([54]). *If $k = k(n) \in \mathbb{N}$ is such that*

$$k < \frac{2^n}{\sum_{i=0}^{d} \binom{n}{i}},$$

then we have that

$$\text{capt}_k(Q_n) > d.$$

In particular, the desired lower bound for the capture time holds, provided that

$$k < \frac{1}{2} \cdot 2^n \left(\frac{d}{en}\right)^d,$$

for some $d \leq n/3$.

Proof. In d steps, any cop can reach $\sum_{i=0}^{d}\binom{n}{i}$ vertices. Therefore, regardless of how cops are initially distributed, they can reach at most $k\sum_{i=0}^{d}\binom{n}{i} < 2^n$ vertices in d steps. Hence, the robber can pick an initial vertex that is at distance at least $d+1$ from any cop and stay put. She clearly survives for more than d rounds.

The second part follows from the fact that for any $d \leq n/3$ we have

$$\frac{2^n}{\sum_{i=0}^{d}\binom{n}{i}} \geq \frac{2^n}{2\binom{n}{d}} \geq \frac{2^n}{2(en/d)^d}. \qquad \square$$

We now summarize temporal speed up results for hypercubes in one corollary.

Corollary 10.2.5 ([54]). *Let $\varepsilon > 0$, and define*

$$\begin{aligned}
g(x) &= 2x\log_2(2x) + (1-2x)\log_2(1-2x) - x\log_2 x \\
&\quad -(1-x)\log_2(1-x), \\
c &= 1/2 - \sqrt{2}/4 \approx 0.1464, \quad and \\
b &= -g(c) \approx 0.2716.
\end{aligned}$$

Suppose $k \geq c(Q_n)$. The following hold for large enough n.

(i) If $k \leq 2^{n^\alpha}$ for some $\alpha < 1$, then $\operatorname{capt}_k(Q_n) = \Theta(n\log n)$;

(ii) If $k \leq 2^{n(1-b+\varepsilon)}$, then $\Omega(n) = \operatorname{capt}_k(Q_n) = O(n\log n)$;

(iii) If $2^{n(1-b+\varepsilon)} < k \leq 2^{n(1-\varepsilon)}$, then $\operatorname{capt}_k(Q_n) = \Theta(n)$;

(iv) If $k = 2^{n-f(n)}$ with $\log n \ll f(n) = o(n)$, then

$$\operatorname{capt}_k(Q_n) = \Theta\left(\frac{f(n)}{\log(n/f(n))}\right) = \Theta\left(\frac{n}{\omega\log\omega}\right)$$

where $\omega = \omega(n) = n/f(n)$ (note that $\omega \to \infty$, so $1 \ll \operatorname{capt}_k(Q_n) = o(n)$);

(v) if $k = 2^{n-f(n)}$ with $f(n) = O(\log n)$ (which is equivalent to $k \geq 2^n/n^{O(1)}$), then $\operatorname{capt}_k(Q_n) = O(1)$.

Proof. Recall from Theorem 10.1.1 that the capture time of Q_n with $c(Q_n)$ cops is $\Theta(n \log n)$. Since $\mathrm{capt}_k(G)$ is monotone nonincreasing in k, this establishes the upper bounds in parts (i) and (ii). The lower bound in part (i) follows immediately from Theorem 10.2.3. The lower bound in part (iii) (and hence, by monotonicity also in (ii)) follows from Theorem 10.2.4, with $d = \alpha n$ chosen so that $\varepsilon > \log_2((e/\alpha)^\alpha) + 1/n$. For the upper bound in part (iii) note that if $d = cn - 2$, then using Stirling's formula we have that

$$36 \cdot 2^n n \frac{(2d+1)_{d+1}}{(n-d)_{d+1}} = O(1) \cdot 2^n n \frac{(2cn)^{2cn-1}((1-2c)n)^{(1-2c)n}}{(cn)^{cn}((1-c)n)^{(1-c)n}}$$

$$= O(1) \cdot 2^n \left(\frac{(2c)^{2c}(1-2c)^{1-2c}}{c^c(1-c)^{1-c}} \right)^n$$

$$= 2^{n(1+g(c)+o(1))} < k,$$

(when n is large enough to make the $o(1)$ term less than ε) and Theorem 10.2.2 yields the linear upper bound. To obtain the upper bound in part (iv) we apply Theorem 10.2.2 again, this time with $d = 2n/(\omega \log \omega)$. Since $1 \ll \omega = o(n/\log n)$, we have that

$$36 \cdot 2^n n \left(\frac{3d}{n(1-c)} \right)^{d+1} \leq 2^{n+O(\log n)+\frac{2n}{\omega \log \omega} \log\left(\frac{8}{\omega \log \omega}\right)}$$

$$= 2^{n - \frac{(2+o(1))n \log \log \omega}{\omega}}$$

$$< 2^{n-f(n)} = k$$

and the desired upper bound holds. To derive the matching lower bound we will use Theorem 10.2.4 with $d = n/(2\omega \log \omega)$. This time we need to verify that

$$\frac{1}{2} \cdot 2^n \left(\frac{d}{en} \right)^d = 2^{n-1+\frac{n}{2\omega \log \omega} \log\left(\frac{1}{2e\omega \log \omega}\right)}$$

$$= 2^{n-(1+o(1))\frac{n}{2\omega}}$$

$$> 2^{n-f(n)} = k,$$

and the desired lower bound holds. Finally, part (v) follows immediately from Theorem 10.2.2 with d constant. $\qquad\square$

Binomial Random Graphs

Using ideas from [136, 167], we may obtain bounds for the capture time of $\mathbb{G}(n,p)$. For simplicity, we restrict ourselves to dense random graphs $(d = p(n-1) \geq \log^3 n)$ and for a large number of cops $(k = k(n) \geq C\sqrt{n\log n})$.

Theorem 10.2.6 ([54]). *Suppose that $d = p(n-1) \geq \log^3 n$ and $C\sqrt{n\log n} \leq k = k(n) \leq n$ for some sufficiently large constant C, and let $r = r(d,k)$ be the smallest positive integer such that $d^{r+1} \geq Cn\log n/k$. If $G = (V,E) \in \mathbb{G}(n,p)$, then a.a.s. $\mathrm{capt}_k(G) = \Theta(r)$.*

Proof. We sketch the proof only. A.a.s. $\mathbb{G}(n,p)$ is a good expander. Let $N(v,j)$ be the set of vertices at distance at most j from vertex v. We can show that a.a.s. for any vertex v and every j such that $d^j = o(n)$, $N(v,j) \sim d^j$. It is well known that any graph G with minimum degree $\delta = \delta(G) > 2$ has a dominating set of cardinality $O(n\log\delta/\delta)$. Hence, we may assume that $d < \sqrt{n\log n}$ as for denser graphs we immediately obtain that a.a.s. $\mathrm{capt}_k(G) = \Theta(1)$ for any $C\sqrt{n\log n} \leq k < n$. Finally, we may assume that $d < Cn\log n/k$. Indeed, if k is too large such that $d \geq Cn\log n/k$, then the result for $k' = C\sqrt{n\log n}$ implies that a.a.s. $\mathrm{capt}_k(G) \leq \mathrm{capt}_{k'}(G) = \Theta(1)$.

We place k cops at random as in the proof of Theorem 10.2.2. The robber appears at some vertex $v \in V$. Note that it follows from the definition of d that

$$d^r < \frac{Cn\log n}{k} \leq \sqrt{n\log n} \leq \frac{k}{C}.$$

The main difficulty is to show that with probability $1 - o(n^{-1})$, it is possible to assign distinct cops to all vertices u in $N(v,r) \setminus N(v,r-1)$ such that a cop assigned to u is within distance $(r+1)$ of u. (Note that here, the probability refers to the randomness in distributing the cops; the random graph is fixed.) If this can be done, then after the robber appears these cops can begin moving straight to their assigned destinations in $N(v,r) \setminus N(v,r-1)$. Since the first move belongs to the cops, they have $r+1$ steps to do so, after which the robber must still be inside $N(v,r)$, while $N(v,r) \setminus N(v,r-1)$ is fully occupied by cops. Then in at most

r additional steps, the cops can "tighten the net" around v and capture the robber. Hence, the cops will win after at most $2r + 1$ steps with probability $1 - o(n^{-1})$, for each possible starting vertex $v \in V$. Hence, this strategy gives a win for the cops a.a.s.

We will use Hall's theorem for bipartite graphs to show that the desired assignment exists. We need to verify that, for any set $S \subseteq N(v,r)\backslash N(v,r-1)$, there are at least $|S|$ cops lying on vertices within distance $r + 1$ from some vertex in S. One thing to make sure of is that Hall's condition holds for $S = N(v,r)\backslash N(v,r-1)$. It follows from expansion properties that $|N(v,r)| < 2d^r \le 2k/C$, so there are enough cops to achieve this goal. The main bottleneck is to satisfy the condition for sets with $|S| = 1$. Since for any vertex u, the expected number of cops in $N(u, r + 1)$ is asymptotic to $kd^{r+1}/n \ge C \log n$, the condition holds provided that C is large enough (for more on this approach, see the proofs of Lemma 5.2.2 or Theorem 10.2.2; see also [167, 168]).

To derive the lower bound, we need to use expansion properties again. It is possible to show that a.a.s. for any starting position of k cops, the number of vertices at distance at most $r - 1$ from them is asymptotic to kd^{r-1}, since $kd^{r-1} < Cn \log n/d = o(n)$. The robber can start at distance at least r from any cop and wait there.   □

10.3 Deterministic Burning

Internet memes seem to appear at random, and spread quickly across social networks like Facebook and Instagram. The burning number of a graph was introduced as a simple model of spreading memes or other kinds of social influence [46, 173]. The smaller the burning number is, the faster an influence can be spread in the network. We may add randomness to the problem by burning vertices via a probabilistic process, or by studying burning on random graphs.

We introduce the burning process as described first in the doctoral thesis of Roshanbin [46, 173]. Given a finite, simple, undirected graph G, the burning process on G is a discrete-time process defined as follows. Initially, at time $t = 0$ all vertices are unburned. At each time step $t \geq 1$, one new unburned vertex is chosen to burn (if such a vertex is available); if a vertex is burned, then it remains in that state until the end of the process. Once a vertex is burned in round t, in round $t + 1$ each of its unburned neighbors becomes burned. The process ends when all vertices of G are burned (that is, let T be the smallest positive integer such that there is at least one vertex not burning in round $T - 1$ and all vertices are burned in round T). The *burning number* of a graph G, denoted by $b(G)$, is the minimum number of rounds needed for the process to end. Note that with our notation, $b(G) = T$. The vertices that are chosen to be burned are referred to as a *burning sequence*; a shortest such sequence is called *optimal*. Note that optimal burning sequences have length $b(G)$.

For example, for the path P_4 with vertices $\{v_1, v_2, v_3, v_4\}$, the sequence (v_2, v_4) is an optimal burning sequence; see Figure 10.1.

FIGURE 10.1: Burning the path P_4 (the open circles represent burned vertices).

It is straightforward to see that burning the graph G in k steps is equivalent to covering $V(G)$ with k balls (closed neighborhoods) of radii $0, 1, \ldots, k - 1$, respectively; see [46, 173]. Further, for every connected graph G we have that $b(G) \leq \operatorname{diam}(G) + 1$. This trivial bound is sometimes tight, as shown below for the example of random graphs. However, the bound can also be far from being tight, as it is known that for a path on n vertices we have $\operatorname{diam}(P_n) = n - 1$, whereas $b(P_n) = \lceil \sqrt{n} \rceil$ (we leave this as an exercise, or see [46]). For more information on the graph burning and the burning number of graphs we refer the reader to [24, 25, 129, 173]. The results of this section are to be found in [148]. We note that graph burning is analogous to but distinct from Firefighter, discussed in Chapter 8.

Binomial Random Graphs

We first consider burning for $\mathbb{G}(n,p)$, with results summarized in the following theorem.

Theorem 10.3.1 ([148]). *Let* $G \in \mathbb{G}(n,p)$, $\varepsilon > 0$, *and* $\omega = \omega(n) \to \infty$ *as* $n \to \infty$ *but* $\omega = o(\log \log n)$. *Suppose first that*

$$d = d(n) = (n-1)p \gg \log n$$

and

$$p \leq 1 - (\log n + \log \log n + \omega)/n.$$

Let $i \geq 2$ *be the smallest integer such that*

$$d^i/n - 2\log n \to \infty.$$

Then the following property holds a.a.s.

$$b(G) = \begin{cases} i & \text{if} \quad d^{i-1}/n \geq (1+\varepsilon)\log n \\ i \ \text{or} \ i+1 & \text{if} \quad (1-\varepsilon)\log d \leq d^{i-1}/n < (1+\varepsilon)\log n \\ i+1 & \text{if} \quad d^{i-1}/n < (1-\varepsilon)\log d. \end{cases} \quad (10.2)$$

If

$$1 - (\log n + \log \log n + \omega)/n < p \leq 1 - (\log n + \log \log n - \omega)/n,$$

then a.a.s.

$$b(G) = 2 \ \text{or} \ 3. \quad (10.3)$$

Finally, if

$$p > 1 - (\log n + \log \log n - \omega)/n,$$

then a.a.s.

$$b(G) = 2. \quad (10.4)$$

For upper bounds on the burning number of G, we will make use of the following result for random graphs, which follows easily from [35, Theorem 10.10].

Lemma 10.3.2 ([148]). *If* $d = (n-1)p \gg \log^3 n$, $p < 1 - \varepsilon$ *for some* $\varepsilon > 0$, *and*

$$d^i/n - 2\log n \to \infty \quad \text{and} \quad d^{i-1}/n - 2\log n \to -\infty,$$

then the diameter of $G \in \mathbb{G}(n,p)$ *is equal to* i *a.a.s.*

Proof. As required in Theorem 10.10 of [35], we have $d \gg \log^3 n$. By our assumptions on i, we have $d^i/n = \log(n^2/c)$ for $c \to 0$, and hence, by Theorem 10.10 of [35], a.a.s. the diameter of $G \in \mathbb{G}(n, p)$ is equal to i. $\qquad\qquad\square$

From the proof of this result, we have the following corollary.

Corollary 10.3.3 ([148]). *If $d = (n-1)p \gg \log^3 n$ and*

$$d^i/n - 2\log n \to \infty,$$

then the diameter of $G \in \mathbb{G}(n, p)$ is at most i a.a.s.

To obtain lower bounds and the upper bound in the first case of (10.2), we will need the following expansion lemma investigating the shape of typical neighborhoods of vertices. Before we state the lemma we need a few definitions. For any $j \geq 0$, let us denote by $N(v, j)$ the set of vertices at distance at most j from v, and by $S(v, j)$ the set of vertices at distance exactly j from v (note that $S(v, 0) = \{v\}$).

Lemma 10.3.4 ([148]). *If $G = (V, E) \in \mathbb{G}(n, p)$ and $\varepsilon > 0$, and suppose that $d = (n-1)p$ is such that $\log n \ll d = o(n)$, and let $i \geq 1$ be the largest integer such that $d^i = o(n)$, then a.a.s. the following properties hold.*

(i) *For all $j = 1, 2, \ldots, i$ and all $v \in V$,*

$$|N(v, j)| = |S(v, j)|(1 + O(1/d)) \sim d^j.$$

(ii) *There exists $v \in V$ such that $N(v, i+1) = V$, provided that $d^{i+1}/n \geq (1 + \varepsilon) \log n$; otherwise, $N(v, i+2) = V$.*

(iii) *For all $v \in V$,*

$$|V \setminus N(v, i+1)| = e^{-c(1+o(1))}n,$$

provided that $c = c(n) = d^{i+1}/n \leq (1 - \varepsilon) \log n$.

Proof. Let $v \in V$ and consider the random variable $U = U(v) = |S(v,1)|$. It is clear that $U \in \text{Bin}(n-1, p)$ so we obtain that $\mathbb{E}U = d$. A consequence of the Chernoff bounds is that

$$\mathbb{P}\left(|U - \mathbb{E}U| \geq \varepsilon U\right) \leq 2\exp\left(-\frac{\varepsilon^2 \mathbb{E}U}{3}\right) \tag{10.5}$$

for $0 < \varepsilon < 3/2$. Hence, after taking $\varepsilon = 2\sqrt{\log n / d}$, we find that with probability $1 - o(n^{-1})$ we have

$$U = \mathbb{E}U(1 + O(\varepsilon)) \sim d.$$

We will continue expanding neighborhoods of v using the BFS procedure. Suppose that for some $j \geq 1$ we have $|N(v,j)| = s$, and our goal is to estimate $|N(v, j+1)|$. Consider the random variable $X = X(N(v,j))$ counting the number of vertices outside of $N(v,j)$ with at least one neighbor in $N(v,j)$. (Note that X is only a lower bound for $|N(v, j+1)|$ since $X = |N(v, j+1) \setminus N(v,j)|$. We will show below that $|N(v, j+1)| \leq (1 + O(1/d))X$.) We will bound X in a stochastic sense. There are two things that need to be estimated: the expected value of X, and the concentration of X around its expectation.

It is evident that

$$\begin{aligned}
\mathbb{E}X &= \left(1 - \left(1 - \frac{d}{n-1}\right)^s\right)(n - s) \\
&= \left(1 - \exp\left(-\frac{ds}{n}(1 + O(d/n))\right)\right)(n - s) \\
&= \frac{ds}{n}(1 + O(ds/n))(n - s) = ds(1 + O(ds/n)).
\end{aligned}$$

It follows that $\mathbb{E}X = ds(1 + O(\log^{-1} n))$ provided $ds \leq n/\log n$, and $\mathbb{E}X \sim ds$ for $ds = o(n)$. Next, we use the Chernoff bounds which implies that with probability $1 - o(n^{-2})$ we have $|X - ds| \leq \varepsilon ds$ for $\varepsilon = 3\sqrt{\log n/(ds)}$. In particular, we derive that with probability $1 - o(n^{-2})$ we have $X = ds(1 + O(\log^{-1} n))$, provided $\log^3 n < ds < n/\log n$, and $X \sim ds$ for $ds = o(n)$. We consider the BFS procedure up to the i'th neighborhood provided that $d^i = o(n)$. Note that this implies that $i = O(\log n/\log\log n)$. Then the cumulative multiplicative error term is

$$(1 + o(1))^3 (1 + O(\log^{-1} n))^i \sim (1 + O(i \log^{-1} n)) \sim 1.$$

(Note that it might take up to two iterations to reach at least $\log^3 n$ vertices to be able to use the error of $(1+O(\log^{-1} n))$ and possibly one more iteration when the number of vertices reached is $o(n)$ but is larger than $n/\log n$.) Hence, by a union bound over $1 \le j \le i$, with probability $1 - o(n^{-1})$, we have $|N(v,j) \setminus N(v,j-1)| \sim d^j$ for all $1 \le j \le i$, provided that $d^i = o(n)$. By taking a union bound one more time (this time over all vertices), a.a.s. the same bound holds for all $v \in V$. Therefore, a.a.s., for all $v \in V$ and all $1 \le j \le i$, $|N(v,j)| = |S(v,j)|(1+O(1/d))$ and also $|N(v,j)| \sim d^j$, and Part (i) follows.

We continue investigating the neighborhood of a given vertex $v \in V$. It follows from the previous part that $s = |N(v,i)| \sim d^i$ with probability $1 - o(n^{-1})$. Let $c = c(n) = d^{i+1}/n$, and note that it follows from the definition of i that $c \ge c'$ for some $c' > 0$. In this case, it is easier to focus on the random variable Y counting the number of vertices outside of $N(v,i)$ that have no neighbor in $N(v,i)$. It follows that

$$
\begin{aligned}
\mathbb{E}Y &= \left(1 - \frac{d}{n-1}\right)^s (n-s) \\
&= \exp\left(-\frac{ds}{n}(1+O(d/n))\right) n(1+o(1)) = e^{-c(1+o(1))}n.
\end{aligned}
$$

If $c \ge (1+\varepsilon)\log n$ for some $\varepsilon > 0$, then $\mathbb{E}Y \le n^{-\varepsilon+o(1)} = o(1)$ and so a.a.s. $Y = 0$ by Markov's inequality. Otherwise, $\mathbb{E}Y \le (1+o(1))e^{-c'}n \le (1-2\varepsilon')n$ for some $\varepsilon' > 0$ and so a.a.s. $Y \le (1-\varepsilon')n$ by Markov's inequality. It follows that a.a.s. $s' = |N(v,i+1)| \ge \varepsilon'n$ and repeating the same argument one more time we easily find that a.a.s. $|N(v,i+2)| = V$, and (ii) holds.

Finally, if $c \le (1-\varepsilon)\log n$ for some $\varepsilon > 0$, then $\mathbb{E}Y \ge n^{\varepsilon+o(1)}$ and so, using (10.5) one more time, we obtain that with probability $1 - o(n^{-1})$, $Y = (1+o(1))e^{-c(1+o(1))}n$. Part (iii) holds by taking a union bound over all $v \in V$. $\qquad\square$

The proof of Theorem 10.3.1 follows by Lemma 10.3.4 together with some well-known results. We first consider the case $d = o(n)$. Before we start considering the three cases of (10.2), let us notice that it follows from the definition of i that $d^{i-2}/n = O(\log n/d) = o(1)$. Since we aim for a result that holds a.a.s., we may assume that G satisfies (deterministically) the properties from Lemma 10.3.4 and Corollary 10.3.3.

Suppose that $d^{i-1}/n \geq (1 + \varepsilon)\log n$. It follows from Lemma 10.3.4(ii) that there exists vertex $v \in V$ such that $N(v, i-1) = V$. To show that $b(G) \leq i$, it suffices to start burning the graph from v. By Lemma 10.3.4(i), regardless of the strategy used, the number of vertices burning after $i - 1$ steps is at most

$$\sum_{j=0}^{i-2} d^j (1 + o(1)) = d^{i-2}(1 + o(1)) = o(n).$$

Hence, $b(G) \geq i$, and the first case is done for $d = o(n)$.

Suppose now that $(1-\varepsilon)\log d \leq d^{i-1}/n < (1+\varepsilon)\log n$. Exactly the same argument as in the first case gives $b(G) \geq i$; the number of vertices burning after $i-1$ steps is at most $d^{i-2}(1+o(1)) = o(n)$. It follows from Lemma 10.3.4(ii) that there exists $v \in V$ such that $N(v, i) = V$, and so $b(G) \leq i + 1$. The second case is done for $d = o(n)$.

Finally, suppose that $c = d^{i-1}/n < (1-\varepsilon)\log d$. It follows from Lemma 10.3.4(i) and (iii) that, no matter which burning sequence is used, after i steps, the number of vertices not burning is at least

$$e^{-c(1+o(1))}n - \sum_{j=0}^{i-2} d^j (1 + o(1))$$

$$\geq \exp\left(-(1 - \varepsilon + o(1))\log d\right)n - d^{i-2}(1 + o(1))$$

$$= \frac{nd^{\varepsilon+o(1)}}{d} - \frac{cn}{d}(1 + o(1))$$

$$\geq \frac{nd^{\varepsilon+o(1)}}{d} - \frac{n\log d}{d} \geq \frac{nd^{\varepsilon+o(1)}}{2d} \to \infty.$$

Hence, $b(G) \geq i+1$ and, since the diameter is at most i, we have $b(G) = i+1$ and the third case is completed for $d = o(n)$.

We now consider $p = \Omega(1)$ and $p \leq 1 - (\log n + \log \log n + \omega)/n$. For this range of the parameter p, by Corollary 10.3.3, the diameter of G is a.a.s. at most 2. We also have $i = 2$, and we are still in the third case of (1). Since $b(G) = 1$ if and only if the graph consists of one vertex, it is clear that for this range of p the only two possibilities for the burning number are 2 and 3. Note that $b(G) = 2$ if and only if there exists $v \in V$ such that $N(v, 1)$ covers all but perhaps one vertex. Indeed, if we start burning the graph from v, all but at most one vertex is burning in the next round and the remaining vertex (in case it exists) can be selected as the next one to burn. However, if no such $v \in V$ exists, then no matter which vertex is selected as the starting one, there is at least one vertex not burning in the next round. This sufficient and necessary condition for having $b(G) = 2$ is equivalent to the property that the complement of G has a vertex of degree at most one (that is, the minimum degree of the complement of G is at most one). It is well known that the threshold for having minimum degree at least 2 is equal to $p_0 = (\log n + \log \log n)/n$. Hence, if $1 - p \geq (\log n + \log \log n + \omega)/n$, then a.a.s. the minimum degree in the complement of G is at least two, and so $b(G) = 3$ a.a.s. This finishes the proof of (10.2).

For the range of p given in (10.3), note that a.a.s. the minimum degree of the complement of G is equal to one or two (recall that $\omega = o(\log \log n)$, and so the complement of G is a.a.s. connected). Hence, $b(G) \in \{2, 3\}$ a.a.s. Finally, the range of p given in (10.4)) is below the critical window for having minimum degree 2 in the complement of G, and hence, it follows that a.a.s. the minimum degree of the complement of G is at most one, and so $b(G) = 2$ a.a.s. The proof of Theorem 10.3.1 follows.

The following corollary is now immediate.

Corollary 10.3.5 ([148]). *Let $d = d(n) = (n-1)p \gg \log n$ and let $G \in \mathcal{G}(n, p)$. Then a.a.s. we have that*

$$b(G) - D(G) \in \{0, 1\}.$$

10.4   Random Burning

We consider next the effect of *random burning* of graphs, where vertices chosen to burn at each time-step are chosen via a random process. In particular, at time i of the process, x_i is selected uniformly at random from V; that is, for each $v \in V$, $\mathbb{P}(x_i = v) = 1/n$. Note that it might happen that v is already burning or maybe even was selected earlier in the process. Note that now, the burning number is a random variable. Let $b_R(G)$ be the random variable associated with the first time all vertices of G are burning. That is, $b_R(G)$ is the minimum number of rounds needed until all vertices of G are burning with random burning.

We consider random burning on P_n; suppose that

$$V(P_n) = \{v_1, v_2, \ldots, v_n\},$$

with v_i being adjacent to v_{i-1} for $2 \le i \le n$. As mentioned above, we have that $b(P_n) = \lceil \sqrt{n} \rceil$ (see [46]; the lower bound will actually be proved again here below in the proof of the lower bound for $b(P_m \square P_n)$, and the upper bound follows by covering greedily the path with balls of radii $0, 1, \ldots, \lceil \sqrt{n} \rceil - 1$. The following result shows that $b_R(P_n)$ is much larger than $b(P_n)$.

Theorem 10.4.1 ([148]). *A.a.s. we have that*

$$b_R(P_n) \sim \sqrt{n \log n / 2}.$$

Proof. We start with the proof of the upper bound for $b_R(P_n)$. Let $k = \sqrt{n(\log n/2 + \omega)}$, where $\omega = \omega(n) = (\log n)^{2/3} = o(\log n)$. We will use the first moment method to show that at time k a.a.s. all vertices are burning. It will be convenient to think of covering the path P_n with k balls of radii $0, 1, \ldots, k - 1$, where the radii are measured in terms of the graph distance; that is, the ball of radius i around a vertex v is $N(v, i)$. Partition P_n into $\sqrt{n}$ subpaths, $p_1, p_2, \ldots, p_{\sqrt{n}}$, each of length $\sqrt{n}$. (For expressions such as $\sqrt{n}$ that clearly have to be an integer, we round up or down but do not specify which: the choice of which does not affect the argument.)

For $1 \le i \le \sqrt{n}$, let X_i be the indicator random variable for the event that no ball contains the whole p_i. In particular, $X_i = 0$ if there exists $j \in [k]$ such that the ball centered at x_j and of radius $j - 1$ contains the path p_i; otherwise, $X_i = 1$. Let $X = \sum_{i=1}^{\sqrt{n}} X_i$. It is straightforward to see that if $X = 0$, then all vertices are burning at time k, and so our goal is to show that $\mathbb{P}(X \ge 1) = o(1)$.

We consider a subpath p_i with $\sqrt{\log n} \le i \le \sqrt{n} - \sqrt{\log n}$. Note that all vertices of p_i are at distance at least $(\sqrt{\log n} - 1) \cdot \sqrt{n} > k$ from the endpoints of P_n. As a result, p_i is sufficiently far from them to be affected by boundary effects. It is evident that

$$\mathbb{E}X_i = \mathbb{P}(X_i = 1) = \prod_{j=\sqrt{n}/2}^{k} \left(1 - \frac{2j - \sqrt{n}}{n}\right)$$

$$= \exp\left(-\left(1 + O\left(\frac{k}{n}\right)\right) \sum_{j=\sqrt{n}/2}^{k} \frac{2j - \sqrt{n}}{n}\right)$$

$$= \exp\left(-\left(1 + O\left(\frac{k}{n}\right)\right) \frac{(k - \sqrt{n}/2)^2}{n}\right)$$

$$= \exp\left(-\left(1 + O\left(\frac{\sqrt{n}}{k}\right)\right) \frac{k^2}{n}\right).$$

Let $x = \frac{1}{2}\log n + \omega$, where $\omega = (\log n)^{2/3}$. Using the definition of k, we derive that

$$\mathbb{E}X_i = \mathbb{P}(X_i = 1) = \exp\left(-\left(1 + O\left(\frac{1}{\sqrt{\log n}}\right)\right) x\right)$$

$$= \exp\left(-x + O(\sqrt{\log n})\right)$$

$$\le \exp\left(-\frac{1}{2}\log n - \frac{1}{2}\omega\right). \tag{10.6}$$

For any $i < \sqrt{\log n}$ or $i > \sqrt{n} - \sqrt{\log n}$, p_i is close to one of the endpoints of P_n, but we can nevertheless estimate the probability of $X_i = 1$ as follows:

$$\mathbb{E}X_i = \mathbb{P}(X_i = 1) \le \prod_{j=\sqrt{n}}^{k} \left(1 - \frac{j}{n}\right)$$

$$= \exp\left(-\left(1 + O\left(\frac{k}{n}\right)\right) \sum_{j=\sqrt{n}}^{k} \frac{j}{n}\right)$$

$$= \exp\left(-\left(1 + O\left(\frac{k}{n}\right)\right) \frac{k^2\left(1 + O(n/k^2)\right)}{2n}\right)$$

$$= \exp\left(-\left(1 + O\left(\frac{n}{k^2}\right)\right) \frac{k^2}{2n}\right)$$

$$\le \exp\left(-\frac{1}{4}\log n - \frac{1}{4}\omega\right). \tag{10.7}$$

From (10.6) and (10.7) we have that

$$\mathbb{E}X \le \left(\sqrt{n} - 2\sqrt{\log n}\right) \exp\left(-\frac{1}{2}\log n - \frac{1}{2}\omega\right)$$

$$+ 2\sqrt{\log n} \exp\left(-\frac{1}{4}\log n - \frac{1}{4}\omega\right)$$

$$= o(1).$$

The upper bound holds a.a.s. by Markov's inequality.

Now, we will show an asymptotically almost sure matching lower bound for $b_R(P_n)$. For this, we introduce the parameter $b_{R'}$. At time i of the process, suppose that in the burning process, x_i is selected uniformly at random from those vertices that were not selected before; that is, for each $v \in V$ that was not selected earlier in the process, $\mathbb{P}(x_i = v) = 1/(n - i + 1)$. However, it still might happen that v is already burning. Let $b_{R'}(G)$ be the burning parameter with these rules. Since $b_{R'}(P_n) \le b_R(P_n)$, the proof of the lower bound will follow if we lower bound $b_{R'}(P_n)$.

Let $k = \sqrt{n(\log n/2 - \log\log n/2 - \omega)}$, where now $\omega = \omega(n) = o(\log\log n)$ is any function tending to infinity as $n \to \infty$, arbitrarily slowly. This time, we will use the second moment method to show that at time k a.a.s. at least one vertex is not burning. To avoid highly dependent events, we focus only on vertices that are at distance at least $2k$ from each other.

For $1 \le i \le \ell = n/(2k) - 1 = \Theta(\sqrt{n/\log n})$, let Y_i be the indicator random variable for the event that vertex v_{2ki} is not burning at time k. Let $Y = \sum_{i=1}^{\ell} Y_i$. In particular, if $Y \ge 1$, then at least one vertex is not burning at time k, and so our goal is to show that $\mathbb{P}(Y = 0) = o(1)$. Recall that, since we investigate $b_{R'}(P_n)$, at time j of the process, only vertices not selected earlier have a chance to be selected as the next vertex x_j to be burned. Recall also that the ball centered at x_j will have radius $k - j$ at time k. Hence, for any $i \in [\ell]$,

$$\mathbb{P}(Y_i = 1) = \prod_{j=1}^{k} \left(1 - \frac{2k - 2j + 1}{n - j + 1}\right)$$

$$= \exp\left(-\left(1 + O\left(\frac{k}{n}\right)\right) \sum_{j=1}^{k} \frac{2k - 2j + 1}{n}\right)$$

$$\sim \exp\left(-\frac{k^2}{n}\right)$$

$$= \exp\left(-\frac{1}{2}\log n + \frac{1}{2}\log\log n + \omega\right),$$

and so

$$\mathbb{E}Y \sim \ell \cdot \exp\left(-\frac{1}{2}\log n + \frac{1}{2}\log\log n + \omega\right)$$

$$= \Theta(e^{\omega}) \to \infty,$$

as $n \to \infty$. Now, we estimate the variance of Y as follows:

$$\mathrm{Var}[Y] = \sum_{i,i'} \mathrm{Cov}(Y_i, Y_i')$$

$$\le \sum_{i \ne i'} \left(\mathbb{P}(Y_i = Y_{i'} = 1) - (\mathbb{P}(Y_i = 1))^2\right) + \mathbb{E}Y.$$

Let $i, i' \in [\ell]$ be such that $i \ne i'$. Since the vertices v_{2ki} and $v_{2ki'}$

are far away from each other, by performing similar calculations as before, we have that

$$\mathbb{P}(Y_i = Y_{i'} = 1) = \prod_{j=1}^{k} \left(1 - 2 \cdot \frac{2k - 2j + 1}{n - j + 1} \right)$$
$$\sim \exp\left(-\log n + \log\log n + 2\omega \right)$$
$$\sim \left(\mathbb{P}(Y_i) \right)^2.$$

Therefore,

$$\mathrm{Var}[Y] \leq O(\ell^2) \cdot o\left(\exp\left(-\log n + \log\log n + 2\omega \right) \right) + \Theta(e^\omega)$$
$$= o\left(e^{2\omega} \right) = o\left((\mathbb{E}Y)^2 \right).$$

The lower bound holds a.a.s. by Chebyshev's inequality. $\square$

10.5   Cops and Drunk (but Visible) Robbers

In this section, we assume that the cops have access to all of their strategies but that the robber is *drunk*. More specifically, we assume that the robber performs a random walk on G. We distinguish a drunk robber from an adversarial one, who has access to all of their strategies (as in the usual game of Cops and Robbers).

Given that the robber is at vertex $v \in V$ at time t, the robber moves to $u \in N(v)$ at time $(t + 1)$ with probability equal to $1/\deg(v)$. Observe that the robber probability distribution does not depend on the current position of the cops; in particular, it can happen that the robber moves to a vertex occupied by a cop (something an adversarial robber would never do). Since the robber is drunk, we cannot expect the most suitable vertex is chosen to start with; instead, the robber chooses an initial vertex uniformly at random. The cops are, of course, aware of this and so they try to choose an initial configuration so that the expected length of the game is as short as possible.

Let dct (G, k) be the expected capture time, provided that k cops play the game. In particular, let $\mathrm{dct}(G) = \mathrm{dct}(G, c(G))$. We define the *cost of drunkenness* by

$$F(G) = \frac{\mathrm{capt}(G)}{\mathrm{dct}(G)}$$

and we obviously have $F(G) \geq 1$. This graph parameter was introduced in [118] from which all results presented here were taken.

While we concentrate on the case $k = c(G)$, it is also natural to consider the expected capture time $\mathrm{dct}(G, k)$ for $k \neq c(G)$. The next theorem shows that this is well-defined for any $k \geq 1$ (in particular, even for $k < c(G)$).

Theorem 10.5.1 ([118]). *For any connected graph G and $k \geq 1$ we have* $\mathrm{dct}(G, k) < \infty$.

Proof. Let G be any connected graph, let $D = \mathrm{diam}(G)$ be the diameter of G, and $\Delta = \Delta(G)$ be the maximum degree of G. Fix any vertex $v \in V$, and place k cops on v. The cops will stay on v until the end of the game (this is clearly a suboptimal strategy). For a given vertex $y \in V$ occupied by the drunk robber, the probability that the robber uses a shortest path from y to v to move straight to v is at least $(1/\Delta)^D$. This discussion implies that, regardless of the current position of the robber at time t, the probability that the robber will be caught after at most D further rounds is at least $\varepsilon = (1/\Delta)^D$. The corresponding events for times $t+iD$, $i \in \mathbb{N} \cup \{0\}$ are pairwise independent. Thus, denoting by T the length of the game, we derive immediately that

$$\mathbb{E}[T] = \sum_{t \geq 0} \mathbb{P}(T > t) \leq \sum_{t \geq 0} \mathbb{P}\left(T > \left\lfloor \frac{t}{D} \right\rfloor D\right)$$

$$= \sum_{i \geq 0} D \cdot \mathbb{P}(T > iD) \leq D \sum_{i \geq 0} (1 - \varepsilon)^i$$

$$= \frac{D}{\varepsilon} = D\Delta^D < \infty,$$

and the proof follows. □

Sharper bounds can be obtained for the capture time of a drunk robber, even in the case that the cops are also drunk; for example, see [71]. However, Theorem 10.5.1 will be sufficient for our needs.

We will show that the cost of drunkenness for a Cartesian grid is asymptotic to 8/3. As we already mentioned, it is known that $\text{capt}(P_n \square P_n) = n - 1$ (see [142]), so it remains to investigate $\text{dct}(P_n \square P_n)$. Recall that $c(P_n \square P_n) = 2$.

Theorem 10.5.2 ([118]).

$$\text{dct}(P_n \square P_n) \sim \frac{3}{8} n,$$

and the cost of drunkenness is $F(P_n \square P_n) \sim 8/3$.

Proof. Suppose that the drunk robber occupies a vertex $(u, v) \in [n]^2$ that is not on the boundary of the grid. The decision of where to go from there can be made in the following way: toss a coin to decide whether to modify the first coordinate (u) or the second one (v); independently, toss another coin to decide whether to increase or decrease the value. Hence, the robber will move with probability 1/4 to any of the four neighbors of (u, v). Note that, if we restrict our attention to one dimension only (for example, we call it the North/South direction) we see the robber going North with probability 1/4, going South with the same probability, and staying in place with probability 1/2. Hence, the robber performs the so-called *lazy random walk* on the path move with probability 1/2; otherwise, do nothing. Let us note that this is precisely the case for $C_n \square C_n$, the grid on the torus. For $P_n \square P_n$, one needs to deal with the boundary effect, which is negligible and straightforward. It is a simple application of the Chernoff bounds to obtain that, say, with probability $1 - o(n^{-1})$, the robber stays within distance $O(\sqrt{n \log n}) = o(n)$ of the initial vertex. Hence, if we look at the grid from a large distance, the drunk robber is not moving at all.

Therefore, since we would like to investigate the asymptotic behavior, the problem reduces to finding a set S consisting of two vertices such that the average distance to S is as small as possible.

Cops should start on S to achieve the best outcome. It is evident that, due to the symmetry of $P_n \square P_n$, there are two symmetric optimal configurations for set S:

$$S = \{(n/2 + O(1), n/4 + O(1)), (n/2 + O(1), 3n/4 + O(1))\},$$
$$S = \{(n/4 + O(1), n/2 + O(1)), (3n/4 + O(1), n/2 + O(1))\}.$$

In any case, the average distance is

$$\sum_{u=0}^{n-1} \sum_{v=0}^{n-1} d((u,v), S) \sim 8n \int_{x=0}^{1/2} \int_{y=0}^{1/4} (x+y) dy\, dx = \frac{3}{8}n.$$

The result follows. $\qquad\qquad\qquad\qquad\qquad\qquad\qquad\qquad\quad\Box$

Now, we show that the cost of drunkenness can be arbitrarily close to any real number $c \in [1, \infty)$. To do this, we introduce two families of graphs, barbells and lollipops.

Let $n \in \mathbb{N}$ and $c \geq 0$. The *barbell* $B(n, c)$ is a graph that is obtained from two complete graphs $K_{\lfloor cn \rfloor}$ connected by a path P_n (that is, one end of the path belongs to the first clique whereas the other end belongs to the second one). The number of vertices of $B(n, c)$ is $(1 + 2c)n + O(1)$, and $c(B(n, c)) = 1$. To capture (either the adversarial or the drunk) robber, the cop should start at the center of the path and move toward the robber; $\text{capt}(B(n, c)) = n/2 + O(1)$. This family can be used to obtain any cost of drunkenness from $(1, 2]$.

Theorem 10.5.3 ([118]). *If $c \geq 0$, then*

$$\text{dct}(B(n, c)) \sim \frac{n}{2} \cdot \frac{1 + 4c}{2 + 4c},$$

and the cost of drunkenness is

$$F(B(n, c)) = \frac{\text{capt}(B(n, c))}{\text{dct}(B(n, c))} \sim 1 + \frac{1}{1 + 4c}.$$

Proof. Using the Chernoff bounds as in the previous proof, we derive that with probability, say, $1 - o(n^{-1})$, the robber does not move too far from their original position. The drunk robber starts on a clique with probability $(2c)/(1+2c) + o(1)$. In this case, the capture occurs at time $n/2 + O(\sqrt{n \log n})$ with probability $1 - o(n^{-1})$. If instead the robber chooses a vertex at distance k from the cop, then the robber is captured after $k + O(\sqrt{n \log n})$ steps, again with probability $1 - o(n^{-1})$. Hence, the expected capture time is

$$(1 + o(1)) \left(\frac{2c}{1 + 2c} \cdot \frac{n}{2} + \frac{1}{1 + 2c} \cdot \frac{n}{4} \right) \sim \frac{n}{2} \cdot \frac{1 + 4c}{2 + 4c},$$

and the result follows. □

Let $n \in \mathbb{N}$ and $c \geq 0$. The *lollipop* $L(n, c)$ is a graph that is obtained from a complete graph $K_{\lfloor cn \rfloor}$ connected to a path P_n (that is, one end of the path belongs to the clique). See Figure 10.2. The number of vertices of $L(n, c)$ is $(1 + c)n + O(1)$, and the cop number $c(L(n, c))$ is 1. To capture an adversarial robber, the cop should start at the center of the path and move toward the robber; $\text{capt}(L(n, c)) = n/2 + O(1)$. However, it is not clear what the optimal strategy is to capture the drunk robber. The larger the clique is, the closer to the clique the cop should start the game.

FIGURE 10.2: A lollipop graph, $L(7, 1/2)$

Theorem 10.5.4 ([118]). *If $c \geq 0$, then*

$$\text{dct}(L(n, c)) \sim \begin{cases} \frac{n}{4} \cdot \frac{(\sqrt{2} - 1 + c)(\sqrt{2} + 1 - c)}{1 + c}, & \text{for } c \in [0, 1], \\ \frac{n}{2(1 + c)}, & \text{for } c > 1. \end{cases}$$

The cost of drunkenness is

$$F(L(n, c)) = \frac{\text{capt}(L(n, c))}{\text{dct}(L(n, c))}$$

$$\sim \begin{cases} \frac{2(1 + c)}{(\sqrt{2} - 1 + c)(\sqrt{2} + 1 - c)}, & \text{for } c \in [0, 1], \\ 1 + c, & \text{for } c > 1. \end{cases}$$

Before we move to the proof of this result, we mention that the cost of drunkenness (as a function of the parameter c) has an interesting behavior. For $c = 0$ it is 2 (we play on the path), but then decreases to hit its minimum of $1 + \sqrt{2}/2$ for $c = \sqrt{2} - 1$. After that it increases back to 2 for $c = 1$, and goes to infinity together with c. Therefore, this family can be used to obtain any ratio at least $1 + \sqrt{2}/2 \approx 1.71$.

Proof of Theorem 10.5.4. Let the cop start on the vertex v at distance $(1 + o(1))bn$ from the clique ($b \in [0, 1]$ will be chosen to minimize the expected capture time). The drunk robber starts on a vertex of the clique with probability $c/(1 + c) + o(1)$. In this case, the capture occurs at time $bn + O(\sqrt{n \log n})$ with probability $1 - o(n^{-1})$. If instead the robber chooses a vertex at distance k from the cop, then the robber is captured, again with probability $1 - o(n^{-1})$, after $k + O(\sqrt{n \log n})$ rounds. The robber starts between the cop and the clique with probability $b/(1 + c) + o(1)$ and on the other side with the remaining probability. Hence, the expected capture time is equal to

$$(1 + o(1)) \left(\frac{c}{1 + c} \cdot bn + \frac{b}{1 + c} \cdot \frac{bn}{2} + \frac{1 - b}{1 + c} \cdot \frac{(1 - b)n}{2} \right)$$
$$\sim \frac{n}{1 + c} \left(b^2 + (c - 1)b + 1/2 \right).$$

The above expression is a function of b (that is, a function of the starting vertex v for the cop) and is minimized at $b = \max\{\frac{1-c}{2}, 0\}$, and the theorem holds. $\square$

It follows immediately from Theorems 10.5.3 and 10.5.4 that the cost of drunkenness can be arbitrarily close to any constant $c \geq 1$.

Corollary 10.5.5 ([118]). *For every real constant $c \geq 1$, there exists a sequence of graphs $(G_n : n \geq 1)$ such that*

$$\lim_{n \to \infty} F(G_n) = \lim_{n \to \infty} \frac{\text{capt}(G_n)}{\text{dct}(G_n)} = c.$$

10.6   Cops and Drunk (and Invisible) Robbers

In this section, we consider a variant of the game played on G in which the robber is invisible to the cops. It is clear that, given enough cops, the expected capture time will be finite. This is obviously true for $|V(G)|$ cops, but $c(G)$-many cops suffice, as demonstrated in the following theorem from [118].

Theorem 10.6.1 ([118]). *Suppose that $c(G)$ cops perform a lazy random walk on a connected graph G, starting from any initial position. The robber, playing perfectly, tries to avoid being captured. Let the random variable T be the length of the game. Then $\mathbb{E}[T] < \infty$.*

Proof. Let G be any connected graph, and let $\Delta = \Delta(G)$ be the maximum degree of G. Put $k = c(G)$. For any vector $x \in V^k$ consisting of vertices occupied by k cops and any vertex $y \in V$ occupied by the robber, there exists a winning strategy $S_{x,y}$ that guarantees that the robber is caught after at most some number $t_{x,y}$ of rounds. It is clear that the cops will follow $S_{x,y}$ with probability at least $(1/(2\Delta))^{kt_{x,y}}$. Now, we define

$$\varepsilon = \min_{x \in V^k, y \in V} (1/(2\Delta))^{kt_{x,y}} = (1/(2\Delta))^{kT_0} > 0,$$

where $T_0 = \max_{x \in V^k, y \in V} t_{x,y}$. We find that, regardless of the current positions of the players at time t, the probability that the robber will be caught after at most T_0 further rounds is at least ε. Observe that the corresponding events for times $t, t+T_0, t+2T_0, \ldots$ are pairwise independent. Thus, we immediately derive that

$$\mathbb{E}[T] = \sum_{t \geq 0} \mathbb{P}(T > t) \leq \sum_{t \geq 0} \mathbb{P}\left(T > \left\lfloor \frac{t}{T_0} \right\rfloor T_0\right)$$

$$= \sum_{i \geq 0} T_0 \mathbb{P}(T > iT_0) \leq T_0 \sum_{i \geq 0} (1 - \varepsilon)^i = \frac{T_0}{\varepsilon} < \infty,$$

and we are done. $\square$

Hence, $c(G)$ is the *minimum* number of cops required to capture an adversarial invisible robber in finite expected time, since this task is at least as hard as capturing an adversarial *visible* robber. Of course, generally it will take longer, compared to the visible robber case, to capture the invisible robber. For the invisible robber, define $\mathrm{ict}_{x,y}(G, k)$, for $x \in V^k$ and $y \in V$, to be the expected capture time when the initial cops/robber configuration is (x, y) and both the k cops and the robber play optimally; we also define

$$\mathrm{ict}(G, k) = \min_{x \in V^k} \max_{y \in V} \mathrm{ict}_{x,y}(G, k)$$

and, finally, $\mathrm{ict}(G) = \mathrm{ict}(G, c(G))$.

We now turn to the *drunk* invisible robber. The robber chooses a starting vertex uniformly at random and performs a random walk, as before. For a given starting position $x \in V^k$ for k cops, there is a strategy that yields the smallest expected capture time $\mathrm{idct}_x(G, k)$. The cops aim to minimize this by selecting a good starting position:

$$\mathrm{idct}(G, k) = \min_{x \in V^k} \mathrm{idct}_x(G, k).$$

As usual, $\mathrm{idct}(G) = \mathrm{idct}(G, c(G))$, but this time it makes sense to consider any $k \geq 1$. The proof of the next theorem has a flavor similar to Theorem 10.5.1 and so it is omitted.

Theorem 10.6.2 ([118]). *For any connected graph G and $k \geq 1$, we have* $\mathrm{idct}(G, k) < \infty$.

Finally, the *cost of drunkenness* for the invisible robber game is

$$F_i(G) = \frac{\mathrm{ict}(G)}{\mathrm{idct}(G)}.$$

It follows from the last theorem that this graph parameter is well-defined.

We will show now that the (invisible) cost of drunkenness can be arbitrarily close to any value from the interval $[2, \infty)$. No example of a graph G with $|V(G)| \geq 2$ and $F_i(G) < 2$ has been found, but it remains an open problem to determine the range of possible values the cost of drunkenness can take.

We consider a family of graphs that we call *brooms*. The broom $B(c, n)$ on n vertices with parameter c $(0 < c \leq 1)$ consists of a path with cn vertices, with one endpoint (the *center* of the broom) identified with the center of a star with $(1 - c)n$ end-vertices. The *end* of the broom is the other endpoint of the path. Bounding $\text{ict}(B(c, n))$ and $\text{idct}(B(c, n))$ is interesting in itself as an illustration of the game-theoretic flavor of this variant. The following results are from [117].

Theorem 10.6.3 ([117]). $\text{idct}(B(c, n)) \sim \frac{c^2 n}{2}$.

Theorem 10.6.4 ([117]). $\text{ict}(B(c, n)) \sim n$.

Corollary 10.6.5 ([117]). *For any* $0 < c \leq 1$,

$$F_i(B(c, n)) = \frac{\text{ict}(B(c, n))}{\text{idct}(B(c, n))} \sim \frac{2}{c^2}.$$

Hence, for every $a \in [2, \infty)$ *there exists* $c \in (0, 1]$ *such that*

$$F_i(B(c, n)) \sim a.$$

Proof of Theorem 10.6.3. One possible strategy for the cop is the following: start at the center of the broom, wait there for one step, and then go to the end of the broom. Since the robber's initial position is distributed uniformly at random and the robber has to move in every step, we have the following cases.

(i) With probability $1 - c$, the robber starts on an end-vertex of the star and is caught in $O(1)$ steps.

(ii) With probability c, the robber starts on some vertex of the path and (starting on the second round) the cop starts moving toward the end. In this case, the problem is reduced to catching a drunk robber on a path of cn vertices, and this takes $(1 + o(1)) \frac{cn}{2}$ steps in expectation.

Hence, we have that

$$\text{idct}(B(c,n)) \leq (1-c) \cdot O(1) + c \cdot (1+o(1))\frac{cn}{2} \sim \frac{c^2 n}{2}.$$

However, this is an asymptotically optimal strategy: with probability c the robber is on one of the vertices of the path and, since the vertices of the star do not help, in this case the expected capture time is $(1+o(1))cn/2$. Hence, $idct(B(c,n)) \sim \frac{c^2 n}{2}$. $\qquad\square$

Proof of Theorem 10.6.4. For an upper bound, we consider the following strategy for the cop. The cop starts at the end of the broom, moves to the center, and, once there, checks all end-vertices in a random order without repetitions. The robber starts at a randomly selected end-vertex and remains there until caught (actually the robber could move into a different end-vertex or even outside of the star as long as the cop is sufficiently distant, but this would not change the expected capture time). The expected capture time under these strategies is the sum of two terms:

(i) the cop needs cn steps to go from the end to the center;

(ii) once there, the cop will on average need to check $(1-c)\,n/2$ end-vertices before they capture the robber, and every such check requires two steps (one from the center to the end-vertex and one back), except the very last move, which requires only one step.

Hence,

$$\text{ict}\,(B\,(c,n)) \leq cn + 2\,(1-c)\,n/2 = n.$$

A lower bound for $\text{ict}(B(c,n))$ can be established by describing a robber's strategy and proving that it is optimal. Before we describe such a strategy, we need some additional notation. Let us assign the "coordinate" 0 to the end of the broom and the "coordinate" cn to the center. The cop's initial position X_0 can then be written as bn, where $b \in [0,c]$. Note that b is a parameter of the cop's strategy; also, the robber will observe b before they make their first move.

Before the game starts, the robber announces that the following strategy will be used: the robber will go to the end of the broom with probability $q = q(b) = b$, and to a randomly chosen end-vertex with probability $1 - q$. The cop is aware of this, and their only reasonable responses (after having started at bn) are the following.

(i) With probability p, the cop moves to the end of the broom, and then back to the center to sweep all end-vertices.

(ii) With probability $1 - p$, the cop moves to the center, sweeps a randomly chosen x-fraction of the end-vertices (for some $x \in [0, 1]$), then moves to the end of the broom, then goes back to the center, sweeping all end-vertices. Of course, for $x = 1$, the cop will have captured the robber by the time they reach the end, at the latest, so they will not need to revisit the center.

We do not consider the possibility that the cop starts at an end-vertex because this does not influence the asymptotic behavior of the expected capture time; this justifies our previous claim that the cop's initial position can be parametrized by bn.

It is easily seen that the above *family* of strategies (parametrized by (b, p, x)) guarantees capture (of course capture may take place before the full schedule is executed). Note that if c is exactly 1, then we have a path, and $\operatorname{ict}(B(c, n)) \sim n$. Also, if b is exactly 0, then the cop starts at the end of the broom, and the robber's only reasonable strategy is to hide at a randomly chosen end-vertex. In this case, the expected capture time is also $(1 + o(1))n$. Excluding these cases from the following analysis, we can break down the expected capture time into the following cases.

The case in which the robber hides in the end-vertices and the cop starts by moving toward the center is split into two subcases (and hence, requires two rows in the above table).

(i) In the first subcase, the cop checks an x-fraction of the end-vertices and captures the robber *before* visiting the end of the broom. This happens with probability x, and the average number of end-vertices checked is $x \cdot (1 - c) \cdot n/2$.

(ii) In the second subcase, the cop checks an x-fraction of the

TABLE 10.1: The possible ways in which the robber can be captured.

R starts at	C moves to	Prob.	Distance Traveled
end	end	bp	bn
an end-vertex	end	$(1-b)p$	$(b+c+(1-c))n$
end	center	$b(1-p)$	$((c-b)+2x(1-c)+c)n$
an end-vertex	center	$(1-b)(1-p)x$	$((c-b)+x(1-c))n$
an end-vertex	center	$(1-b)(1-p)(1-x)$	$((c-b)+(2x(1-c)+2c+(1-c)))n$

end-vertices, fails to capture the robber, visits the end of the broom, and then returns to check all the end-vertices (and so capture the robber with certainty). This happens with probability $1-x$, and the number of end-vertices checked is $x \cdot (1-c) \cdot n$ during the first visit and $(1-c) \cdot n/2$ (on the average) during the second visit.

The expected capture time for a given value of c is

$$\mathbb{E}[T] \sim f_c(p, b, x) \cdot n,$$

where $f_c(p, b, x) \cdot n$ is obtained by multiplying the entries of the last two columns of Table 10.1, adding, and performing some algebra. We finally derive

$$\begin{aligned}
f_c(b, p, x) = {} & (-1 + b + c + p - bc - bp - cp + bcp)\, x^2 \\
& + (1 + b - 3c - p + bc - bp + 3cp - bcp)\, x \\
& + (2c - 2b + 2bp - 2cp + 1).
\end{aligned}$$

Hence, $f_c(b, p, x)$ is a quadratic function in x, and can be rewritten as follows: $f_c(b, p, x) = a_2 x^2 + a_1 x + a_0$, with

$$\begin{aligned}
a_2 &= -(1-p)(1-b)(1-c) \\
a_1 &= (1-p)(1-3c+bc+b) \\
a_0 &= 2(1-p)(c-b) + 1.
\end{aligned}$$

Since a_2 is negative (unless $p = 1$, in which case the whole function is zero), the parabola is downward concave, and thus, it achieves its minimum either at $x = 0$ or at $x = 1$. If $x = 0$, then $f_c(b, p, x) =$

$a_0 = 1+2(1-p)(c-b) \geq 1$. If $x = 1$, then $f_c(b, p, x) = a_2+a_1+a_0 = 1$, and hence, the minimum is achieved there. Thus, $\text{ict}(B(c, n)) \geq (1 - o(1))n$; this, together with the upper bound, shows that the robber strategy is optimal and that $\text{ict}(B(c, n)) \sim n$. $\quad\square$

Chapter 11

Miscellaneous topics

As we have seen in previous chapters, there are a wealth of vertex pursuit games and processes that have been actively investigated. While a full survey of all of these games would be impossible, we focus on four relatively new games and processes that may garner interest in the future. We gather together in this chapter various topics ranging from the Toppling Game, Revolutionary and Spies, Robot Crawler, to Seepage. For each of the games and processes we consider, we focus on their behavior on random graphs such as $\mathbb{G}(n, p)$, and use probabilistic methods to analyze them. As we give an overview of various results, the proofs of technical lemmas are usually omitted.

11.1 Toppling Number

The first game we consider in this chapter was introduced by Cranston and West [73]. A *configuration* of a graph G is a placement of chips on the vertices of G. We represent a configuration by a function $c : V(G) \to \mathbb{N} \cup \{0\}$, where $c(v)$ indicates the number of chips on vertex v. Let c be a configuration of a graph G. We may *fire* a vertex v provided that $c(v) \geq \deg(v)$; when v is fired, $\deg(v)$ chips are removed from v, and one chip is added to each of its neighbors.

A configuration is *volatile* if there is some infinite sequence $v_1, v_2, \ldots$ of vertices that may be fired in order. A *stable* configuration is one in which no vertex may be fired; that is, $c(v) < \deg(v)$ for all $v \in V(G)$. See Figure 11.1.

Björner, Lovász, and Shor [28] proved that for any configuration of any graph, the order of vertex firings does not matter.

FIGURE 11.1: A volatile and a stable configuration on P_5.

More precisely, they showed that if c is a volatile configuration, then after any list of vertex firings, the resulting configuration remains volatile; if instead c is not volatile, then any two maximal lists of vertex firings yield the same stable configuration. They also showed that every volatile configuration of a graph G has at least $|E(G)|$ chips, and that every configuration having at least $2|E(G)| - |V(G)| + 1$ chips is volatile.

The *toppling game* is a two-player game played on a graph G. At the beginning of the game there are no chips on any vertices. The players, *Max* and *Min*, alternate turns as per usual. On each turn, a player adds one chip to a vertex of their choosing. The game ends when a volatile configuration is reached. Max aims to maximize the number of chips played before this point, while Min aims to minimize it. If Max starts and both players play optimally, then the length of the game on G is the *toppling number* of G, denoted by $t(G)$. A *turn* of the game is the placement of a single chip; a *round* is a pair of consecutive turns, one by each player. After each turn of the game, vertices are fired until a stable configuration arises (unless of course the configuration is volatile, in which case the game is over). We occasionally postpone firing vertices until it is convenient, or stop firing before reaching a stable configuration. As a result, we may end up firing vertices in a different order than if we had always fired immediately; however, by [28] this does not affect whether or not the current configuration is volatile.

We may consider the play of the toppling game on random graphs, which was first analyzed in [48]. Asymptotic bounds were given there on the toppling number of complete graphs as noted in the following theorem.

Theorem 11.1.1 ([48]). *For sufficiently large n we have that*

$$0.596400n^2 < t(K_n) < 0.637152n^2.$$

The toppling number of binomial random graphs and complete graphs are closely connected, and this relationship is made clear with the following theorem.

Theorem 11.1.2 ([48]). *If p is such that $pn \geq n^{2/\sqrt{\log n}}$ and $G \in \mathbb{G}(n,p)$, then a.a.s. we have that*

$$\mathrm{t}(G) \sim p\,\mathrm{t}(K_n).$$

Our goal in this section will be to prove Theorem 11.1.2. We begin by introducing a variant of the toppling game that facilitates the connection between the complete graph and the random graph.

Given $p \in (0,1)$, the *fractional toppling game* on a graph G is played similarly to the ordinary toppling game, but with different vertex firing rules. In the fractional game, a vertex v fires once the number of chips on v is at least $p \deg(v)$; when v fires, $p \deg(v)$ chips are removed from v, and p chips are added to each neighbor of v. Since p may be any real number, vertices need not contain whole numbers of chips (although each player still places exactly one chip on each turn). If Max plays first and both players play optimally, the length of the game is denoted by $t_p(G)$.

Despite their differences, the ordinary toppling game and fractional toppling game are related in a strong sense. We make this explicit in the following theorem (with proof omitted).

Theorem 11.1.3 ([48]). *For every n-vertex graph G, we have that $t_p(G) = p\,\mathrm{t}(G) + O(n)$.*

The central idea behind our main result is that the toppling game on $\mathbb{G}(n,p)$ behaves analogously to the fractional game on K_n, so long as p tends to zero slowly enough. We need the following lemmas, whose proofs are omitted.

Lemma 11.1.4 ([48]). *Let c_1 and c_2 be configurations of a graph G with c_1 dominating c_2. If c_2 is volatile, then so is c_1.*

Lemma 11.1.5 ([48]). *Let c be a configuration of a graph G. If c admits some list of firings such that every vertex of G fires at least once, then c is volatile.*

We describe a useful change to the rules of the fractional game that does not asymptotically affect the length of the game on K_n.

Lemma 11.1.6 ([48]). *Fix any function $\omega = \omega(n)$ tending to infinity with n. Consider the fractional game on K_n. If we forbid either player from placing more than ωpn chips on the same vertex, then the length of the game (assuming both players play optimally) is $(1 + o(1)) \operatorname{t}_p(K_n)$.*

Proof. For the upper bound it suffices to restrict only Min, since restricting Max cannot increase the length of the game. For the lower bound it suffices to restrict only Max. Both arguments proceed analogously.

Denote the players as "A" and "B." We consider two games: the *ordinary game*, in which there are no restrictions on moves, and the *restricted game*, in which A can play no more than ωpn chips on any one vertex. When restricting A, we consider only chips placed by A, and ignore chips placed by B. We give a strategy for A to ensure that the length of the restricted game is asymptotically the same as that of the ordinary game. To simplify analysis, we postpone all vertex firing until the end of the game.

Player A plays as follows. At the beginning of the game, A chooses an arbitrary indexing $v_1, v_2, \ldots, v_n$ of the vertices of K_n. Whenever B plays in the restricted game, A imagines that B played identically in the ordinary game. On A's turn, they first choose their move in the ordinary game according to some optimal strategy for that game; suppose they chose to play at vertex v. If A has placed fewer than $\lfloor (\omega - 1)pn \rfloor$ chips on v in the restricted game, then A places a chip on v. Otherwise, A places a chip on the least-indexed vertex on which they have not already placed $\lfloor (\omega - 1)pn \rfloor$ chips. (There must be some such vertex given sufficiently large n, since A places fewer than pn^2 chips before the game ends, and for large n we have that $n \lfloor (\omega - 1)pn \rfloor \geq pn^2$.)

Suppose now that A is Min. Min plays optimally in the ordinary game, so it finishes after at most $\operatorname{t}_p(K_n)$ turns; the restricted game may last longer. Once the ordinary game finishes, let c and c_r be configurations of the ordinary and restricted games, respectively. Let $S = \{v \in V(G) : c(v) \neq c_r(v)\}$. Each vertex in S, except perhaps for one, contains $\lfloor (\omega - 1)pn \rfloor$ chips in the restricted game; hence, $|S| \leq \frac{pn^2}{\lfloor(\omega-1)pn\rfloor} = o(n)$. Allow the restricted game to continue for another $|S|\,pn$ rounds; on these extra turns, Min places another pn chips on each vertex of S. Let $\hat{c}_r$ be the resulting configuration of the restricted game.

We claim that $\hat{c}_r$ is volatile. Let $u_1, u_2, \ldots$ be an infinite firing sequence for c; we construct an analogous firing sequence for $\hat{c}_r$. At all times we maintain the invariants that

$$\sum_{v \in S}(\hat{c}_r(v) - c(v)) \geq pn\,|S|$$

and that $\hat{c}_r(v) \geq c(v)$ for $v \notin S$. Both invariants clearly hold before any vertices are fired, and are maintained under the firing (in both games) of any vertex not in S. Suppose we fire, in the original game, some vertex v in S. If v may be fired in the restricted game, then doing so preserves both invariants. Otherwise, since $\sum_{v \in S}(\hat{c}_r(v) - c(v)) \geq pn\,|S|$, there is some vertex v' in S such that $\hat{c}_r(v') - c(v') \geq pn$. Since $\hat{c}_r(v') \geq p(n-1)$ we may fire v' in the restricted game, and doing so maintains both invariants. We may repeat the process indefinitely; hence, $\hat{c}_r$ is volatile. Therefore, the total number of chips played in the restricted game is at most $t_p(K_n) + 2pn\,|S| = t_p(K_n) + o(pn^2) \sim t_p(K_n)$, as desired.

Now suppose instead that A is Max. By Max's strategy, the ordinary game lasts for at least $t_p(K_n)$ turns, but the restricted game may finish earlier. By an argument similar to that in the preceding paragraph, once the restricted game ends, Min can reach a volatile configuration for the ordinary game after another $o(pn^2)$ turns. Letting t_r denote the length of the restricted game, we have that $t_p(K_n) \leq t_r + o(pn^2)$, so $t_r \geq t_p(K_n) - o(pn^2) \sim t_p(K_n)$, as claimed. □

We also need the following well-known observation (see for example, [37] Lemma 2.2). For a given vertex v and integer i, let $N(v, i)$ denote the set of vertices at distance i from v.

Lemma 11.1.7 ([48]). *Let $d = p(n-1)$ and suppose that $\log n \ll d \ll n$. For $G \in \mathbb{G}(n, p)$, a.a.s. for every $v \in V(G)$ and $i \in \mathbb{N}$ such that $d^i = o(n)$ we have that*

$$|N(v, i)| \sim d^i.$$

In particular, a.a.s. for every $v \in V(G)$, we have that $\deg(v) \sim d$.

Our next lemma is an analogue of Lemma 11.1.6 for the ordinary game on $\mathbb{G}(n,p)$.

Lemma 11.1.8 ([48]). *Fix any function* $\omega = \omega(n)$ *tending to* ∞ *with* n. *Consider the toppling game on* $G \in \mathbb{G}(n,p)$, *where* $p \gg \log n/n$. *If we forbid either or both of the players from placing more than* ωpn *chips on the same vertex, then a.a.s. the length of the game (assuming both players play optimally) is* $(1+o(1))\,\mathrm{t}(G)$.

Proof. Since we aim to show that the specified bound on $\mathrm{t}(G)$ holds a.a.s., we may assume that the property stated in Lemma 11.1.7 holds deterministically for G.

As in the proof of Lemma 11.1.6, we bound the length of the game above and below; for the upper bound it suffices to restrict only Min, and for the lower bound it suffices to restrict only Max. The two players' strategies are very similar.

Denote the players as "A" and "B." We consider two games: the *restricted game*, in which A can play no more than ωpn chips on any one vertex, and the *ordinary game*, in which there are no restrictions on moves. Player A uses the ordinary game to guide their play in the restricted game. We give a strategy for A to ensure that the length of the restricted game is asymptotically the same as that of the real game. To simplify analysis, we postpone firing vertices until it is convenient.

When B makes a move in the ordinary game, A simply imagines that B made the same move in the restricted game. When A plays, more care is needed. Call a vertex *saturated* if A has already placed $(\omega-3)pn$ chips there in the restricted game. Player A chooses their move for the ordinary game according to some optimal strategy for that game; suppose they play at vertex v. If v is not saturated, then A plays at v in the restricted game as well. If v is saturated, then A instead attempts to play (in the restricted game) on some vertex u in $N(v)$. This may require recursion: if u is saturated, then A attempts to play on some neighbor of u, and so on until they reach an unsaturated vertex. On subsequent moves by A at v in the ordinary game, A chooses different neighbors on which to play in the restricted game, until all neighbors have been used. At this point we fire v in the ordinary game, which resolves the discrepancy between the two games due to placement of excess chips at v. Player A handles subsequent moves at v similarly— by playing instead at each neighbor of v in turn, firing v, and repeating the process.

Suppose A is Min. Since Min plays optimally in the ordinary game, that game finishes in at most $t(G)$ turns. If at that point the restricted game has already finished, then Min has enforced the desired upper bound, and we are done. Otherwise, it suffices to show that Min can cause the restricted game to end in $o(pn^2)$ additional rounds, since $t(G) \geq |E(G)| = \Theta(pn^2)$. Let c and c_r be the current configurations for the ordinary and restricted games, respectively. We show that adding $o(pn^2)$ chips to c_r yields a configuration that dominates c; since c is volatile, the claim then follows by Lemma 11.1.4. However, this is straightforward. There is only one possible reason for discrepancy between c and c_r: for each saturated vertex v in the restricted game, it may be that Min attempted to play at some, but not all, neighbors of v. To resolve this discrepancy, it suffices to add, to each such vertex, $\deg(v)$ chips, and then to fire. Since the game lasts for $\Theta(pn^2)$ turns, the number of saturated vertices is $o(n)$; by Lemma 11.1.7, the number of chips added is thus $o(pn^2)$. Observe that no saturated vertex v gains more than $2\deg(v)$ chips during this process, so in total Min plays no more than $(\omega - 3)pn + 2\deg(v)$ chips on v; for sufficiently large n, this number of chips is less than ωpn.

Suppose instead that A is Max. This time, the ordinary game finishes in at least $t(G)$ turns. If at this point the restricted game has not yet finished, then Max has enforced the desired lower bound. Suppose instead that the restricted game finishes first. We claim that once the restricted game ends, Min can cause the ordinary game to end within $o(pn^2)$ additional rounds. Min can do this using a strategy similar to that in the last paragraph: to each saturated vertex v, add $\deg(v)$ chips in the ordinary game, and then fire. This produces a configuration in the ordinary game that dominates the configuration of the restricted game, and hence, is volatile. □

Before proving the main result, we need three technical lemmas about the structure of binomial random graphs (with proofs omitted).

Lemma 11.1.9 ([48]). *Let* $d = p(n-1)$ *and suppose that* $n^{2/\sqrt{\log n}} \leq d \ll n$. *Fix* $G \in \mathbb{G}(n,p)$, *let* $m \in \mathbb{N}$ *be such that* $d^m = o(n)$, *and let* $\psi = \psi(n) = n/d^m$. *A.a.s. for every* $v \in V(G)$ *the following properties hold.*

(i) *Fix* $j < m$. *For every* $u \in N(v, j-1) \cup N(v, j)$, *we have that*

$$|N(v, j-1) \cap N(u)| < 1.5\sqrt{\log n}/(m-j).$$

(ii) *For every* $u \in N(v, m-1) \cup N(v, m)$, *we have that*

$$|N(v, m-1) \cap N(u)| = O(\log n).$$

(iii) *For every* $u \in V \setminus \bigcup_{j \leq m-1} N(v, j)$, *we have that*

$$|N(v, m) \cap N(u)| = \begin{cases} O(d/\psi) & \text{if } \psi \leq d/\log n, \\ O(\log n) & \text{if } \psi > d/\log n. \end{cases}$$

The second lemma we need is the following.

Lemma 11.1.10 ([48]). *Let* p *be such that* $pn \geq n^{2/\sqrt{\log n}}$. *Fix* $G \in \mathbb{G}(n,p)$ *and let* $\omega = \log\log n$. *A.a.s. for every configuration* c *of* G *such that* $c(v) \leq 2\sqrt{\omega}pn$ *for all vertices* v, *and every legal firing sequence* $F = (u_1, u_2, \ldots, u_n)$ *under* c, *every vertex of* G *appears in* F *only* $O(pn/\log^2 n)$ *times.*

The final lemma is about properties of random graphs.

Lemma 11.1.11 ([48]). *Let* p *be such that* $pn \geq n^{2/\sqrt{\log n}}$. *Fix* $G \in \mathbb{G}(n,p)$ *and let* $\omega = \log\log n$. *A.a.s. for all sets* S *such that* $|S| \leq 2n/\omega$, *we have either*

$$\sum_{v \in V(G)} |\text{disc}(v)| = o(pn)\,|S|$$

or

$$\sum_{v \in V(G)} |\text{disc}(v)| = O(n\log n),$$

where $\text{disc}(v) = |N(v) \cap S| - p\,|S|$.

We are now finally ready to prove our main result of the section on the toppling number of random graphs.

Proof of Theorem 11.1.2. Since we aim for a property that holds a.a.s., we may assume that the properties in Lemmas 11.1.7, 11.1.10, and 11.1.11 hold deterministically for G. Let $\omega = \log \log n$.

We show that $t(G) \sim t_p(K_n)$, from which the result follows by Theorem 11.1.3. We denote the players as "A" and "B," and we provide a strategy for player A. We play two games, the ordinary game on G and the fractional game on K_n. The ordinary game is the "real" game on which both players play, while the fractional game is "imagined" by A to guide their strategy for the ordinary game. Except where otherwise specified, we postpone firing vertices until the end of the game.

Player A plays as follows. On A's turns, they first play according to some optimal strategy for the fractional game, and then make the same move in the ordinary game. When B plays in the ordinary game, A simply duplicates this move in the fractional game. Since $\sqrt{\omega} \to \infty$, by Lemma 11.1.6 we may assume (without affecting the asymptotic length of the game) that A plays no more than $\sqrt{\omega}pn$ chips on any one vertex in the fractional game (and hence, also in the ordinary game). Likewise, we may assume by Lemma 11.1.8 that B plays no more than $\sqrt{\omega}pn$ chips on any one vertex in the ordinary game (and hence, also in the fractional game). Thus, all vertices have at most $2\sqrt{\omega}pn$ chips at all times in both games.

Assume first that A is Min. We aim to show that $t(G) \leq (1 + o(1))p\,t(K_n) + o(pn^2)$ (which suffices since $t(K_n) = \Theta(n^2)$). By Lemmas 11.1.3 and 11.1.6, the fractional game ends after at most $(1+o(1))p\,t(K_n)$ turns. If the ordinary game finishes first, then we are done, so suppose otherwise. It suffices to show that Min can force the ordinary game to end after $o(pn^2)$ additional rounds or, equivalently, that the ordinary game can be brought to a volatile configuration by adding another $o(pn^2)$ chips.

Let c and c_f denote the configurations of the ordinary and fractional games, respectively. We would like to fire vertices while maintaining the property that $c(v) \geq c_f(v)$ for all vertices v. (Initially, we have equality for all vertices.) Since the fractional game has ended, c_f is volatile. Let $F = (u_1, u_2, \ldots, u_n)$ be a firing sequence of length n for c_f (the v_i need not be distinct). For $i \geq 1$, define $S_i = \{v : v$ appears at least i times in S$\}$. Since each vertex u_i gains at most $p(n-1)$ chips from predecessors in S, we must have $c_f(v) > (i-1)p(n-1)$ whenever $v \in S_i$. Let $k = \max\{i : S_i \neq \emptyset\}$. Construct a sequence F' by listing all elements of S_k, followed by all elements of S_{k-1}, and so on down to S_1, with the restriction that when listing the elements of S_1 we do so in order of their final appearances in S. We claim that F' is a legal firing sequence for c_f: since each chip in S_i can fire $i-1$ times without assistance from earlier vertices, the only potential problems come from the firings in S_1, but each vertex receives at least as many chips before firing as it did under F.

We aim to show that F' is also a legal firing sequence in the ordinary game (after adding $o(pn^2)$ more chips). We do this by firing vertices in large groups. With the S_i defined as in the previous paragraph, let

$$\ell = \min\{i : |S_i| \geq n/\omega\}.$$

Let F^* be the portion of F' consisting of those vertices in $S_k, S_{k-1}, \ldots, S_{\ell+1}$. In the fractional game, each vertex has at most $2\sqrt{\omega}pn$ chips under c_f, and gains at most $p(n-1)$ more as F' is fired; consequently, $k \leq (2 + o(1))\sqrt{\omega}$. Thus, $|F^*| \leq (2 + o(1))\sqrt{\omega} \cdot \frac{n}{\omega} = o(n)$. For all i such that $i \in \{1, 2, \ldots, \ell\}$, we divide the portion of F' corresponding to S_i into consecutive blocks with sizes between n/ω and $2n/\omega$.

We have expressed F' in the form $(F^*, F_1, \ldots, F_m)$, where $|F^*| = o(n)$, no vertex appears more than once in any F_i, and each F_i has size between n/ω and $2n/\omega$. We aim to fire each subsequence in both games, possibly after adding a few extra chips in the ordinary game. At all times we maintain the property that each vertex has at least as many chips in the ordinary game as in the fractional game. Each time we attempt to fire a vertex v in the ordinary game, we first add $\lceil \deg(v) - p(n-1) \rceil$ chips to v. This ensures that v loses exactly $p(n-1)$ chips in the course of firing; it also guarantees that whenever we may fire v in the fractional game, we may also fire it in the ordinary game. By Lemma 11.1.7, we add only $o(pn^2)$ chips across all n firings.

We first fire all of F^*. In the fractional game, each vertex gains at most $p\,|F^*|$ chips from these firings. This need not be the case in the ordinary game. To compensate, before firing any vertices in the ordinary game, we add $\lceil p\,|F^*|\rceil$ chips to each vertex; this ensures that, after firing, each vertex still has at least as many chips in the ordinary game as in the fractional game. Since $|F^*| = o(n)$, we only add $o(pn^2)$ extra chips.

We next fire the F_i in order. Let $F_i = (v_1, v_2, \ldots, v_m)$, and recall from the construction of F_i that the v_j must be distinct. When firing F_i, some vertices may receive more chips in the fractional game than in the ordinary game. As before, we compensate for this discrepancy before firing. To each v_j we add $p\,|F_i|$ chips, to ensure that v_j will have enough chips to fire when it is reached in the firing sequence; the total number of chips added is $p\,|F_i|^2 = o(pn^2/\omega)$. Each vertex v outside F_i receives exactly $p\,|F_i|$ chips from the firing in the fractional game, but only $|F_i \cap N_G(v)|$ chips in the ordinary game. Thus, to each vertex v we add $\max\{|F_i \cap N_G(v)| - p\,|F_i|, 0\}$ chips; by Lemma 11.1.11, the number of extra chips needed is only $o(pn^2/\omega)$. The number of chips added to compensate for the discrepancy due to each F_i is $o(pn^2/\omega)$, so the total number added throughout the full firing sequence is $o(pn^2)$.

Let U be the set of vertices not appearing in F'. All vertices not in U have been fired at least once. Since we have fired n vertices, each vertex in U has at least pn chips in the fractional game, and hence, also in the ordinary game. By Lemma 11.1.7, we may now add another $o(pn)$ chips to each vertex in U to ready it for firing. Lemma 11.1.5 now implies that we have reached a volatile configuration, and it follows that $t(G) \leq (1+o(1))p\,t(K_n)+o(pn^2)$ as claimed.

Suppose now that A is Max. We now aim to show that

$$t(G) \geq (1 + o(1))p\,t(K_n) - o(pn^2) \sim p\,t(K_n).$$

Play both games until the ordinary game ends. It follows from Lemma 11.1.8 that the ordinary game lasts for at least $(1 + o(1))\,t(G)$ turns. If the fractional game finishes first, then we are done, so suppose otherwise. It suffices to show that Min can force the fractional game to end after $o(pn^2)$ additional rounds or, equivalently, that the fractional game can be brought to a volatile configuration by adding another $o(pn^2)$ chips. Such a property would give that $(1 + o(1))\, t_p(K_n) \le (1 + o(1))\, t(G) + o(pn^2)$, which, by Lemma 11.1.3, is equivalent to the desired lower bound on $t(G)$.

Let c and c_f be the current configurations of the ordinary and fractional games, respectively. Initially $c_f(v) = c(v)$ for all vertices v; we aim to fire vertices while maintaining $c_f(v) \ge c(v)$ for all v. During this process we may need to add $o(pn^2)$ extra chips to the fractional game. Let $F = (u_1, u_2, \ldots, u_n)$ be a legal firing sequence under c. Partition F into contiguous subsequences $F_1, F_2, \ldots, F_k$, where each F_i has size between n/ω and $2n/\omega$, and $k \le \omega$. We show how to fire the F_i, in order, in both games.

Fix i; we aim to fire F_i in both games and reestablish $c_f(v) \ge c(v)$ for all v, while adding only $o(pn^2/\omega)$ chips to the fractional game. For sufficiently large n, Lemma 11.1.10 ensures that each vertex of G appears at most $Cpn/\log^2 n$ times in F_i, for some constant C. For $1 \le j \le Cpn/\log^2 n$, let S_j be the set of vertices appearing at least j times in F_i. Define $\mathrm{disc}_j(v) = |N(v) \cap S_j| - p|S_j|$. When F_i is fired in both games, vertex v receives $\sum_j (\mathrm{disc}_j(v) + p\,|S_j|)$ chips in the ordinary game, but only $\sum_j p\,|S_j|$ chips in the fractional game. To compensate for this discrepancy, before firing F_i, we add $\max\{\sum_j \mathrm{disc}_j(v), 0\}$ chips to each vertex v in the fractional game. By Lemma 11.1.11, we have that

$$\sum_v \sum_j \mathrm{disc}_j(v) \; \le \; \sum_j o(pn)\,|S_j| + \sum_j O(n\log n)$$

$$= \; o(pn^2/\omega) + O\left(n\log n \frac{Cpn}{\log^2 n}\right)$$

$$= \; o(pn^2/\omega).$$

We must exercise caution when firing, as vertices in F_i may receive chips "sooner" in the ordinary game than in the fractional game. To compensate for this, to each vertex v appearing in F_i, we add an additional $p\,|F_i|$ chips; consequently, v receives $\sum_j |N(v) \cap S_j|$ chips in the fractional game before any vertices of F_i are fired, which ensures that each vertex in F_i receives in advance, in the fractional game, all chips it receives from firing F_i in the ordinary game. Thus, when v may be fired in the ordinary game, it may also be fired in the fractional game. In total, this costs only $O(pn^2/\omega^2) = o(pn^2/\omega)$ extra chips. Finally, when a vertex is fired in the ordinary game, it may lose $o(pn)$ fewer chips than in the fractional game; we compensate for this by adding extra chips to each vertex in the fractional game before it is fired. Again, this requires adding only $o(pn^2/\omega)$ extra chips in total.

We now fire all F_i in order. For each F_i fired, we add $o(pn^2/\omega)$ chips to the fractional game, so in total we add $o(pn^2)$ chips. Since we have fired n vertices in the fractional game, every vertex has either fired or has received pn chips and hence, may now fire. Thus, by Lemma 11.1.5, the fractional game has ended. Since we have added only $o(pn^2)$ chips to the fractional game, the desired lower bound on $t(G)$ follows. $\qquad\square$

11.2 Revolutionary and Spies

In this section, we study the game of Revolutionaries and Spies first considered by József Beck in the mid-1990s (as referenced in [74]). We play on a fixed graph G, and there are two players: a team of r *revolutionaries* and a team of s *spies*. The revolutionaries want to arrange a one-time meeting consisting of m revolutionaries free of oversight by spies—the spies want to prevent this from happening. The revolutionaries start by occupying some vertices as their initial positions; more than one revolutionary is allowed to occupy some vertex. After that the spies do the same: they can start from some vertices already occupied by revolutionaries or choose brand new vertices. In each subsequent round, each revolutionary may move to an adjacent vertex or choose to stay where they are, and then each spy has the same option. Note that, as in the game of Cops and Robbers, the game is played with perfect information.

A *meeting* is a set of at least m revolutionaries occupying one vertex; a meeting is *unguarded* if there is no spy at that vertex. The revolutionaries win if at the end of some round there is an unguarded meeting. The spies win if they have a strategy to prevent this forever. See Figure 11.2.

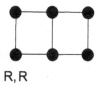

R, R

FIGURE 11.2: An unguarded meeting of two revolutionaries. Three spies can prevent unguarded meetings.

For given r and m, the minimum number of spies required to win on a graph G is the *spy number* $\sigma(G, r, m)$. Since $\min\{|V(G)|, \lfloor r/m \rfloor\}$ meetings can be initially formed, at least that many spies are needed to have a chance to win. Note that $r - m + 1$ spies can create a matching with $r - m + 1$ distinct revolutionaries and follow them during the whole game, preventing any unguarded meeting from taking place. If $|V(G)| < r - m + 1$, then this can be improved since occupying all vertices clearly does the job as well. We thus obtain the following immediate bounds on the spy number of a graph:

$$\min\{|V(G)|, \lfloor r/m \rfloor\} \leq \sigma(G, r, m) \leq \min\{|V(G)|, r - m + 1\}.$$

It is known that the lower bound is sufficient when G is a tree, and at most one additional spy is needed to win on any unicyclic graph [74]. The upper bound can be obtained for hypercubes, for example; see [60]. Grid-like graphs were studied in [113].

We focus on the spy number in binomial random graphs, and we first recall some notation. For $p \in (0, 1)$ or $p = p(n) > 0$ tending to 0 with n, define $\mathbb{L}n = \log_{\frac{1}{1-p}} n$. For constant p, clearly $\mathbb{L}n = \Theta(\log n)$, but for $p = o(1)$ we have that

$$\mathbb{L}n = \frac{\log n}{-\log(1 - p)} \sim \frac{\log n}{p}.$$

Preliminary results for random graphs have been proved in [60], where it is shown that for constant $p \in (0,1)$ and for $r < c\log\frac{1}{\min\{p,1-p\}} n$ with $c < 1$, and also for constant r and $p \gg n^{-1/r}$, $\sigma(G,r,m) = r - m + 1$ a.a.s. (the required condition is that $pn^r \to \infty$). The following bounds improve on these results.

Theorem 11.2.1 ([147]). *Let $\varepsilon > 0$, $r = r(n) \in \mathbb{N}$, and $m = m(n) \in \mathbb{N}$. If $G \in \mathbb{G}(n,p)$ with $p = p(n) < 1 - \varepsilon$, then a.a.s.*

$$\sigma(G,r,m) \leq \frac{r}{m} + 2(2 + \sqrt{2} + \varepsilon)\mathbb{L}n.$$

In particular, it follows from Theorem 11.2.1 that $\sigma(G,r,m) \sim r/m$ whenever $r/m \gg \mathbb{L}n$, and $\sigma(G,r,m) = \Theta(r/m)$ if $r/m = \Theta(\mathbb{L}n)$. The next theorem provides a lower bound.

Theorem 11.2.2 ([147]). *Let $\varepsilon > 0$, $\eta \in (0, 1/3]$, $r = r(n) \in \mathbb{N}$, and $m = m(n) \in \mathbb{N}$. Consider a random graph $G \in \mathbb{G}(n,p)$ with $p = n^{-1/3+\eta+o(1)} < 1 - \varepsilon$. Then a.a.s. we have that*

$$\sigma(G,r,m) \geq \min\{r - m, 2.99\eta\mathbb{L}n\} + 1.$$

Theorems 11.2.1 and 11.2.2 give an asymptotic behavior (up to a constant factor) of the spy number of random graphs with average degree $n^{2/3+\eta+o(1)}$ for some $\eta \in (0, 1/3]$. Therefore, we obtain a comprehensive set of results for such dense random graphs.

Corollary 11.2.3 ([147]). *Let $\varepsilon > 0$, $\eta \in (0, 1/3]$, $r = r(n) \in \mathbb{N}$, and $m = m(n) \in \mathbb{N}$. Consider a random graph $G \in \mathbb{G}(n,p)$ with $p = n^{-1/3+\eta+o(1)} < 1 - \varepsilon$. Then a.a.s. we have that*

$$\sigma(G,r,m) = \begin{cases} r - m + 1 & \text{if } r - m \leq 2.99\eta\mathbb{L}n, \\ \Theta(\mathbb{L}n) & \text{if } r - m > 2.99\eta\mathbb{L}n \\ & \text{and } r/m = O(\mathbb{L}n), \\ (1 + o(1))r/m & \text{if } r/m \gg \mathbb{L}n. \end{cases}$$

To prove these results, we first need some notation. Let $\mathcal{S}(v,i)$ denote the set of vertices whose distance from v is precisely i, and $\mathcal{N}(v,i)$ the set of vertices whose distance from v is at most i. (In particular, $\mathcal{N}(v) = \mathcal{S}(v,1)$, the neighborhood of v.) Also, $\mathcal{N}[S,i] =$

328   *Graph Searching Games and Probabilistic Methods*

$\bigcup_{v \in S} \mathcal{N}(v, i)$, $\mathcal{N}[S]$ denotes $\mathcal{N}[S, 1]$, the closed neighborhood of S, and $\mathcal{N}(S) = \mathcal{N}[S] \setminus S$, the (open) neighborhood of S. Finally, $\mathcal{N}^c(v)$ denotes $V(G) \setminus \mathcal{N}(v, 1)$, the nonneighborhood of v.

We need the following lemma and its corollary (with proofs omitted).

Lemma 11.2.4 ([147]). *Let $\alpha > 0$, $\beta > \frac{1+\alpha}{\alpha}$, and $\varepsilon > 0$. Consider a random graph $G \in \mathbb{G}(n, p)$ with $p < 1 - \varepsilon$. Then a.a.s. for every set $S \subseteq V(G)$ of cardinality $\beta \mathbb{L}n$ we have that*

$$\left| \bigcap_{v \in S} \mathcal{N}^c(v) \right| \leq \alpha \beta \mathbb{L}n.$$

Corollary 11.2.5 ([147]). *Let $\beta > 1$ and $\varepsilon > 0$. Consider a random graph $G \in \mathbb{G}(n, p)$ with $p < 1 - \varepsilon$. Then a.a.s. for every set $S \subseteq V(G)$ of cardinality $\beta \mathbb{L}n$ we have that*

$$\left| \bigcap_{v \in S} \mathcal{N}^c(v) \right| \leq (1 + \varepsilon) \frac{\beta}{\beta - 1} \mathbb{L}n.$$

It is not difficult to show that a.a.s. a random graph $\mathbb{G}(n, p)$ has a dominating set of size $(1 + o(1))\mathbb{L}n$. However, we can show that a larger set can not only dominate the rest of the graph, but can also create a matching with any set of cardinality $O(\mathbb{L}n)$.

Lemma 11.2.6 ([147]). *Let $\varepsilon > 0$ and consider a random graph $G \in \mathbb{G}(n, p)$ with $p = n^{-\eta + o(1)} < 1 - \varepsilon$ for some $\eta \in [0, 1]$. Let $\gamma, \delta > 0$ be such that $\gamma - \delta > 1 + \eta$. Then a.a.s. there exists a set $S \subseteq V(G)$ of cardinality $\gamma \mathbb{L}n$ such that for all sets $T \subseteq V(G) \setminus S$ of size at most $\delta \mathbb{L}n$ there is a perfect matching from T to some subset S' of S.*

Now, we are ready to prove the main result of this section, which gives the upper bound on the spy number on random graphs.

Proof of Theorem 11.2.1. Consider a random graph $G \in \mathbb{G}(n, p)$ with $p = n^{-\eta + o(1)} < 1 - \varepsilon$ for some $\eta \in [0, 1]$. Let

$$\delta = \frac{1 - \eta + \sqrt{\eta^2 + 2\eta + 5}}{2} > 1$$

and

$$\gamma = 1 + \eta + \delta + \varepsilon = \frac{3 + \eta + \sqrt{\eta^2 + 2\eta + 5}}{2} + \varepsilon \le 2 + \sqrt{2} + \varepsilon.$$

It follows from Lemma 11.2.6 that a.a.s. there exists a set $A \subseteq V(G)$ of cardinality $\gamma \mathbb{L}n$ such that for all sets $T \subseteq V(G) \setminus A$ of size at most $\delta \mathbb{L}n$ there is a perfect matching from T to some set $A' \subseteq A$.

We will split spies into three groups: the first two groups, τ_1 and τ_2, consist each of $\gamma \mathbb{L}n$ *super-spies*, and the third group τ_3 consists of $\lfloor r/m \rfloor$ *regular-spies*. super-spies from team τ_1 will occupy the entire whole set A at odd times but some of them might be sent to a mission at even times. If this is the case, then they will be back to the set A in the next round. Similarly, super-spies from team τ_2 will occupy A at even times but might be used to protect some other vertices at odd times. In particular, the set A will be constantly protected and so no unguarded meeting can take place there. Regular-spies (team τ_3) will occupy a set $B_t \subseteq V(G) \setminus A$ at time t. Moreover, no two regular-spies will occupy the same vertex, so $|B_t| = \lfloor r/m \rfloor$ for all t.

The revolutionaries start the game by occupying a set R_1 and forming at most $\lfloor r/m \rfloor$ meetings. The super-spies (from both teams τ_1 and τ_2) go to the set A. The regular-spies can easily protect the vertices of $V(G) \setminus A$ in which meetings take place by placing a spy on each vertex where there are at least m revolutionaries. The remaining regular-spies go to arbitrary vertices of $V(G) \setminus A$ not occupied by another spy. As a result, no unguarded meeting takes place in the first round.

Suppose that no unguarded meeting takes place at time $t - 1$ and that the regular-spies occupy a set $B_{t-1} \subseteq V(G) \setminus A$ of cardinality $\lfloor r/m \rfloor$. At the beginning of round t, the revolutionaries might form at most $\lfloor r/m \rfloor$ meetings at vertices of $M_t \subseteq V(G) \setminus A$ (as we already pointed out, meetings at A are constantly protected by super-spies, so we do not have to worry about them). Let $B = M_t \cap B_{t-1}$ be the set of vertices in which meetings are already guarded by regular-spies. It remains to show that there exists a perfect matching between $M_t \setminus B$ and some subset S of

$A \cup (B_{t-1} \setminus B)$. Indeed, if this is the case, then the regular-spies that do not protect any meeting as well as the super-spies from the team protecting A in the previous round, move from S to $M_t \setminus B$. The super-spies from another team come back to A to guard this set and prepare themselves for another mission. No unguarded meeting takes place at time t, $B_t \subseteq V(G) \setminus A$, and no two regular-spies occupy the same vertex. The result will follow by induction on t.

To show that a perfect matching can be formed, we use Hall's theorem. For a given $T \subseteq M_t \setminus B$, we need to show that $|\mathcal{N}(T) \cap (A \cup (B_{t-1} \setminus B))| \geq |T|$. If $|T| \leq \delta \mathbb{L}n$, then a perfect matching from T to some $A' \subseteq A$ exists and so

$$|\mathcal{N}(T) \cap (A \cup (B_{t-1} \setminus B))| \geq |\mathcal{N}(T) \cap A| \geq |T|.$$

Hall's condition then holds for such small sets. Suppose then that $|T| > \delta \mathbb{L}n$. It follows from Corollary 11.2.5 that all but at most

$$\left(1 + \frac{\varepsilon}{4}\right) \frac{\delta}{\delta - 1} \mathbb{L}n = \left(1 + \frac{\varepsilon}{4}\right) \frac{3 + \eta + \sqrt{\eta^2 + 2\eta + 5}}{2} \mathbb{L}n$$
$$< \left(\frac{3 + \eta + \sqrt{\eta^2 + 2\eta + 5}}{2} + \varepsilon\right) \mathbb{L}n$$
$$= \gamma \mathbb{L}n$$

vertices of $V(G) \setminus T$ have at least one neighbor in T. Hence,

$$|\mathcal{N}(T) \cap (A \cup (B_{t-1} \setminus B))| > |A| + \lfloor r/m \rfloor - |B| - \gamma \mathbb{L}n$$
$$= \lfloor r/m \rfloor - |B|$$
$$\geq |M_t| - |B| \geq |T|.$$

Thus, Hall's condition holds for larger sets as well and the proof follows. $\qquad \square$

For the proofs of the lower bounds in Theorem 11.2.2, we employ the following adjacency property (also used earlier for some other games such as Cops and Robbers) and its generalizations. For fixed positive integers k and l, we say that a graph G is (l, k)-*existentially closed* (or (l, k)-e.c.) if for any two disjoint subsets of $V(G)$, $A \subseteq V(G), B \subseteq V(G)$, with $|A| = l$ and $|B| = k$, there exists a vertex $z \in V(G) \setminus (A \cup B)$ not adjacent to any vertex in B and adjacent to every vertex in A. We will use the following simple observation.

Theorem 11.2.7 ([147]). *Let* r, m, s *be positive integers such that* $s \leq r - m$, *and let* G *be any* $(2, s)$-*e.c. graph. Then*

$$\sigma(G, r, m) \geq s + 1.$$

In particular, if G *is* $(2, r - m)$-*e.c., then* $\sigma(G, r, m) = r - m + 1$.

Proof. Suppose that r revolutionaries play the game on a graph G against s spies. The revolutionaries start by occupying r distinct vertices. No matter what s spies do, there will be $r - s \geq m$ unguarded revolutionaries. Since G is $(2, s)$-e.c., any two of them can meet in the next round and stay unguarded. In the following round, another revolutionary can join the two forming a group of three unguarded revolutionaries. This argument can be repeated (each time at least one more revolutionary joins the group) until an unguarded meeting of m revolutionaries is formed, and the game ends. The result holds. □

It remains to investigate for which values of s a random graph is $(2, s)$-e.c. a.a.s. Since we would like to match an upper bound of $O(\mathbb{L}n)$, our goal is to obtain a lower bound of $\Omega(\mathbb{L}n)$. Hence, the graph must be dense enough.

Lemma 11.2.8 ([147]). *Consider a random graph* $G \in \mathbb{G}(n, p)$ *with* $p = n^{-1/3 + \eta + o(1)} < 1 - \varepsilon$ *for some* $\eta \in (0, 1/3]$ *and* $\varepsilon > 0$. *Then for*

$$s = 2.99 \eta \mathbb{L}n$$

we have that a.a.s. G *is* $(2, s)$-*e.c.*

Proof. Fix any two disjoint subsets of $V(G)$, A, B, with $|A| = 2$ and $|B| = s$. For a vertex $z \in V(G) \setminus (A \cup B)$, the probability that z is adjacent to both vertices of A and no vertex of B is $p^2(1-p)^s$. Since the edges are chosen independently, the probability that no suitable vertex can be found for this particular choice of A and B is $(1 - p^2(1-p)^s)^{n-s-2}$.

Let X be the random variable counting the number of pairs of A, B for which no suitable z can be found. We have that

$$
\begin{aligned}
\mathbb{E}[X] &= \binom{n}{2}\binom{n-2}{s}(1 - p^2(1-p)^s)^{n-s-2} \\
&\leq n^{s+2} \exp\left(-p^2(1-p)^s(n-s-2)\right) \\
&= \exp\left((s+2)\log n - p^2(1-p)^s n(1+o(1))\right).
\end{aligned}
$$

If $p = \Theta(1)$, then $s = \Theta(\log n)$ and so it follows that

$$
\begin{aligned}
\mathbb{E}[X] &\leq \exp\left(O(\log^2 n) - \Omega(n^{1-2.99\eta})\right) \\
&= \exp\left(O(\log^2 n) - \Omega(n^{0.003})\right) \\
&= o(1).
\end{aligned}
$$

For $p = o(1)$ we have $s \sim 2.99\eta(\log n)/p = n^{1/3-\eta+o(1)}$ and thus, we derive that

$$
\begin{aligned}
\mathbb{E}[X] &\leq \exp\left(n^{1/3-\eta+o(1)} - n^{2(-1/3+\eta+o(1))-2.99\eta+1}\right) \\
&= \exp\left(n^{1/3-\eta+o(1)} - n^{1/3-0.99\eta+o(1)}\right) \\
&= o(1).
\end{aligned}
$$

The result follows by Markov's inequality.   □

Theorem 11.2.2 now follows from Theorem 11.2.7 and Lemma 11.2.8.

Proof of Theorem 11.2.2. It follows from Lemma 11.2.8 that a.a.s. G is $(2, s_0)$-e.c. for $s_0 = 2.99\eta\mathbb{L}n$. If $r - m \geq s_0$, then we find that $\sigma(G, r, m) \geq s_0 + 1$ by Theorem 11.2.7 (applied with $s = s_0$). Suppose then that $r - m < s_0$. Note that Theorem 11.2.7 cannot be applied with $s = s_0$ anymore. However, it is clear that any $(2, s_0)$-e.c. graph is also $(2, r - m)$-e.c. Using Theorem 11.2.7 with $s = r - m$, we derive that $\sigma(G, r, m) = r - m + 1$.   □

11.3   Robot Crawler

A *crawler* is a software application designed to gather information from web pages. Crawlers visit web pages and then traverse links as they explore the network. Information gathered by crawlers is then stored and indexed, as part of the anatomy of a search engine such as Google or Bing. Walks in graph theory have long been studied, beginning with Euler's study of the Königsberg bridges problem in 1736, and including the traveling salesperson problem [11] and the sizeable literature on Hamiltonicity problems (see, for example, [180]).

A generalization of Eulerian walks was introduced by Messinger and Nowakowski in [143], as a variant of graph cleaning processes (see, for example, [7, 144]). In the model of [143], called the *robot vacuum*, a building with dirty corridors is cleaned by an autonomous robot. The robot cleans these corridors in a greedy fashion, such that the next corridor cleaned is always the dirtiest to which it is adjacent. This is modeled as a walk in a graph. Li and Vetta [132] gave an interesting example where the robot vacuum takes exponential time to clean the graph.

We consider a recent model of [42], where the crawler cleans *vertices* rather than edges. The *robot crawler* $\mathcal{RC}(G, \omega_0) = \left((\omega_t, v_t) \right)_{t=1}^{L}$ of a connected graph $G = (V, E)$ on n vertices with an *initial weighting* $\omega_0 : V \to B_n$, that is a bijection from the vertex set to B_n, is defined as follows.

(i) Initially, set v_1 to be the vertex in V with weight $\omega_0(v_1) = -n + 1$.

(ii) Set $\omega_1(v_1) = 1$; the other values of ω_1 remain the same as in ω_0.

(iii) Set $t = 1$.

(iv) If all the weights are positive (that is, $\min_{v \in V} \omega_t(v) > 0$), then set $L = t$, stop the process, and return L and $\mathcal{RC}(G, \omega_0) = \left((\omega_t, v_t) \right)_{t=1}^{L}$.

(v) Let v_{t+1} be the dirtiest neighbor of v_t. More precisely, let v_{t+1} be such that

$$\omega_t(v_{t+1}) = \min\{\omega_t(v) : v \in N(v_t)\}.$$

(vi) $\omega_{t+1}(v_{t+1}) = t + 1$; the other values of ω_{t+1} remain the same as in ω_t.

(vii) Increment to time $t + 1$ (that is, increase t by 1) and return to 4.

If the process terminates, then define

$$\mathrm{rc}(G, \omega_0) = L.$$

In particular, $\mathrm{rc}(G, \omega_0)$ is equal to the number of steps in the *crawling sequence* $(v_1, v_2, \ldots, v_L)$ (including the initial state) taken by the robot crawler until all vertices are clean; otherwise $\mathrm{rc}(G, \omega_0) = \infty$. For a given ω_0, all steps of the process are deterministic; further, at each point of the process, the weighting ω_t is an injective function. Hence, there is always a unique vertex v_{t+1}, neighbor of v_t of minimum weight (see step (4) of the process). We refer to a vertex as *dirty* if it has a nonpositive weight (that is, it has not been yet visited by the robot crawler), and *clean*, otherwise.

The process always terminates in a finite number of steps by a theorem from [42] (with proof omitted).

Theorem 11.3.1 ([42]). *For a connected graph $G = (V, E)$ on n vertices and a bijection $\omega_0 : V \to B_n$, we have that $\mathcal{RC}(G, \omega_0)$ terminates after a finite number of steps; that is, $\mathrm{rc}(G, \omega_0) < \infty$.*

Given Theorem 11.3.1, we may consider optimizing the robot crawler process. Let G be any connected graph on n vertices. Let Ω_n be the family of all initial weightings $\omega_0 : V \to B_n$. Then

$$\mathrm{rc}(G) = \min_{\omega_0 \in \Omega_n} \mathrm{rc}(G, \omega_0) \quad \text{and} \quad \mathrm{RC}(G) = \max_{\omega_0 \in \Omega_n} \mathrm{rc}(G, \omega_0).$$

Hence, $\mathrm{rc}(G)$ and $\mathrm{RC}(G)$ are the smallest and largest number of time-steps, respectively, needed to crawl G, over all choices of initial weightings.

We will use the following result.

Theorem 11.3.2 ([42]). *If $T = (V, E)$ is a tree of order $n \geq 2$, then we have that*

$$\mathrm{rc}(T) = 2n - 1 - \mathit{diam}(T) \quad \text{and} \quad \mathrm{RC}(T) = 2n - 2,$$

where $\mathit{diam}(T)$ is the diameter of T.

Binomial Random Graphs

It is known (see, for example, [124]) that a.a.s. $\mathbb{G}(n,p)$ has a Hamiltonian cycle (and so also a Hamiltonian path) provided that $pn \geq \log n + \log \log n + \omega$, where $\omega = \bar{\omega}(n)$ is any function tending to infinity together with n. A.a.s. $\mathbb{G}(n,p)$ has no Hamiltonian cycle if $pn \leq \log n + \log \log n - \omega$. It is straightforward to show that in this case a.a.s. there are more than two vertices of degree at most 1 and so a.a.s. there is no Hamiltonian path. Combining these observations, we derive the following result.

Corollary 11.3.3 ([42]). *If $\omega = \omega(n)$ is any function tending to infinity together with n, then the following hold a.a.s.*

(i) *If $pn \geq \log n + \log \log n + \omega$, then $\mathrm{rc}(\mathbb{G}(n,p)) = n$.*

(ii) *If $pn \leq \log n + \log \log n - \omega$, then $\mathrm{rc}(\mathbb{G}(n,p)) > n$.*

The next result provides an upper bound on $\mathrm{RC}(\mathbb{G}(n,p))$ for the stated range of p.

Corollary 11.3.4 ([42]). *If $pn \geq C\sqrt{n \log n}$, for a sufficiently large constant $C > 0$, then a.a.s. we have that*

$$\mathrm{RC}(\mathbb{G}(n,p)) \leq n^3.$$

Proof. Given two different vertices u, v, the number of common neighbors is distributed as $\mathrm{Bin}(n-2, p^2)$, which by the Chernoff bounds is at least 1 with probability $1 - o(n^{-2})$ (by taking C large enough). Therefore, a.a.s. every pair of vertices of $\mathbb{G}(n,p)$ has some common neighbor and hence, the diameter is at most 2. The maximum degree is $\Delta \leq n - 1$. Note that for a connected graph G of order n, maximum degree Δ, and diameter d, it is straightforward to show that

$$\mathrm{RC}(G) \leq n(\Delta + 1)^d. \tag{11.1}$$

Therefore, by (11.1) a.a.s. $\mathrm{RC}(\mathbb{G}(n,p)) \leq n^3$. $\qquad\square$

We now consider the lower bound on $\mathrm{RC}(\mathbb{G}(n,p))$.

Theorem 11.3.5 ([42]). *If $C\sqrt{n\log n} \leq pn \leq (1-\varepsilon)n$, for constants $C > 1$ and $\varepsilon > 0$, then a.a.s. we have that*

$$\mathrm{RC}(\mathbb{G}(n,p)) \geq (2 - p + o(p))n.$$

Proof. Fix a vertex v of $G \in \mathbb{G}(n,p)$, and expose its neighborhood $K = N(v)$. Let $k = |K|$, $M = V \setminus K$, and $m = |M|$. We will show that the following properties hold a.a.s.

(a) $k \sim pn$ and $m = \Theta(n)$.

(b) The subgraphs of G induced by K and by M (which we denote $G[K]$ and $G[M]$, respectively) are both Hamiltonian.

(c) There is a pair $C_K = (u_1, u_2, \ldots, u_k)$ and $C_M = (v_1, v_2, \ldots, v_m)$ of Hamiltonian cycles of $G[K]$ and $G[M]$, respectively, such that $u_1 v_{m-1}$ and $u_k v_m$ are edges in G (that is, edges v_{m-1}, v_m and u_k, u_1 are contained in a 4-cycle).

If all of these properties hold, then we assign the initial weightings to vertices in the following order from dirtiest to cleanest:

$$v_1, v_2, \ldots, v_{m-1}, u_1, u_2, \ldots, u_k, v_m, v.$$

The robot crawler is then forced to first clean vertices $v_1, v_2, \ldots, v_{m-1}$, next $u_1, u_2, \ldots, u_k$, then $v_m v_1, v_2, \ldots, v_{m-1}$ and finally u_1 and v. Note that the only way of cleaning v is coming from a vertex in K at the previous time-step. This takes

$$(m - 1) + k + m + 2 = 2n - k - 1 = (2 - p + o(p))n$$

steps, as required.

We show items (a), (b), and (c) hold a.a.s. The first statement in (a) follows directly from the Chernoff bounds, without exposing any edges other than those incident to v. The second one uses the fact that $p \leq 1-\varepsilon$. For (b), note that $pk \sim p^2 n \geq C\log n \geq C\log k$, so the average degree in $G[K]$ is at least $C\log k$ and it is well known that $G[K]$ is a.a.s. Hamiltonian, provided that $C > 1$ (recall the discussion before Corollary 11.3.3). For an analogous reason (with even more room to spare), $G[H]$ is also a.a.s. Hamiltonian. Finally, for (c), fix any Hamiltonian cycles C_K and C_M and any edge $e =$

(x, y) in C_K. Let $\{e_i = (v_{2i-1}, v_{2i}) : 1 \leq i \leq \lfloor m/2 \rfloor\}$ be a set of independent edges of C_M. An important observation is that the previous arguments did not expose edges between K and M. For each i, let A_i be the event that x is adjacent to x_i and y is adjacent to y_i. Events $A_1, A_2, \ldots, A_{\lfloor m/2 \rfloor}$ are independent and each one has probability of holding equal to p^2. The number of successful events is distributed as $\text{Bin}(\lfloor m/2 \rfloor, p^2)$ and has expectation $\Omega(np^2) = \Omega(\log n)$. By the Chernoff bounds, we conclude a.a.s. that A_i holds for some i, and therefore, property (c) holds after conveniently relabeling the vertices of C_K and C_M. (Note that there is room to spare as one needs the expected value to tend to infinity at any rate.) □

Preferential Attachment Model

The *preferential attachment model*, introduced by Barabási and Albert [18], was an early stochastic model of complex networks. We will use the following precise definition of the model, as considered by Bollobás and Riordan in [38] as well as Bollobás, Riordan, Spencer, and Tusnády [39].

Let G_1^0 be the null graph with no vertices (or let G_1^1 be the graph with one vertex, v_1, and one loop). The random graph process $(G_1^t)_{t>0}$ is defined inductively as follows. Given G_1^{t-1}, we form G_1^t by adding vertex v_t together with a single edge between v_t and v_i, where i is selected randomly with the following probability distribution:

$$\mathbb{P}\,(i = s) = \begin{cases} \deg(v_s, t-1)/(2t-1) & 1 \leq s \leq t-1, \\ 1/(2t-1) & s = t, \end{cases}$$

where $\deg(v_s, t-1)$ denotes the degree of v_s in G_1^{t-1}.

Hence, in the model, we send an edge e from v_t to a random vertex v_i, where the probability that a vertex is chosen as v_i is proportional to its degree at the time, counting e as already contributing one to the degree of v_t.

For $m \in \mathbb{N} \setminus \{1\}$, the process $(G_m^t)_{t \geq 0}$ is defined analogously with the only difference that m edges are added to G_m^{t-1} to form G_m^t (one at a time), counting previous edges as already contributing to the degree distribution. Equivalently, we can define the process

$(G_m^t)_{t\geq 0}$ by considering the process $(G_1^t)_{t\geq 0}$ on a sequence $v_1', v_2', \ldots$ of vertices; the graph G_m^t if formed from G_1^{tm} by identifying vertices $v_1', v_2', \ldots, v_m'$ to form v_1, identifying vertices $v_{m+1}', v_{m+2}', \ldots, v_{2m}'$ to form v_2, and so on. Note that in this model G_m^t is in general a multigraph, possibly with multiple edges between two vertices (if $m \geq 2$) and self-loops. For the purpose of the robot crawler, loops can be ignored and multiple edges between two vertices can be treated as a single edge.

It was shown in [39] that for any $m \in \mathbb{N}$ a.a.s. the degree distribution of G_m^n follows a power law: the number of vertices with degree at least k falls off as $(1 + o(1))ck^{-2}n$ for some explicit constant $c = c(m)$ and large $k \leq n^{1/15}$. Let us start with the case $m = 1$, where each G_1^n is a forest. Each vertex sends an edge either to itself or to an earlier vertex, so the graph consists of components which are trees, each with a loop attached. The expected number of components is then $\sum_{t=1}^n 1/(2t-1) \sim (1/2)\log n$ and, since events are independent, we derive that a.a.s. there are $(1/2 + o(1))\log n$ components in G_1^n by the Chernoff bounds. Pittel [157] essentially showed that a.a.s. the largest distance between two vertices in the same component of G_1^n is $(\gamma^{-1} + o(1))\log n$, where γ is the solution of $\gamma e^{1+\gamma} = 1$ (see Theorem 13 in [38]). The following results derive from these facts; see [42] for more information.

Theorem 11.3.6 ([42]). *The following properties hold a.a.s. for any connected component G of G_1^n:*

$$
\begin{aligned}
\mathrm{rc}(G) &= 2|V(G)| - 1 - \mathrm{diam}(G) = 2|V(G)| - O(\log n), \\
\mathrm{RC}(G) &= 2|V(G)| - 2.
\end{aligned}
$$

We may modify the definition of the model to ensure G_1^n is a tree on n vertices, by starting from G_1^2 being an isolated edge and not allowing loops to be created in the process (this is the original model in [18]). For such variant, we would have that a.a.s. $\mathrm{rc}(G_1^n) \sim \mathrm{RC}(G_1^n) \sim 2n$, as the diameter would be negligible comparing to the order of the graph.

The case $m \geq 2$ is more complex, and what is known is summarized in the following theorem. The case $m \geq 3$ remains open. A.a.s. G_m^n is connected and its diameter is asymptotic to $\log n / \log \log n$, as shown in [38], and in contrast to the result for $m = 1$ presented above.

Theorem 11.3.7 ([42]). *A.a.s.* $\mathrm{rc}(G_2^n) \geq (1 + \xi + o(1))n$, *where*

$$\xi = \max_{c \in (0,1/2)} \left(\frac{2\sqrt{c}}{3} - c - \frac{c^2}{6} \right) \approx 0.10919.$$

Proof. Consider the process $(G_m^t)_{t \geq 0}$ on the sequence of vertices $(v_t)_{t \geq 0}$. We will call vertex v_i *lonely* if $\deg(v_i, n) = m$; that is, no loop is created at the time v_i is introduced and no other vertex is connected to v_i later in the process. The vertex v_i is called *old* if $i \leq cn$ for some constant $c \in (0, 1)$ that will be optimized at the end of the argument; otherwise, v_i is called *young*. Finally, v_i is called *j-good* if v_i is lonely and exactly j of its neighbors are old.

Suppose that *an* vertices are young and 1-good, *bn* vertices are young and 2-good, and *dn* vertices are old and lonely (which implies that they are 2-good). The robot crawler needs to visit all young vertices and all old and lonely ones, which takes at least $(1 - c)n + dn$ steps. Observe that each time a young and 2-good vertex is visited, the crawler must come from an old but not-lonely vertex and move to another such one right after. Similarly, each time the crawler visits a young and 1-good vertex, it must come from or move to some vertex that is old but not lonely. It follows that vertices that are old but not lonely must be visited at least $an/2 + bn + O(1)$ times. Hence, the process must take at least $(1 - c + d + a/2 + b + o(1))n$ steps, and goal is to show this gives a nontrivial bound for some value of $c \in (0, 1)$.

The probability that v_i is lonely is elementary to estimate from the equivalent definition of G_m^n obtained in terms of G_1^{mn}. For $i \gg 1$, we derive that

$$\mathbb{P}(v_i \text{ is lonely}) = \mathbb{P}(\deg(v_i, i) = m) \prod_{t=im+1}^{nm} \left(1 - \frac{m}{2t - 1} \right)$$

$$\sim \exp\left(-\sum_{t=im+1}^{nm} \frac{m}{2t - 1} + O\left(\sum_{t=im+1}^{nm} t^{-2} \right) \right)$$

$$\sim \; \exp\left(-\frac{m}{2}\sum_{t=im+1}^{nm} t^{-1}\right)$$

$$\sim \; \exp\left(-\frac{m}{2}\log\left(\frac{nm}{im}\right)\right) = \left(\frac{i}{n}\right)^{m/2}.$$

We consider the behavior of the following random variable: for $\lfloor cn \rfloor \le t \le n$, let

$$Y_t = \sum_{j \le cn} \deg(v_j, t).$$

In view of the identification between the models G_m^n and G_1^{mn}, it will be useful to investigate the following random variable instead: for $m\lfloor cn \rfloor \le t \le mn$, let

$$X_t = \sum_{j \le cmn} \deg_{G_1^t}(v_j', t).$$

We have that $Y_t = X_{tm}$. It follows that $X_{m\lfloor cn \rfloor} = Y_{\lfloor cn \rfloor} = 2m\lfloor cn \rfloor$. Further, for $m\lfloor cn \rfloor < t \le mn$,

$$X_t = \begin{cases} X_{t-1} + 1 & \text{with probability } \frac{X_{t-1}}{2t-1}, \\ X_{t-1} & \text{otherwise.} \end{cases}$$

The conditional expectation is given by

$$\begin{aligned} \mathbb{E}(X_t | X_{t-1}) &= (X_{t-1} + 1) \cdot \frac{X_{t-1}}{2t-1} + X_{t-1}\left(1 - \frac{X_{t-1}}{2t-1}\right) \\ &= X_{t-1}\left(1 + \frac{1}{2t-1}\right). \end{aligned}$$

Taking the expectation again, we derive that

$$\mathbb{E}X_t = \mathbb{E}X_{t-1}\left(1 + \frac{1}{2t-1}\right).$$

Hence, arguing as before, it follows that

$$\begin{aligned} \mathbb{E}Y_t &= \mathbb{E}X_{tm} = 2m\lfloor cn \rfloor \prod_{s=m\lfloor cn \rfloor+1}^{tm}\left(1 + \frac{1}{2s-1}\right) \\ &\sim 2cmn\left(\frac{tm}{cmn}\right)^{1/2} \\ &= 2mn\sqrt{ct/n}. \end{aligned}$$

Noting that $\mathbb{E}Y_t = \Theta(n)$ for any $\lfloor cn \rfloor \leq t \leq n$, and that Y_t increases by at most m each time (X_t increases by at most one), we obtain that with probability $1 - o(n^{-1})$, $Y_t = \mathbb{E}Y_t + O(\sqrt{n \log n}) \sim \mathbb{E}Y_t$ (using a standard martingale argument; for the Azuma-Hoeffding inequality that is used here see, for example, [115]). Hence, we may assume that $Y_t \sim 2mn\sqrt{ct/n}$ for any $\lfloor cn \rfloor \leq t \leq n$.

Note that for a given $t = xn$ with $c \leq x \leq 1$, the probability that an edge generated at this point of the process goes to an old vertex is asymptotic to $(2mn\sqrt{ct/n})/(2mt) = \sqrt{cn/t} = \sqrt{c/x}$. Recall that v_t is lonely with probability asymptotic to $(t/n)^{m/2} = x$ for the case $m = 2$. It follows that

$$a \sim \int_c^1 2\sqrt{c/x}(1 - \sqrt{c/x})x\,dx = \frac{4\sqrt{c}}{3} - 2c + \frac{2c^2}{3},$$

$$b \sim \int_c^1 (\sqrt{c/x})^2 x\,dx = c - c^2,$$

$$d \sim \int_0^c x\,dx = \frac{c^2}{2}.$$

The proof follows since

$$1 - c + d + a/2 + b \sim 1 + \frac{2\sqrt{c}}{3} - c - \frac{c^2}{6}$$

is maximized at $c = \dfrac{\left(\left(4+4\sqrt{5}\right)^{2/3}-4\right)^2}{4\left(4+4\sqrt{5}\right)^{2/3}} \approx 0.10380.$ $\quad\square$

11.4 Seepage

Our final vertex pursuit game is motivated by the one-way flow of information in certain real-world networks. In *hierarchical social networks* such as Twitter or the organizational structure of a company, users are organized on ranked levels below the source, with links (and as such, information) flowing from the source downwards to sinks. We may view hierarchical social networks as *directed acyclic graphs*, or *DAGs* for short. In such networks, information flows downwards from the source to sinks. Disrupting the flow of information may correspond to halting the spread of news or gossip in an online social network, or intercepting a message sent in a terrorist network.

How do we disrupt this flow of information while minimizing the resources used? We consider a simple model in the form of a vertex-pursuit game called Seepage introduced in [68]. Seepage is motivated by the 1973 eruption of the Eldfell volcano in Iceland. To protect the harbor, the inhabitants poured water on the lava in order to solidify it and thus, halt its progress.

The game of Seepage has two players, the *sludge* and a set of *greens* (note that one player controls all the greens), a DAG with one source (corresponding to the top of the volcano) and many sinks (representing the lake). The greens are analogous to the cops and the sludge to the robber. The players take turns, with the sludge going first by contaminating the top vertex (source). Then it is the greens' turn, and they choose some nonprotected, non-contaminated vertices to protect. On subsequent rounds the sludge moves a nonprotected vertex that is adjacent (that is, downhill) to the vertex the sludge is currently occupying and contaminates it; note that the sludge is located at a single vertex in each turn. The greens, on their turn, proceed as before; that is, choose some non-protected, noncontaminated vertices to protect. Once protected or contaminated, a vertex stays in that state to the end of the game. The sludge wins if some sink is contaminated; otherwise the greens win, that is, if they erect a cut set of vertices which separates the contaminated vertices from the sinks. The smallest number of greens needed to guarantee a win on a DAG is its *green number*. We make this more precise below. The game of Seepage has some elements in common with the game of Cops and Falling Robbers; see Section 3.4.

For an example, see the DAG in Figure 11.3. We assume all edges point from higher vertices to lower ones.

We analyze Seepage and the green number when played on a random DAG as a model of disrupting a given hierarchical social network. We give a precise formulation of our random DAG model below; the model includes as a parameter the total degree distribution of vertices in the DAG. This has some similarities to the $G(\mathbf{w})$

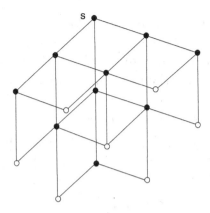

FIGURE 11.3: A DAG where 2 greens are needed to win. The white vertices are the sinks.

model of random graphs with expected degree sequences (see [66]) or the pairing model (see [184]).

Now, we more formally introduce Seepage. Fix $v \in V(G)$ a vertex of G, and we name v the *source*. For $i \in \mathbb{N}$ let

$$L_i = L_i(G, v) = \{u \in V(G) : \text{dist}(u, v) = i\},$$

where $\text{dist}(u, v)$ is the distance between u and v in G. In particular, $L_0 = \{v\}$. For a given $j \in \mathbb{N}^+$ and $c \in \mathbb{R}^+$, let $\mathbb{G}(G, v, j, c)$ be the game played on graph G with the source v and the *sinks* L_j. The game proceeds over a sequence of discrete time-steps. Exactly

$$c_t = \lfloor ct \rfloor - \lfloor c(t-1) \rfloor$$

new vertices are protected at time-step t. In particular, at most ct vertices are protected by the time t. Note that if c is an integer, then exactly c vertices are protected at each time-step, so this is a natural generalization of Seepage. To avoid trivialities, we assume that $L_j \neq \emptyset$.

The *sludge* starts the game on the vertex $v_1 = v$. The second player, the *greens*, can protect $c_1 = \lfloor c \rfloor$ vertices of $G \setminus \{v\}$. Once vertices are protected they will stay protected to the end of the game. At time $t \geq 2$, the sludge makes the first move by sliding along an edge from v_{t-1} to v_t, which is an out-neighbor of v_{t-1}.

After that the greens have a chance to protect another c_t vertices. Since the graph is finite and acyclic, the sludge will be forced to stop moving, and so the game will eventually terminate. If they reach any vertex of L_j, then the sludge wins; otherwise, the greens win.

If $c = \Delta(G)$ (the maximum out-degree of G), then the game $\mathbb{G}(G, v, j, c)$ can be easily won by the greens by protecting of all neighbors of the source. Therefore, the following graph parameter, *the green number*, is well-defined:

$$g_j(G, v) = \inf\{c \in \mathbb{R}^+ : \mathbb{G}(G, v, j, c) \text{ is won by the greens}\}.$$

It is clear that for any $j \in \mathbb{N}_+$ we have that $g_{j+1}(G, v) \le g_j(G, v)$.

There are two parameters of the model: $n \in \mathbb{N}^+$ and an infinite sequence

$$\mathbf{w} = (w_1, w_2, \ldots)$$

of nonnegative integers. Note that the w_i's may be functions of n. The first layer (that is, the source) consists of one vertex: $L_0 = \{v\}$. The next layers are recursively defined. For the inductive hypothesis, suppose that all layers up to and including the layer k are created, and we label all vertices of those layers. In particular,

$$L_k = \{v_{d_{k-1}+1}, v_{d_{k-1}+2}, \ldots, v_{d_k}\},$$

where $d_k = \sum_{i=0}^{k} |L_i|$. We would like the vertices of L_k to have a total degree with the following distribution $(w_{d_{k-1}+1}, w_{d_{k-1}+2}, \ldots, w_{d_k})$. However, it can happen that some vertex $v_i \in L_k$ has an in-degree $\deg^-(v_i)$ already larger than w_i, and so there is no hope for the total degree of w_i. If this is not the case, then the requirement can be easily fulfilled. As a result, $\mathbf{w}$, the desired degree distribution, will serve as a deterministic lower bound for the actual degree distribution we obtain during the (random) process.

Let S be a new set of vertices of cardinality n. All directed edges that are created at this time-step will be from the layer L_k to a random subset of S that will form a new layer L_{k+1}. Each vertex $v_i \in L_k$ generates $\max\{w_i - \deg^-(v_i), 0\}$ random directed edges from v_i to S. Therefore, we generate

$$e_k = \sum_{v_i \in L_k} \max\{w_i - \deg^-(v_i), 0\}$$

random edges at this time-step. The destination of each edge is chosen uniformly at random from S. All edges are generated independently, and so we perform e_k independent experiments. The set of vertices of S that were chosen at least once forms a new layer L_{k+1}. Note that it can happen that two parallel edges are created during this process. However, this is a rare situation for the sparse random graphs we are going to investigate in this paper. Hence, our results on the green number will also hold for a modified process that excludes parallel edges.

Random Power Law DAGs

We have three parameters in this model: $\beta > 2$, $d > 0$, and $0 < \alpha < 1$. For a given set of parameters, let

$$M = M(n) = n^\alpha, \quad i_0 = i_0(n) = n \left(\frac{d}{M} \frac{\beta - 2}{\beta - 1} \right)^{\beta - 1},$$

and

$$c = \left(\frac{\beta - 2}{\beta - 1} \right) dn^{\frac{1}{\beta - 1}}.$$

Finally, for $i \geq 1$ let

$$w_i = c(i_0 + i - 1)^{-\frac{1}{\beta - 1}}.$$

In this case, we refer to the model as *random power law DAGs*.

We note that the sequence $\mathbf{w}$ is decreasing (in particular, the source has the largest expected degree). The number of coordinates that are at least k is equal to

$$n \left(\frac{\beta - 2}{\beta - 1} \frac{d}{k} \right)^{\beta - 1} - i_0 \sim n \left(\frac{\beta - 2}{\beta - 1} \frac{d}{k} \right)^{\beta - 1} = \Theta(nk^{-\beta + 1}),$$

and hence, the sequence follows a power-law with exponent β. From the same observation it follows that the maximum value is

$$w_1 = ci_0^{-\frac{1}{\beta - 1}} = M.$$

Finally, the average of the first n values is

$$\frac{c}{n}\sum_{i=i_0}^{i_0+n-1} i^{-\frac{1}{\beta-1}} \sim \frac{c}{n}\left(\frac{\beta-1}{\beta-2}\right) n^{1-\frac{1}{\beta-1}} \sim d,$$

since $M = o(n)$.

The main result on the green number $g_j = g_j(G, v)$ in the case of power law sequences is the following. We include this result without proof.

Theorem 11.4.1 ([51]). *Let*

$$\gamma = d^{\beta-1}\left(\frac{\beta-2}{\beta-1}\right)^{\beta-2}\left(\left(1+\left(d\frac{\beta-2}{\beta-1}\right)^{1-\beta}\right)^{\frac{\beta-2}{\beta-1}} - 1\right)$$

if $\frac{1}{\alpha}-\beta+3 \in \mathbb{N}^+\backslash\{1,2\}$, and $\gamma = 1$, otherwise. Let j_1 be the largest integer satisfying $j_1 \leq \max\{\frac{1}{\alpha}-\beta+3, 2\}$. Let $j_2 = O(\log\log n)$ be the largest integer such that

$$d^{\beta-1}\left(\frac{\gamma}{d^{\beta-1}}n^{\alpha(j_1-1)-1}\right)^{\left(\frac{\beta-2}{\beta-1}\right)^{j_2-j_1}} \leq (\omega\log\log n)^{-\max\{2,(\beta-1)^2\}}.$$

Finally, let

$$\xi = \left(\frac{\beta-2}{\beta-1}\right)d\left(\left(\frac{d(\beta-2)}{\beta-1}\right)^{\beta-1}+1\right)^{-\frac{1}{\beta-1}}.$$

Then for $1 \leq j \leq j_2 - 1$ we have that a.a.s.

$$(1+o(1))\bar{w}_j \leq g_j \leq (1+o(1))\bar{w}_{j-1}, \tag{11.2}$$

where $\bar{w}_0 = \bar{w}_1 = M$, for $2 \leq j < \frac{1}{\alpha}-\beta+3$,

$$\bar{w}_j = \begin{cases} n^\alpha & \text{if } 2 \leq j < \frac{1}{\alpha}-\beta+2, \\ \xi n^\alpha & \text{if } 2 \leq j = \frac{1}{\alpha}-\beta+2, \\ \left(\frac{\beta-2}{\beta-1}\right)dn^{\frac{1-\alpha(j-1)}{\beta-1}} & \text{if } \frac{1}{\alpha}-\beta+2 < j < \frac{1}{\alpha}-\beta+3 \\ & \text{and } j \geq 2, \end{cases}$$

and for $j_1 \leq j \leq j_2 - 1$,

$$\bar{w}_j = \left(\frac{\beta-2}{\beta-1}\right)\left(\frac{\gamma}{d^{\beta-1}}n^{\alpha(j_1-1)-1}\right)^{-\left(\frac{\beta-2}{\beta-1}\right)^{j-j_1}/(\beta-1)}.$$

In the power law case, Theorem 11.4.1 tells us that the green number is smaller for large j.

Random Regular DAGs

We consider in this subsection a constant sequence; that is, for $i \in \mathbb{N}^+$ we set $w_i = d$, where $d \geq 3$ is a constant. In this case, we refer to the stochastic model as *random d-regular DAGs*. Since $w_i = d$, observe that $|L_j| \leq d(d-1)^{j-1}$ for any j, since at most $d(d-1)^{j-1}$ random edges are generated when L_j is created. We will write g_j for $g_j(G,v)$ since the graph G is understood to be a d-regular random graph, and $L_0 = \{v\} = \{v_1\}$.

Theorem 11.4.2 ([51]). *Let $\omega = \omega(n)$ be any function that grows (arbitrarily slowly) as n tends to infinity. For the random d-regular DAGs, we have the following.*

(i) *A.a.s. $g_1 = d$.*

(ii) *If $2 \leq j = O(1)$, then a.a.s.*

$$g_j = d - 2 + \frac{1}{j}.$$

(iii) *If $\omega \leq j \leq \log_{d-1} n - \omega \log \log n$, then a.a.s.*

$$g_j = d - 2.$$

(iv) *If $\log_{d-1} n - \omega \log \log n \leq j \leq \log_{d-1} n - \frac{5}{2}s \log_2 \log n + \log_{d-1} \log n - O(1)$ for some $s \in \mathbb{N}^+$, then a.a.s.*

$$d - 2 - \frac{1}{s} \leq g_j \leq d - 2.$$

(v) *Let $s \in \mathbb{N}^+$, $s \geq 4$. There exists a constant $C_s > 0$ such that if $j \geq \log_{d-1} n + C_s$, then a.a.s.*

$$g_j \leq d - 2 - \frac{1}{s}.$$

We give a snapshot of the proof of Theorem 11.4.2, proving all but the final item. The theorem tells us that the green number is bigger than $d-2$ if the sinks are located near the source, and then it is $d-2$ for a large interval of j. Later, it might decrease since an increasing number of vertices have already in-degree 2 or more, but only for large j (part (v)) we can prove better upper bounds than $d-2$. One interpretation of this fact is that the resources needed to disrupt the flow of information is in a typical regular DAG is (almost) independent of j, and relatively low (as a function of j).

The following lemma (proof omitted) gives the threshold for appearance of vertices of in-degree k.

Lemma 11.4.3 ([51]). *Let $\omega = \omega(n)$ be any function that grows (arbitrarily slowly) as n tends to infinity. Then a.a.s. the following properties hold.*

(i) $|L_j| \sim d(d-1)^{j-1}$ *for any* $1 \le j \le \log_{d-1} n - \omega$.

(ii) *For all $k \ge 2$, let $j_k = \frac{k-1}{k} \log_{d-1} n$. For every $v \in L_j$, we have that $\deg^-(v) < k$ if $j < j_k - \omega$, and $\deg^-(v) = k$ for some $v \in L_{j_k+\omega}$. In particular, the threshold for the appearance of vertices of in-degree k is j_k.*

(iii) $|L_j| = d(d-1)^{j-1}$ *for* $1 \le j \le \frac{1}{2} \log_{d-1} n - \omega$.

The lemma is enough to prove the first two parts of the main theorem.

Proof of Theorem 11.4.2. As discussed previously, we prove only the first four items. By Lemma 11.4.3 (iii) for $j \le \frac{1}{2} \log_{d-1} n - \omega$ the game is played on a tree. Part (i) is immediate, since the greens have to protect all vertices in L_1, or they lose.

To derive the upper bound of (ii), note that for $c = d - 2 + \frac{1}{j}$ we have that $c_j = d - 1$ ($c_i = d - 2$ for $1 \le i \le j - 1$). The greens can play arbitrarily during the first $j - 1$ steps, and then block the sludge on level j. If $j \ge \omega$, then we have that $d - 2$ is an upper bound of g_j, and the upper bound of (iii) holds.

To derive the lower bound of (ii), note that if $d - 2 \le c <$

$d-2-\frac{1}{j}$, then exactly $d-2$ new vertices are protected at each time-step. Without loss of generality, we may assume that the greens always protect vertices adjacent to the sludge (since the game is played on the tree, there is no advantage to play differently). No matter how the greens play, there is always at least one vertex not protected and the sludge can reach L_j. $\qquad\square$

For a given vertex $v \in L_t$ and integer j, let us denote by $S(v,j)$ the subset of L_{t+j} consisting of vertices at distance j from v (that is, those that are in the j-th level of the subgraph whose root is v). Let $N(v,j) = \sum_{i=1}^{j} S(v,i)$ be the subgraph of all vertices of depth j pending at v. A vertex $u \in S$ is *bad* if $u \in N(v,j)$ and u has in-degree at least 2 (recall that S is a set of n vertices used in the process of generating a random graph). Let $X(v,j)$ be the total number of bad vertices in $N(v,j)$. In the next lemma, we estimate $X(v,j)$.

Lemma 11.4.4 ([51]). *A.a.s. the following holds for some large enough constant $C' > 0$. For any $v \in L_t$, where $t \le \log_{d-1} n - \omega$, and any j such that $\frac{(d-1)^{t+2j}}{n} \le \log n$,*

$$X(v,j) \le C' \log n.$$

Proof. Fix $v \in L_t$ and let j be the maximum integer satisfying $\frac{(d-1)^{t+2j}}{n} \le \log n$. Since there are $O(n)$ possible vertices to consider, it is enough to show that the bound holds with probability $1 - o(n^{-1})$.

For $u \in S$, let $I_u(v,i)$ $(1 \le i \le j)$ be the event that $u \in S(v,i)$ and u is bad. In order for u to be in $S(v,i)$, u must receive at least one edge from a vertex in $S(v,i-1)$, and in order to be bad it must have at least one more edge from either $S(v,i-1)$ or from another vertex at layer L_{t+i-1}. Thus,

$$\mathbb{P}(I_u(v,i)) = \frac{O((d-1)^i)}{n}\frac{O((d-1)^{t+i})}{n} = O\left(\frac{(d-1)^{t+2i}}{n^2}\right),$$

since there are $O((d-1)^i)$ edges emanating from $S(v,i-1)$, and there are $O((d-1)^{t+i})$ edges coming from L_{t+i-1}. Letting $I_u = I_u(v, i)$ be the corresponding indicator variable, we have that

$$\mathbb{E}\left(\sum_{u \in S} I_u\right) = O\left(\frac{(d-1)^{t+2i}}{n}\right).$$

Note that

$$\mathbb{P}(I_u = 1 | I_{u'} = 1) \le \mathbb{P}(I_u = 1),$$

since

$$\mathbb{P}(I_u = 1 | I_{u'} = 1) \le \mathbb{P}(I_u = 1 | I_{u'} = 0)$$

(for a fixed total number of edges, the probability for u to be bad is smaller if another vertex u' is bad) and, by the law of total probability, at least one of the two conditional probabilities has to be at most $\mathbb{P}(I_u = 1)$. Thus, $\sum_{u \in S} I_u$ is bounded from above by $\sum_{u \in S} I'_u$, where the I'_u are independent indicator random variables with

$$\mathbb{P}(I_u = 1) \le \mathbb{P}(I'_u = 1) = C\frac{(d-1)^{t+2i}}{n^2}$$

for some sufficiently large $C > 0$. The total number of bad vertices in the subgraph of depth j pending at v is

$$X = \sum_{i=1}^{j} \sum_{u \in S} I_u(v, i) \le \sum_{i=1}^{j} \sum_{u \in S} I'_u(v, i).$$

Since $\frac{(d-1)^{t+2j}}{n} \le \log n$,

$$\mathbb{E}(X) \le \sum_{i=1}^{j} C\frac{(d-1)^{t+2i}}{n} = O\left(\frac{(d-1)^{t+2j}}{n}\right) = O(\log n),$$

and by the Chernoff bounds, $X \le C' \log n$ with probability $1 - o(n^{-1})$ for some $C' > 0$ large enough.   □

We need a final lemma. For a given vertex $v \in L_t$ and integer j, a vertex $u \in S$ is called *very bad* if it has at least two incoming edges from vertices in $S(v, i-1)$. In particular, every very bad vertex is bad. Let $Z(v, j)$ be the number of very bad vertices in $N(v, j)$.

For a given $T = \Theta(\log \log n)$, and any $\hat{L}_T \subseteq L_T$ such that $|\hat{L}_T| = o(|L_T| / \log^2 n)$, we will consider the subgraph $G(\hat{L}_T)$ consisting of all vertices to which there is a directed path from some vertex in $\hat{L}_T$. For any $t > T$, let $\hat{L}_t$ be a subset of L_t that is in $G(\hat{L}_T)$.

Lemma 11.4.5 ([51]). *Let $\hat{L}_T \subseteq L_T$ for some $T = \Theta(\log \log n)$ be such that $|\hat{L}_T| = o(|L_T| / \log^2 n)$. Then a.a.s. for any $v \in \hat{L}_t$, where $T \leq t \leq \log_{d-1} n - \omega$, and any integer j with $\frac{(d-1)^{t+2j}}{n} \leq \log n$, we have that $Z(v, j) = 0$.*

Proof. Fix any $v \in \hat{L}_t$ for some $T \leq t \leq \log_{d-1} n - \omega$. As in Lemma 11.4.4, by letting $H_u(v, i)$ be the event that $u \in S$ is very bad, we have

$$\mathbb{P}(H_u(v, i)) = \frac{O((d-1)^i)}{n} \frac{O((d-1)^i)}{n} = O\left(\frac{(d-1)^{2i}}{n^2}\right).$$

Letting $H_u = H_u(v, i)$ be the corresponding indicator variable, we have that

$$\mathbb{E}\left(\sum_{u \in S} H_u\right) = O\left(\frac{(d-1)^{2i}}{n}\right).$$

Define independent indicator random variables H'_u with $\mathbb{P}(H_u = 1) \leq \mathbb{P}(H'_u = 1) = C\frac{(d-1)^{2i}}{n^2}$. We have

$$Z(v, j) = \sum_{i=1}^{j} \sum_{u \in S} H_u(v, i) \leq \sum_{i=1}^{j} \sum_{u \in S} H'_u(v, i),$$

and so

$$\mathbb{E}(Z(v, j)) \leq \sum_{i=1}^{j} C\frac{(d-1)^{2i}}{n} = O\left(\frac{(d-1)^{2j}}{n}\right)$$
$$= O\left(\frac{\log n}{(d-1)^t}\right),$$

since $\frac{(d-1)^{t+2j}}{n} \leq \log n$.

As $|\hat{L}_T| = o(|L_T|/\log^2 n)$, we have that

$$|\hat{L}_t| \le |\hat{L}_T|(d-1)^{t-T} = o((d-1)^t/\log^2 n),$$

and so the expected number of very bad vertices found in $\hat{L}_t$ is $o(1/\log n)$. Finally, the expected number of very bad vertices in any sublayer $\hat{L}_t$ ($T \le t \le \log_{d-1} n - \omega$) is $o(1)$, and the result holds by Markov's inequality.   □

We now come back to the proof of the main theorem for random regular DAGs.

Proof of Theorem 11.4.2(iii) and (iv). Note that we already proved an upper bound of (iii) (see the proof of parts (i) and (ii)). Since g_j is nonincreasing as a function of j, an upper bound of (iv) also holds.

We will prove a lower bound of (iv) first. The lower bound of (iii) will follow easily from there. Let $s \in \mathbb{N}^+$ and suppose that we play the game with parameter $c = d - 2 - \frac{1}{s}$. If $s \ne 1$, then for every $i \in \mathbb{N}$, we have that $c_{si+1} = d - 3$ and $c_t = d - 2$, otherwise. (For $s = 1$ we find that $c_t = d - 3$ for any t.) Suppose that the greens play *greedily* (that is, they always protect vertices adjacent to the sludge) and the graph is locally a tree. Note that during the time between $si + 2$ and $s(i + 1)$, they can direct the sludge leaving them exactly one vertex to choose from at each time-step. However, at time-step $s(i + 1) + 1$, the sludge has 2 vertices to choose from. The sludge has to use this opportunity wisely, since arriving at a bad vertex (see definition above) when the greens can protect $d - 2$ vertices would result in them losing the game. Our goal is to show that the sludge can avoid bad vertices and, as a result, they have a strategy to reach the sink L_j. Since we aim for a statement that holds a.a.s. we can assume that all properties mentioned in Lemmas 11.4.3, 11.4.4, and 11.4.5 hold.

Before we describe a winning strategy for the sludge, let us discuss the following useful observation. While it is evident that the greens should use a greedy strategy to play on the tree, it is less evident in our situation. Perhaps instead of playing greedily, the greens should protect a vertex far away from the sludge, provided that there are at least two paths from the sludge to this vertex.

However, this implies that the vertex is very bad and we know that very bad vertices are rare. It follows from Lemma 11.4.5 that there is no very bad vertex within distance j, provided that the sludge is at a vertex in L_t, $t = \Omega(\log\log n)$ and $\frac{(d-1)^{t+2j}}{n} \leq \log n$. (For early steps we know that the graph is locally a tree so there are no bad vertices at all.) Therefore, without loss of generality, we can assume that at any time-step t of the game, the greens protect vertices greedily or protect vertices at distance at least j where j is the smallest value such that $\frac{(d-1)^{t+2j}}{n} > \log n$. We call the latter protected vertices *dangerous*. The sludge has to make sure that there are no nearby bad nor dangerous vertices.

Let
$$T = s(\log_2 \log n + C),$$

where the constant $C > 0$ will be determined soon and is sufficiently large such that the sludge is guaranteed to escape from all bad or dangerous vertices which are close to them. Let $\delta = 3/\log_2\left(\frac{d-1}{d-2}\right)$. During the first δT time-steps, the sludge chooses any arbitrary branch. Since they are given this opportunity at least $\delta \log_2 \log n = 3\log_{(d-1)/(d-2)} \log n$ times and each time they cut the number of possible destinations by a factor of $\frac{d-2}{d-1}$, the number of possible vertices the sludge can reach at time δT is $O(|L_{\delta T}|/\log^3 n)$. From that point on, it follows from Lemma 11.4.5 that there are no nearby very bad vertices. At time $t_1 = \delta T$, by Lemma 11.4.3, there are no bad vertices at distance

$$d_1 = \frac{1}{2}\log_{d-1} n - \delta T - \omega \geq T$$

from the sludge, and hence, no dangerous vertices within this distance. It follows from Lemma 11.4.4 that there are $O(\log n)$ bad vertices at distance

$$\bar{d}_1 = \frac{1}{2}\log_{d-1} n + \frac{1}{2}\log_{d-1}\log n - \frac{\delta T}{2}.$$

There are $O(\log n)$ dangerous vertices within this distance (since the total number of protected vertices during the whole game is of this order). Thus, there are $O(\log n)$ bad or dangerous vertices at a distance between d_1 and $\bar{d}_1$ from the sludge.

To derive a lower bound on the length of the game, we provide a strategy for the sludge that allows them to play for at least a certain number of steps, independently of the greens' behavior. In particular, their goal is to avoid these bad or dangerous vertices: as long as the sludge is occupying a vertex that is not bad, there is at least one vertex on the next layer available to choose from. More precisely, it follows from Lemma 11.4.5, that from time δT onwards, locally there are no very bad vertices. Let us call a *round* a sequence of T time-steps. Since all bad vertices are in distinct branches, in every s-th time-step the sludge can half the number of bad vertices. Therefore, after one round the sludge can escape from all $(C'+1)\log n$ bad or dangerous vertices that are under consideration in a given round, provided that $C > 0$ is a large enough constant. Recall the constant C' is defined in Lemma 11.4.4.

Using this strategy, at time $t_2 = (\delta + 1)T$ there are no bad or dangerous vertices at distance

$$d_2 = \bar{d}_1 - T = \frac{1}{2}\log_{d-1} n + \frac{1}{2}\log_{d-1}\log n - \frac{\delta+2}{2}T \geq T.$$

To see this, note that since the sludge escaped from all bad or dangerous vertices, which at time t_1 were at distance $\bar{d}_1$, and they have advanced T steps by now. Using Lemma 11.4.4 again, we find that there are $O(\log n)$ bad or dangerous vertices at distance

$$\bar{d}_2 = \frac{1}{2}\log_{d-1} n + \frac{1}{2}\log_{d-1}\log n - \frac{\delta+1}{2}T.$$

Arguing as before, we find that it takes another T steps to escape from them.

In general, at time $t_i = (\delta + i - 1)T$, there are $O(\log n)$ bad or dangerous vertices at a distance between

$$d_i = \frac{1}{2}\log_{d-1} n + \frac{1}{2}\log_{d-1}\log n - \frac{\delta+i}{2}T$$

and

$$\bar{d}_i = \frac{1}{2}\log_{d-1} n + \frac{1}{2}\log_{d-1}\log n - \frac{\delta+i-1}{2}T.$$

Thus, as long as $d_i \geq T$, the strategy of escaping from bad or dangerous vertices before actually arriving at that level is feasible. Note that we can finish this round, and so the sludge is guaranteed to use this strategy until time t_i, where i is the smallest value such that $d_i \leq T$. Solving this for i we obtain that

$$\frac{(\delta + i + 2)}{2} T \leq \frac{1}{2} \log_{d-1} n + \frac{1}{2} \log_{d-1} \log n,$$

and so

$$t_i = (\delta + i - 1)T \leq \log_{d-1} n + \log_{d-1} \log n - 3T.$$

Hence,

$$t_i = (\delta + i - 1)T \leq \log_{d-1} n - 3s \log_2 \log n + \log_{d-1} \log n - O(1).$$

Finally, note that if i is the smallest value such that $d_i \leq T$, we obtain that $d_{i-1} \geq T$ and so $d_i \geq \frac{T}{2}$. Hence, another $T/2$ steps can be played, and the constant of the second-order term can be improved from $3s \log_2 \log n$ to $\frac{5}{2} s \log_2 \log n$, yielding part (iv). Part (iii) follows by taking s to be a function of n slowly growing to infinity. $\qquad\square$

Bibliography

[1] M. Aigner, M. Fromme, A game of cops and robbers, *Discrete Applied Mathematics* **8** (1984) 1–12.

[2] N. Alon, F.R.K. Chung, Explicit construction of linear sized tolerant networks, *Discrete Mathematics* **72** (1988) 15–19.

[3] N. Alon, F.R.K. Chung, R.L. Graham, Routing permutations on graphs via matchings, *SIAM Journal of Discrete Mathematics* **7** (1994) 513–530.

[4] N. Alon, A. Merhabian, Chasing a fast robber on planar graphs and random graphs, *Journal of Graph Theory* **78** (2015) 81–96.

[5] N. Alon, V.D. Milman, λ_1, Isoperimetric inequalities for graphs and superconcentrators, *Journal of Combinatorial Theory, Series B* **38** (1985) 73–88.

[6] N. Alon, P. Prałat, Chasing robbers on random geometric graphs—an alternative approach, *Discrete Applied Mathematics* **178** (2014) 149–152.

[7] N. Alon, P. Prałat, N. Wormald, Cleaning regular graphs with brushes, *SIAM Journal on Discrete Mathematics* **23** (2008) 233–250.

[8] N. Alon, J. Spencer, *The Probabilistic Method*, Wiley, New York, 2000.

[9] B. Alspach, Sweeping and searching in graphs: a brief survey, *Matematiche* **59** (2006) 5–37.

[10] O. Angel, I. Shinkar, A tight upper bound on acquaintance time of graphs, *Graphs and Combinatorics* **32** (2016) 1–7.

[11] D.L. Applegate, R.E. Bixby, V. Chvátal, W.J. Cook, *The Traveling Salesman Problem*, Princeton University Press, 2007.

[12] K. Azuma, Weighted sums of certain dependent random variables, *Tohoku Mathematical Journal* **19** (1967) 357–367.

[13] W. Baird, A. Bonato, A. Beveridge, P. Codenotti, A. Maurer, J. McCauley, S. Valeva, On the minimum order of k-cop-win graphs, *Contributions to Discrete Mathematics* **9** (2014) 70–84.

[14] W. Baird, A. Bonato, Meyniel's conjecture on the cop number: a survey, *Journal of Combinatorics* **3** (2012) 225–238.

[15] D. Bal, P. Bennett, A. Dudek, P. Prałat, The total acquisition number of random graphs, *Electronic Journal of Combinatorics* **23(2)** (2016) #P2.55.

[16] D. Bal, A. Bonato, W.B. Kinnersley, P. Prałat, Lazy Cops and Robbers played on hypercubes, *Combinatorics, Probability, and Computing* **24** (2015) 829–837.

[17] D. Bal, A. Bonato, W. Kinnersley, P. Prałat, Lazy Cops and Robbers played on random graphs and graphs on surfaces, *Journal of Combinatorics* **7** (2016) 627–642.

[18] A.L. Barabási, R. Albert, Emergence of scaling in random networks, *Science* **286** (1999) 509–512.

[19] Zs. Baranyai, On the factorization of the complete uniform hypergraph, In: *Infinite and Finite Sets*, Colloquia mathematica Societatis János Bolyai, Vol. 10, North-Holland, Amsterdam, 1975, pp. 91–108.

[20] I. Benjamini, I. Shinkar, G. Tsur, Acquaintance time of a graph, *SIAM Journal on Discrete Mathematics*, **28(2)** (2014) 767–785.

[21] E.A. Bender, E.R. Canfield, The asymptotic number of non-negative integer matrices with given row and column sums, *Journal of Combinatorial Theory, Series A* **24** (1978) 296–307.

[22] S. Ben-Shimon, M. Krivelevich, Random regular graphs of non-constant degree: concentration of the chromatic number, *Discrete Mathematics* **309** (2009) 4149–4161.

[23] A. Berarducci, B. Intrigila, On the cop number of a graph, *Advances in Applied Mathematics* **14** (1993) 389–403.

[24] S. Bessy, A. Bonato, J. Janssen, D. Rautenbach, E. Roshanbin, Bounds on the burning number, Preprint 2017.

[25] S. Bessy, A. Bonato, J. Janssen, D. Rautenbach, E. Roshanbin, Burning a graph is hard, accepted to *Discrete Applied Mathematics*.

[26] A. Beveridge, A. Dudek, A. Frieze, T. Müller, Cops and Robbers on geometric graphs, *Combinatorics, Probability and Computing* **21** (2012) 816–834.

[27] A. Beveridge, A. Dudek, A. M. Frieze, T. Müller, M. Stojakovic, Maker-breaker games on random geometric graphs, *Random Structures and Algorithms* **45** (2014) 553–607.

[28] A. Björner, L. Lovász, P.W. Shor, Chip-firing games on graphs, *European Journal of Combinatorics* **12** (1991) 283–291.

[29] B. Bollobás, A probabilistic proof of an asymptotic formula for the number of labelled regular graphs, *European Journal of Combinatorics* **1** (1980) 311–316.

[30] B. Bollobás, Random graphs, in *Combinatorics* (ed. H.N.V. Temperley) London Mathematical Society Lecture Note Series, 52, Cambridge University Press, Cambridge (1981) pp. 80–102.

[31] B. Bollobás, *The Art of Mathematics: Coffee Time in Memphis*, Cambridge University Press, 2006.

[32] B. Bollobás, The evolution of random graphs, *Transactions of the American Mathematical Society* **286** (1984) 257–274.

[33] B. Bollobás, The isoperimetric number of random regular graphs, *European Journal of Combinatorics* **9** (1984) 241–244.

[34] B. Bollobás, *Modern Graph Theory*, Springer-Verlag, New York, 1998.

[35] B. Bollobás, *Random Graphs*, Cambridge University Press, Cambridge, 2001.

[36] B. Bollobás, G. Kun, I. Leader, Cops and robbers in a random graph, *Journal of Combinatorial Theory, Series B* **103** (2013) 226–236.

[37] B. Bollobás, D. Mitsche, P. Prałat, Metric dimension for random graphs, *Electronic Journal of Combinatorics* **20(4)** (2013) #P1.

[38] B. Bollobás, O. Riordan, The diameter of a scale-free random graph, *Combinatorica* **24** (2004) 5–34.

[39] B. Bollobás, O. Riordan, J. Spencer, G. Tusnády, The degree sequence of a scale-free random graph process, *Random Structures and Algorithms* **18** (2001) 279–290.

[40] B. Bollobás, O. Riordan, A simple branching process approach to the phase transition in $G_{n,p}$, *Electronic Journal of Combinatorics* **19** (2012) P21.

[41] A. Bonato, A. Burgess, Cops and Robbers on graphs based on designs, *Journal of Combinatorial Designs* **21** (2013) 359–418.

[42] A. Bonato, R.M. del Río-Chanona, C. MacRury, J. Nicolaidis, X. Pérez-Giménez, P. Prałat, K. Ternovsky, The robot crawler number of a graph, In: *Proceedings of WAW'15*, 2015.

[43] A. Bonato, N.E. Clarke, S. Finbow, S. Fitzpatrick, M.E. Messinger, A note on bounds for the cop number using tree decompositions, *Contributions to Discrete Mathematics* **9** (2014) 50–56.

[44] A. Bonato, P. Gordinowicz, W.B. Kinnersley, P. Prałat, The capture time of the hypercube, *Electronic Journal of Combinatorics* **20**(2) (2013).

[45] A. Bonato, G. Hahn, P.A. Golovach, J. Kratochvíl, The capture time of a graph, *Discrete Mathematics* **309** (2009) 5588–5595.

[46] A. Bonato, J. Janssen, E. Roshanbin, How to burn a graph, *Internet Mathematics* **1-2** (2016) 85–100.

[47] A. Bonato, G. Kemkes, P. Prałat, Almost all cop-win graphs contain a universal vertex, *Discrete Mathematics* **312** (2012) 1652–1657.

[48] A. Bonato, W.B. Kinnersley, P. Prałat, The toppling number of complete and random graphs, *Discrete Mathematics and Theoretical Computer Science* **16** (2014) 229-252.

[49] A. Bonato, M.E. Messinger, P. Prałat, Fighting constrained fires in graphs, *Theoretical Computer Science* **434** (2012) 11–22.

[50] A. Bonato, D. Mitsche, X. Pérez-Giménez, P. Prałat, A probabilistic version of the game of Zombies and Survivors on graphs, *Theoretical Computer Science* **655** (2016) 2–14.

[51] A. Bonato, D. Mitsche, P. Prałat, Vertex-pursuit in random directed acyclic graphs, *SIAM Journal on Discrete Mathematics* **27** (2013) 732–756.

[52] A. Bonato, R.J. Nowakowski, *The Game of Cops and Robbers on Graphs*, American Mathematical Society, Providence, Rhode Island, 2011.

[53] A. Bonato, P. Prałat, C. Wang, Network security in models of complex networks, *Internet Mathematics* **4** (2009) 419–436.

[54] A. Bonato, X. Pérez-Giménez, P. Prałat, B. Reiniger, Over-prescribed Cops and Robbers, *Graphs and Combinatorics* **33** (2017) 801–815.

[55] A. Bonato, P. Prałat, C. Wang, Vertex pursuit games in stochastic network models, *Proceedings of the 4th Workshop on Combinatorial and Algorithmic Aspects of Networking*, Lecture Notes in Computer Science, Springer, 2007, 46–56.

[56] P. Borowiecki, D. Dereniowski, P. Prałat, Brushing with additional cleaning restrictions, *Theoretical Computer Science* **557** (2014) 76–86.

[57] G. Brightwell, P. Winkler, Gibbs measures and dismantlable graphs, *Journal of Combinatorial Theory, Series B* **78** (2000) 141–169.

[58] W.G. Brown, On graphs that do not contain a Thomsen graph, *Canadian Mathematical Bulletin* **9** (1966) 281–285.

[59] D. Bryant, N. Francetic, P. Gordinowicz, D. Pike, P. Prałat, Brushing without capacity restrictions, *Discrete Applied Mathematics* **170** (2014) 33–45.

[60] J.V. Butterfield, D.W. Cranston, G. Puleo, D.B. West, R. Zamani, Revolutionaries and Spies: spy-good and spy-bad graphs, *Theoretical Computer Science* **463** (2012) 35–53.

[61] L. Cai, Y. Cheng, E. Verbin, Y. Zhou, Surviving rates of graphs with bounded treewidth for the firefighter problem, *SIAM Journal of Discrete Mathematics* **24** (2010) 1322–1335.

[62] L. Cai, W. Wang, The surviving rate of a graph for the firefighter problem, *SIAM Journal of Discrete Mathematics* **23** (2009) 1814–1826.

[63] P. Chebyshev, Mémoire sur les nombres premiers, *Mémoires de l'Académie impériale des sciences de St. P'etersbourg* **7** (1850) 17-33.

[64] N. Chen, On the approximability of influence in social networks, *SIAM Journal of Discrete Mathematics* **23** (2009) 1400–1415.

[65] E. Chiniforooshan, A better bound for the cop number of general graphs, *Journal of Graph Theory* **58** (2008) 45–48.

[66] F.R.K. Chung, L. Lu, *Complex graphs and networks*, American Mathematical Society, Providence RI, 2006.

[67] N.E. Clarke, *Constrained Cops and Robber*, Ph.D. Thesis, Dalhousie University, 2002.

[68] N.E. Clarke, S. Finbow, S.L. Fitzpatrick, M.E. Messinger, R.J. Nowakowski, Seepage in directed acyclic graphs, *Australasian Journal of Combinatorics* **43** (2009) 91–102.

[69] N.E. Clarke, G. MacGillivray, Characterizations of k-copwin graphs, *Discrete Mathematics* **312** (2012) 1421–1425.

[70] C. Cooper, A. Frieze, B. Reed, Random regular graphs of non-constant degree: connectivity and Hamilton cycles, *Combinatorics, Probability and Computing* **11** (2002) 249–262.

[71] D. Coppersmith, P. Tetali, P. Winkler, Collisions among random walks on a graph, *SIAM Journal of Discrete Mathematics* **6** (1993) 363–374.

[72] D.W. Cranston, D.B. West, An introduction to the discharging method via graph coloring, *Discrete Mathematics* **340** (2017) 766–793.

[73] D.W. Cranston, D.B. West, On the length of the toppling game, Preprint 2017.

[74] D.W. Cranston, C.D. Smyth, D.B. West, Revolutionaries and spies in trees and unicyclic graphs, *Journal of Combinatorics* **3** (2012) 195–206.

[75] P. Dembowski, *Finite Geometries*, Springer-Verlag, Berlin, 1968.

[76] R. Diestel, *Graph theory*, Springer-Verlag, New York, 2000.

[77] A. Dudek, A. Frieze, Loose Hamiltonian cycles in random uniform hypergraphs, *Electronic Journal of Combinatorics* **18** (2011) #P48.

[78] A. Dudek, A. Frieze, P.-S. Loh, S. Speiss, Optimal divisibility conditions for loose Hamilton cycles in random hypergraphs, *Electronic Journal of Combinatorics* **19** (2012) #P44.

[79] A. Dudek, P. Gordinowicz, P. Prałat, Cops and Robbers playing on edges, *Journal of Combinatorics* **5** (2014) 131–153.

[80] A. Dudek, P. Prałat, Acquaintance time of random graphs near connectivity threshold, *SIAM Journal of Discrete Mathematics* **30** (2016) 555–568.

[81] P. Erdős, Beweis eines Satzes von Tschebyschef, *Acta Universitatis Szegediensis* **5** (1930-32) 194–198.

[82] P. Erdős, A. Rényi, On the evolution of random graphs, *Publications of the Mathematical Institute of the Hungarian Academy of Sciences* **5** (1960) 17–61.

[83] P. Erdős, A. Rényi, V.T. Sós, On a problem of graph theory, *Studia Scientiarum Mathematicarum Hungarica* **1** (1966) 215–235.

[84] M. Fekete, Über die verteilung der wurzeln bei gewissen algebraischen gleichungen mit ganzzahligen koeffizienten, *Mathematische Zeitschrift* **17** (1923) 228–249.

[85] W. Feller, Generalization of a probability limit theorem of Cramér, *Transactions of the American Mathematical Society* **54** (1943) 361–372.

[86] S. Finbow, A. King, G. MacGillivray, R. Rizzi, The firefighter problem for graphs of maximum degree three, *Discrete Mathematics* **307** (2007) 2094–2105.

[87] S. Finbow, G. MacGillivray, The firefighter problem: a survey of results, directions and questions, *Australasian Journal of Combinatorics* **43** (2009) 57–77.

[88] S. Finbow, P. Wang, W. Wang, The surviving rate of an infected network, *Theoretical Computer Science* **411** (2010) 3651–3660.

[89] S.L. Fitzpatrick, J. Howell, M.E. Messinger, D.A. Pike, A deterministic version of the game of zombies and survivors on graphs, *Discrete Applied Mathematics* **213** (2016) 1–12.

[90] F. Fomin, P. Golovach, J. Kratochvíl, On tractability of the Cops and Robbers game, In: Giorgio Ausiello, Juhani Karhumki, Giancarlo Mauri, C.-H. Luke Ong (Eds.): *Fifth IFIP International Conference On Theoretical Computer Science- TCS 2008, IFIP 20th World Computer Congress, TC 1, Foundations of Computer Science*, Milano, Italy, IFIP 273 Springer 2008, 171–185.

[91] P. Frankl, Cops and robbers in graphs with large girth and Cayley graphs, *Discrete Applied Mathematics* **17** (1987) 301–305.

[92] J. Friedman, A proof of Alon's second eigenvalue conjecture, *Memoirs of the American Mathematical Society*, **195** 118 pp.

[93] A. Frieze, Loose Hamilton cycles in random 3-uniform hypergraphs, *Electronic Journal of Combinatorics* **17** (2010) #N283.

[94] A.M. Frieze, M. Karoński, *Introduction to Random Graphs*, Cambridge University Press, 2015.

[95] A. Frieze, M. Krivelevich, P. Loh, Variations on Cops and Robbers, *Journal of Graph Theory* **69** (2012) 383–402.

[96] Z. Füredi, On the number of edges of quadrilateral-free graphs, *Journal of Combinatorial Theory, Series B* **68** (1996) 1–6.

[97] S. Gaspers, M.E. Messinger, R. Nowakowski, P. Prałat, Clean the graph before you draw it!, *Information Processing Letters* **109** (2009) 463–467.

[98] S. Gaspers, M.E. Messinger, R. Nowakowski, P. Prałat, Parallel cleaning of a network with brushes, *Discrete Applied Mathematics* **158** (2010) 467–478.

[99] T. Gavenčiak, *Games on Graphs*, Master's Thesis, Department of Applied Mathematics, Charles University, Prague, 2007.

[100] T. Gavenčiak, Catching a fast robber on interval graphs, *Theory and Applications of Models of Computation*, Lecture Notes in Computer Science, vol. 6648, Springer, Heidelberg, 2011, pp. 353–364.

[101] R. Glebov, A. Liebenau, T. Szabó, On the concentration of the domination number of the random graph, *SIAM Journal on Discrete Mathematics* **29** (2015) 1186–1206.

[102] A. Godbole, E. Kelley, E. Kurtz, P. Prałat, Y. Zhang, The total acquisition number of the randomly weighted path, *Discussiones Mathematicae Graph Theory* **37** (2017), 919–934.

[103] P. Gordinowicz, R. Nowakowski, P. Prałat, POLISH: Let us play the cleaning game, *Theoretical Computer Science* **463** (2012) 123–132.

[104] G.R. Grimmett, D.R. Stirzaker, *Probability and Random Processes*, Oxford University Press, 2001.

[105] G. Hahn, Cops, Robbers and graphs, *Tatra Mountain Mathematical Publications* *36* (2007) 163–176.

[106] G. Hahn, G. MacGillivray, A characterization of k-cop-win graphs and digraphs, *Discrete Mathematics* **306** (2006) 2492–2497.

[107] P. Hall, On representatives of subsets, *Journal of the London Mathematical Society* **10** (1935) 26–30.

[108] B. Hartnell, Firefighter! An application of domination, Presentation at the *25th Manitoba Conference on Combinatorial Mathematics and Computing*, University of Manitoba, Winnipeg, Canada, 1995.

[109] S.T. Hedetniemi, S.M. Hedetniemi, A. Liestman, A survey of gossiping and broadcasting in communication networks, *Networks* **18** (1998) 319–349.

[110] A. Hill, *Cops and Robbers: Theme and Variations*, PhD Thesis, Dalhousie University, 2008.

[111] W. Hoeffding, Probability inequalities for sums of bounded random variables, *Journal of the American Statistical Association* **58** (1963) 13–30.

[112] S. Hoory, N. Linial, A. Wigderson, Expander graphs and their applications, *Bulletin of the American Mathematical Society* **43** (2006) 439–561.

[113] D. Howard, C.D. Smyth, Revolutionaries and spies on grid-like graphs, Preprint 2017.

[114] E. Infeld, D. Mitsche, P. Prałat, The total acquisition number of random geometric graphs, *Electronic Journal of Combinatorics* **24(3)** (2017), #P3.31.

[115] S. Janson, T. Łuczak, A. Ruciński, *Random Graphs*, Wiley, New York, 2000.

[116] G. Joret, M. Kamiński, D.O. Theis, The cops and robber game on graphs with forbidden (induced) subgraphs, *Contributions to Discrete Mathematics* **5** (2010) 40–51.

[117] A. Kehagias, D. Mitsche, P. Prałat, Cops and invisible Robbers: the cost of drunkenness, *Theoretical Computer Science* **481** (2013) 100–120.

[118] A. Kehagias, P. Prałat, Some remarks on Cops and Drunk Robbers, *Theoretical Computer Science* **463** (2012) 133–147.

[119] D. Kempe, J. Kleinberg, E. Tardos, Maximizing the spread of influence through a social network, In: *Proceedings of KKD*, 2003.

[120] W.B. Kinnersley, D. Mitsche, P. Prałat, A note on the acquaintance time of random graphs, *Electronic Journal of Combinatorics* **20** (2013) #P52.

[121] W.B. Kinnersley, P. Prałat, Game brush number, *Discrete Applied Mathematics* **207** (2016) 1–14.

[122] W.B. Kinnersley, P. Prałat, D.B. West, To catch a falling robber, *Theoretical Computer Science* **627** (2016) 107–111.

[123] N. Komarov, J. Mackey, Containment: a variation of Cops and Robbers, Preprint 2017.

[124] J. Komlós, E. Szemerédi, Limit distribution for the existence of Hamiltonian cycles in a random graph, *Discrete Mathematics* **43** (1983) 55–63.

[125] M. Krivelevich, C. Lee, B. Sudakov, Long paths and cycles in random subgraphs of graphs with large minimum degree, *Random Structures and Algorithms* **46** (2015) 320–345.

[126] M. Krivelevich, B. Sudakov, V.H. Vu, N.C. Wormald, Random regular graphs of high degree, *Random Structures and Algorithms* **18** (2001) 346–363.

[127] S. Kopparty, C.V. Ravishankar, A framework for pursuit evasion games in R^n, *Information Processing Letters* **96** (2005) 114–122.

[128] D.E. Lampert, P.J. Slater, The acquisition number of a graph, *Congressus Numerancium* **109** (1995) 203–210.

[129] M. Land, L. Lu, An upper bound on burning number of graphs, *Proceedings of the 13th Workshop on Algorithms and Models for the Web Graph (WAW 2016)*, Lecture Notes in Computer Science **10088**, Springer, 2016, 1–8.

[130] T.D. LeSaulnier, N. Prince, P. Wenger, D.B. West, P. Worah, Total acquisition in graphs, *SIAM Journal of Discrete Mathematics* **27** (2013) 1800–1819.

[131] A. Li, T. Müller, P. Prałat, Chasing robbers on percolated random geometric graphs, *Contributions to Discrete Mathematics* **10** (2015) 134–144.

[132] Z. Li, A. Vetta, Bounds on the cleaning times of robot vacuums, *Operations Research Letters* **38** (2010) 69–71.

[133] L. Lu, X. Peng, On Meyniel's conjecture of the cop number, *Journal of Graph Theory* **71** (2012) 192–205.

[134] T. Łuczak, Sparse random graphs with a given degree sequence, in *Random Graphs Vol. 2* (eds. A. Frieze & T. Łuczak) Wiley, New York (1992) pp. 165–182.

[135] T. Łuczak, Component behavior near the critical point of the random graph process, *Random Structures and Algorithms* **1** (1990) 287–310.

[136] T. Łuczak, P. Prałat, Chasing robbers on random graphs: zigzag theorem, *Random Structures and Algorithms* **37** (2010) 516–524.

[137] M. Maamoun, H. Meyniel, On a game of policemen and robber, *Discrete Applied Mathematics* **17** (1987) 307–309.

[138] B.D. McKay, Asymptotics for symmetric 0-1 matrices with prescribed row sums, *Ars Combinatoria* **19** (1985) 15–25.

[139] B.D. McKay, N.C. Wormald, Asymptotic enumeration by degree sequence of graphs of high degree, *European Journal of Combinatorics* **11** (1990) 565–580.

[140] B.D. McKay, N.C. Wormald, Asymptotic enumeration by degree sequence of graphs with degrees $o(n^{1/2})$, *Combinatorica* **11** (1991) 369–382.

[141] S. McKeil, *Chip Firing Cleaning Processes*, MSc Thesis, Dalhousie University (2007).

[142] A. Merhrabian, The capture time of grids, *Discrete Mathematics* **311** (2011) 102–105.

[143] M.E. Messinger, R.J. Nowakowski, The Robot cleans up, *Journal of Combinatorial Optimization* **18** (2009) 350–361.

[144] M.E. Messinger, R.J. Nowakowski, P. Prałat, Cleaning a network with brushes, *Theoretical Computer Science* **399** (2008) 191–205.

[145] M.E. Messinger, R. Nowakowski, P. Prałat, Cleaning with brooms, *Graphs and Combinatorics* **27** (2011) 251–267.

[146] M.E. Messinger, R. Nowakowski, P. Prałat, Elimination schemes and lattices, *Discrete Mathematics* **328** (2014) 63–70.

[147] D. Mitsche, P. Prałat, Revolutionaries and spies on random graphs, *Combinatorics, Probability, and Computing* **22** (2013) 417–432.

[148] D. Mitsche, P. Prałat, E. Roshanbin, Burning graphs—a probabilistic perspective, *Graphs and Combinatorics* **33** (2017) 449–471.

[149] D. Mubayi, J. Williford, On the independence number of the Erdős-Rényi and projective norm graphs and a related hypergraph, *Journal of Graph Theory* **56** (2007) 113–127.

[150] T. Müller, P. Prałat, The acquaintance time of (percolated) random geometric graphs, *European Journal of Combinatorics* **48** (2015) 198–214.

[151] S. Neufeld, R. Nowakowski, A game of cops and robbers played on products of graphs, *Discrete Mathematics* **186** (1998) 253–268.

[152] R.J. Nowakowski, P. Winkler, Vertex-to-vertex pursuit in a graph, *Discrete Mathematics* **43** (1983) 235–239.

[153] D. Offner, K. Okajian, Variations of Cops and Robber on the hypercube, *Australasian Journal of Combinatorics* **59** (2014) 229–250.

[154] M.D. Penrose, Connectivity of soft random geometric graphs, *The Annals of Applied Probability* **26** (2016) 986–1028.

[155] M.D. Penrose, *Random Geometric Graphs*, Oxford University Press, 2003.

[156] M. Penrose, The longest edge of the random minimal spanning tree, *Annals of Applied Probability* **7** (1997) 340–361.

[157] B. Pittel, Note on the heights of random recursive trees and random m-ary search trees, *Random Structures and Algorithms* **5** (1994) 337–347.

[158] B. Pittel, J. Spencer, N. Wormald, Sudden emergence of a giant k-core in a random graph, *Journal of Combinatorial Theory, Series B* **67** (1996) 111–151.

[159] P. Prałat, Almost all k-cop-win graphs contain a dominating set of cardinality k, *Discrete Mathematics* **338** (2015) 47–52.

[160] P. Prałat, Cleaning random d-regular graphs with brooms, *Graphs and Combinatorics* **27** (2011) 567–584.

[161] P. Prałat, Cleaning random graphs with brushes, *Australasian Journal of Combinatorics* **43** (2009) 237–251.

[162] P. Prałat, Containment game played on random graphs: another zig-zag theorem, *Electronic Journal of Combinatorics* **22(2)** (2015) #P2.32.

[163] P. Prałat, Graphs with average degree smaller than $\frac{30}{11}$ burn slowly, *Graphs and Combinatorics* **30** (2014) 455–470.

[164] P. Prałat, How many zombies are needed to catch the survivor on toroidal grids? Preprint 2017.

[165] P. Prałat, Sparse graphs are not flammable, *SIAM Journal on Discrete Mathematics* **27** (2013) 2157–2166.

[166] P. Prałat, When does a random graph have constant cop number?, *Australasian Journal of Combinatorics* **46** (2010) 285–296.

[167] P. Prałat, N. Wormald, Meyniel's conjecture holds for random graphs, *Random Structures and Algorithms* **48** (2016) 396–421.

[168] P. Prałat, N. Wormald, Meyniel's conjecture holds for random d-regular graphs, Preprint 2017.

[169] A. Quilliot, Jeux et pointes fixes sur les graphes, *Thèse de 3ème cycle*, Université de Paris VI, 1978, 131–145.

[170] A. Quilliot, *Problèmes de jeux, de point Fixe, de connectivité et de represésentation sur des graphes, des ensembles ordonnés et des hypergraphes*, Thèse d'Etat, Université de Paris VI, 1983, 131–145.

[171] D. Reichman, New bounds for contagious sets, *Discrete Mathematics* **312** (2012) 1812–1814.

[172] R.W. Robinson, N.C. Wormald, Almost all cubic graphs are Hamiltonian, *Random Structures and Algorithms* **3** (1992) 117–125.

[173] E. Roshanbin, *Burning a graph as a model of social contagion*, PhD Thesis, Dalhousie University, 2016.

[174] A. Scott, B. Sudakov, A bound for the cops and robbers problem, *SIAM Journal of Discrete Mathematics* **25** (2011) 1438–1442.

[175] J. Sgall, A solution to David Gale's lion and man problem, *Theoretical Computer Science* **259** (2001) 663–670.

[176] P.J. Slater, Y. Wang, Some results on acquisition numbers, *Journal of Combinatorial Mathematics and Combinatorial Computing* **64** (2008) 65–78.

[177] J. Spencer (with L. Florescu), *Asymptopia*, American Mathematical Society, 2014.

[178] Z.A. Wagner, Cops and Robbers on diameter two graphs, *Discrete Mathematics* **338** (2015) 107–109.

[179] A. Wald, Some generalizations of the theory of cumulative sums of random variables, *The Annals of Mathematical Statistics*, **16** (1945) 287–293.

[180] D.B. West, *Introduction to Graph Theory, 2nd edition*, Prentice Hall, 2001.

[181] B. Wieland, A. Godbole, On the domination number of a random graph, *Electronic Journal of Combinatorics* **8** (2001) 37–37.

[182] N.C. Wormald, The asymptotic connectivity of labelled regular graphs, *Journal of Combinatorial Theory, Series B* **31** (1981) 156–167.

[183] N.C. Wormald, The differential equation method for random graph processes and greedy algorithms, *Lectures on Approximation and Randomized Algorithms*, eds. M. Karoński and H. J. Prömel, PWN, Warsaw, pp. 73–155, 1999.

[184] N.C. Wormald, Models of random regular graphs, *Surveys in Combinatorics*, 1999, J.D. Lamb and D.A. Preece, eds. London Mathematical Society Lecture Note Series, vol 276, pp. 239–298. Cambridge University Press, Cambridge, 1999.

[185] N.C. Wormald, *Some Problems in the Enumeration of Labelled Graphs*, PhD thesis, University of Newcastle, 1978.

[186] N.C. Wormald, Analysis of greedy algorithms on graphs with bounded degrees, EuroComb '01 (Barcelona) *Discrete Mathematics* **273** (2003) 235–260.

Index